FUNDAMENTALS OF PHOSPHORS

FUNDAMENTALS OF PHOSPHORS

Edited by

William M. Yen

Shigeo Shionoya (Deceased)

Hajime Yamamoto

CRC Press
Taylor & Francis Group
Boca Raton London New York

CRC Press is an imprint of the
Taylor & Francis Group, an **informa** business

CRC Press
Taylor & Francis Group
6000 Broken Sound Parkway NW, Suite 300
Boca Raton, FL 33487-2742

First issued in paperback 2019

ISBN-13: 978-1-4200-4367-9 (hbk)
ISBN-13: 978-0-367-38964-2 (pbk)

Visit the Taylor & Francis Web site at
http://www.taylorandfrancis.com

and the CRC Press Web site at
http://www.crcpress.com

Dedication

Dr. Shigeo Shionoya 1923–2001

This volume is a testament to the many contributions Dr. Shionoya made to phosphor art and is dedicated to his memory.

In Memoriam

Shigeo Shionoya
Formerly of the University of Tokyo
The Institute for Solid
  State Physics
Tokyo, Japan

Shosaku Tanaka
Tottori University
Department of Electrical & Electronic
  Engineering
Tottori, Japan

The Editors

William M. Yen obtained his B.S. degree from the University of Redlands, Redlands, California in 1956 and his Ph.D. (physics) from Washington University in St. Louis in 1962. He served from 1962–65 as a Research Associate at Stanford University under the tutelage of Professor A.L. Schawlow, following which he accepted an assistant professorship at the University of Wisconsin-Madison. He was promoted to full professorship in 1972 and retired from this position in 1990 to assume the Graham Perdue Chair in Physics at the University of Georgia-Athens.

Dr. Yen has been the recipient of a J.S. Guggenheim Fellowship (1979–80), of an A. von Humboldt Senior U.S. Scientist Award (1985, 1990), and of a Senior Fulbright to Australia (1995). He was recently awarded the Lamar Dodd Creative Research Award by the University of Georgia Research Foundation. He is the recipient of the ICL Prize for Luminescence Research awarded in Beijing in August 2005. He has been appointed to visiting professorships at numerous institutions including the University of Tokyo, the University of Paris (Orsay), and the Australian National University. He was named the first Edwin T. Jaynes Visiting Professor by Washington University in 2004 and has been appointed to an affiliated research professorship at the University of Hawaii (Manoa). He is also an honorary professor at the University San Antonio de Abad in Cusco, Peru and of the Northern Jiatong University, Beijing, China. He has been on the technical staff of Bell Labs (1966) and of the Livermore Laser Fusion Effort (1974–76).

Dr. Yen has been elected to fellowship in the American Physical Society, the Optical Society of America, the American Association for the Advancement of Science and by the U.S. Electrochemical Society.

Professor Shionoya was born on April 30, 1923, in the Hongo area of Tokyo, Japan and passed away in October 2001. He received his baccalaureate in applied chemistry from the faculty of engineering, University of Tokyo, in 1945. He served as a research associate at the University of Tokyo until he moved to the department of electrochemistry, Yokohama National University as an associate professor in 1951. From 1957 to 1959, he was appointed to a visiting position in Professor H.P. Kallman's group in the physics department of New York University. While there, he was awarded a doctorate in engineering from the University of Tokyo in 1958 for work related to the industrial development of solid-state inorganic phosphor materials. In 1959, he joined the Institute for Solid State Physics (ISSP, Busseiken) of the University of Tokyo as an associate professor; he was promoted to full professorship in the Optical Properties Division of the ISSP in 1967. Following a reorganization of ISSP in 1980, he was named head of the High Power Laser Group of the Division of Solid State under Extreme Conditions. He retired from the post in 1984 with the title of emeritus professor. He helped in the establishment of the Tokyo Engineering University in 1986 and served in the administration and as a professor of Physics. On his retirement from the Tokyo Engineering University in 1994, he was also named emeritus professor in that institution.

During his career, he published more than two hundred scientific papers and authored or edited a number of books—the *Handbook on Optical Properties of Solids* (in Japanese, 1984) and the *Phosphor Handbook* (1998).

Professor Shionoya has been recognized for his many contributions to phosphor art. In 1977, he won the Nishina Award for his research on high-density excitation effects in semiconductors using picosecond spectroscopy. He was recognized by the Electrochemical Society in 1979 for his contributions to advances in phosphor research. Finally, in 1984 he was the first recipient of the ICL Prize for Luminescence Research.

Hajime Yamamoto received his B.S. and Ph.D. degrees in applied chemistry from the University of Tokyo in 1962 and 1967. His Ph.D. work was performed at the Institute for Solid State Physics under late Professors Shohji Makishima and Shigeo Shionoya on spectroscopy of rare earth ions in solids. Soon after graduation he joined Central Research Laboratory, Hitachi Ltd., where he worked mainly on phosphors and p-type ZnSe thin films. From 1971 to 1972, he was a visiting fellow at Professor Donald S. McClure's laboratory, Department of Chemistry, Princeton University. In 1991, he retired from Hitachi Ltd. and moved to Tokyo University of Technology as a professor of the faculty of engineering. Since 2003, he has been a professor at the School of Bionics of the same university.

Dr. Yamamoto serves as a chairperson of the Phosphor Research Society and is an organizing committee member of the Workshop on EL Displays, LEDs and Phosphors, International Display Workshops. He was one of the recipients of Tanahashi Memorial Award of the Japanese Electrochemical Society in 1988, and the Phosphor Award of the Phosphor Research Society in 2000 and 2005.

Preface

This volume originated from the *Phosphor Handbook* which has enjoyed a moderate amount of sale success as part of the CRC Laser and Optical Science and Technology Series and which recently went into its second edition. The original *Handbook* was published in Japanese in 1987 through an effort of the Phosphor Research Society of Japan. The late professor Shionoya was largely instrumental in getting us involved in the translation and publication of the English version. Since the English publication in 1998, the *Handbook* has gained wide acceptance by the technical community as a central reference on the basic properties as well as the applied and practical aspects of phosphor materials.

As we had expected, advances in the display and information technologies continue to consume and demand phosphor materials which are more efficient and more targeted to specific uses. These continuing changes in the demand necessitated an update and revision of the *Handbook* and resulted in the publication of the second edition which incorporates almost all additional topics, especially those of current interest such as quantum cutting and LED white lighting phosphor materials.

At the same time, it has also become apparent to some of us that the evolution of recent technologies will continue to place demands on the phosphor art and that research activity in the understanding and development of new phosphor materials will continue to experience increases. For this reason, it has been decided by CRC Press that a series of titles dedicated to Phosphor Properties be inaugurated through the publication of correlated sections of the *Phosphor Handbook* into three separate volumes. Volume I deals with the fundamental properties of luminescence as applied to solid state phosphor materials; the second volume includes the description of the synthesis and optical properties of phosphors used in different applications while the third addresses experimental methods for phosphor evaluation. The division of the Handbook into these sections, will allow us as editors to maintain the currency and timeliness of the volumes by updating only the section(s) which necessitate it.

We hope that this new organization of a technical series continues to serve the purpose of serving as a general reference to all aspects of phosphor properties and applications and as a starting point for further advances and developments in the phosphor art.

William M. Yen
Athens, GA, USA
October, 2006

Hajime Yamamoto
Tokyo, Japan
October, 2006

Contributors

Chihaya Adachi
Kyushu University
Fukuoka, Japan

Pieter Dorenbos
Delft University of Technology
Delft, The Netherlands

Gen-ichi Hatakoshi
Toshiba Research
Consulting Corp.
Kawasaki, Japan

Naoto Hirosaki
National Institute of Materials Science
Tsukuba, Japan

Sumiaki Ibuki
Formerly of Mitsubishi Electric Corp.
Amagasaki, Japan

Kenichi Iga
Formerly of Tokyo Institute of Technology
Yokohama, Japan

Tsuyoshi Kano
Formerly of Hitachi, Ltd.,
Tokyo, Japan

Hiroshi Kobayashi
Tokushima Bunri University
Kagawa, Japan

Hiroshi Kukimoto
Toppan Printing Co., Ltd.
Tokyo, Japan

Yasuaki Masumoto
University of Tsukuba
Ibaraki, Japan

Hiroyuki Matsunami
Kyoto University
Kyoto, Japan

Mamoru Mitomo
National Institute of Materials Science
Tsukuba, Japan

Noboru Miura
Meiji University
Kawasaki, Japan

Makoto Morita
Formerly of Seikei University
Tokyo, Japan

Shuji Nakamura
University of California
Santa Barbara, California

Eiichiro Nakazawa
Formerly of Kogakuin University
Tokyo, Japan

Shigetoshi Nara
Hiroshima University
Hiroshima, Japan

Hiroshi Sasakura
Formerly of Tottori University
Tottori, Japan

Masaaki Tamatani
Toshiba Research Consulting
  Corporation
Kawasaki, Japan

Shinkichi Tanimizu
Formerly of Hitachi, Ltd.
Tokyo, Japan

Tetsuo Tsutsui
Kyushu University
Fukuoka, Japan

Rong-Jun Xie
Advanced Materials Laboratory, National
 Institute of Materials Science
Tsukuba, Japan

Hajime Yamamoto
Tokyo University of Technology
Tokyo, Japan

Toshiya Yokogawa
Matsushita Electric Ind. Co., Ltd.
Kyoto, Japan

Contents

chapter one — section one

Fundamentals of luminescence

Eiichiro Nakazawa

Contents

1.1   Absorption and emission of light

Most phosphors are composed of a transparent microcrystalline host (or a matrix) and an activator, i.e., a small amount of intentionally added impurity atoms distributed in the host crystal. Therefore, the luminescence processes of a phosphor can be divided into two parts: the processes mainly related to the host, and those that occur around and within the activator.

Processes related to optical absorption, reflection, and transmission by the host crystal are discussed, from a macroscopic point of view, in 1.1.1. Other host processes (e.g., excitation by electron bombardment and the migration and transfer of the excitation energy in the host) are discussed in a later section. 1.1.2 deals with phenomena related to the activator atom based on the theory of atomic spectra.

The interaction between the host and the activator is not explicitly discussed in this section; in this sense, the host is treated only as a medium for the activator. The interaction processes such as the transfer of the host excitation energy to the activator will be discussed in detail for each phosphor elsewhere.[1]

1.1.1 *Absorption and reflection of light in crystals*

Since a large number of phosphor host materials are transparent and nonmagnetic, their optical properties can be represented by the optical constants or by a complex dielectric constant.

1.1.1.1 *Optical constant and complex dielectric constant*

The electric and magnetic fields of a light wave, propagating in a uniform matrix with an angular frequency ω (= $2\pi\nu$, ν:frequency) and velocity $v = \omega/k$ are:

$$E = E_0 \exp\left[i\left(\tilde{k} \cdot r - \omega t\right)\right] \tag{1}$$

$$H = H_0 \exp\left[i\left(\tilde{k} \cdot r - \omega t\right)\right], \tag{2}$$

where r is the position vector and $\tilde{k}$ is the complex wave vector.

E and H in a nonmagnetic dielectric material, with a magnetic permeability that is nearly equal to that in a vacuum ($\mu \approx \mu_0$) and with uniform dielectric constant ε and electric conductivity σ, satisfy the next two equations derived from Maxwell's equations.

$$\nabla^2 E = \sigma\mu_0 \frac{\partial E}{\partial t} + \varepsilon\mu_0 \frac{\partial^2 E}{\partial t^2} \tag{3}$$

$$\nabla^2 H = \sigma\mu_0 \frac{\partial H}{\partial t} + \varepsilon\mu_0 \frac{\partial^2 H}{\partial t^2} \tag{4}$$

In order that Eqs. 1 and 2 satisfy Eqs. 3 and 4, the $\tilde{k}$-vector and its length $\tilde{k}$, which is a complex number, should satisfy the following relation:

$$\tilde{k} \cdot \tilde{k} = \tilde{k}^2 = \left(\varepsilon + \frac{i\sigma}{\omega}\right)\mu_0\omega^2 = \tilde{\varepsilon}\mu_0\omega^2 \tag{5}$$

where $\tilde{\varepsilon}$ is the complex dielectric constant defined by:

$$\tilde{\varepsilon} = \varepsilon' + i\varepsilon'' \equiv \varepsilon + i\frac{\sigma}{\omega} \tag{6}$$

Therefore, the refractive index, which is a real number defined as $n \equiv c/v = ck/w$ in a transparent media, is also a complex number:

$$\tilde{n} = n + i\kappa \equiv c\tilde{k}/\omega = \left(\frac{\tilde{\varepsilon}}{\varepsilon_0}\right)^{1/2} \tag{7}$$

where c is the velocity of light in vacuum and is equal to $(\varepsilon_0\mu_0)^{-1/2}$ from Eq. 5. The last term in Eq. 7 is also derived from Eq. 5.

The real and imaginary parts of the complex refractive index, i.e., the real refractive index n and the extinction index κ, are called optical constants, and are the representative constants of the macroscopic optical properties of the material. The optical constants in a nonmagnetic material are related to each other using Eqs. 6 and 7,

$$\frac{e'}{\varepsilon_0} = n^2 - \kappa^2 \tag{8}$$

$$\frac{\varepsilon''}{\varepsilon_0} = 2n\kappa \tag{9}$$

Both of the optical constants, n and κ, are functions of angular frequency ω and, hence, are referred to as dispersion relations. The dispersion relations for a material are obtained by measuring and analyzing the reflection or transmission spectrum of the material over a wide spectral region.

1.1.1.2   Absorption coefficient

The intensity of the light propagating in a media a distance x from the incident surface having been decreased by the optical absorption is given by Lambert's law.

$$I = I_0 \exp(-\alpha x) \tag{10}$$

where I_0 is the incident light intensity minus reflection losses at the surface, and $\alpha(\text{cm}^{-1})$ is the absorption coefficient of the media.

Using Eqs. 5 and 7, Eq. 1 may be rewritten as:

$$E = E_0 \exp(-\omega\kappa x/c)\exp\left[-i\omega(t + nx/c)\right] \tag{11}$$

and, since the intensity of light is proportional to the square of its electric field strength E, the absorption coefficient may be identified as:

$$\alpha = 2\omega\kappa/c \tag{12}$$

Therefore, κ is a factor that represents the extinction of light due to the absorption by the media.

There are several ways to represent the absorption of light by a medium, as described below.

1. Absorption coefficient, $\alpha(\text{cm}^{-1})$: $I/I_0 = e^{-\alpha x}$
2. Absorption cross-section, α/N (cm^2). Here, N is the number of absorption centers per unit volume.
3. Optical density, absorbance, $D = -\log_{10}(I/I_0)$
4. Absorptivity, $(I_0 - I)/I_0 \times 100$, (%)
5. Molar extinction coefficient, $\varepsilon = \alpha\log_{10}e/C$. Here, $C(\text{mol}/l)$ is the molar concentration of absorption centers in a solution or gas.

1.1.1.3   Reflectivity and transmissivity

When a light beam is incident normally on an optically smooth crystal surface, the ratio of the intensities of the reflected light to the incident light, i.e., normal surface reflectivity R_0, can be written in terms of the optical constants, n and κ, by

$$R_0 = \frac{(n-1)^2 + \kappa^2}{(n+1)^2 + \kappa^2} \tag{13}$$

Then, for a sample with an absorption coefficient α and thickness d that is large enough to neglect interference effects, the overall normal reflectivity and transmissivity, i.e., the ratio of the transmitted light to the incident, are; respectively:

$$\overline{R} = R_0\left(1 + \overline{T}\exp(-\alpha d)\right) \tag{14}$$

$$\overline{T} = \frac{(1-R_0)^2\left(1+\kappa^2/n^2\right)\exp(-\alpha d)}{1-R_0^2\exp(-2\alpha d)} \simeq \frac{(1-R_0)^2\exp(-\alpha d)}{1-R_0^2\exp(-2\alpha d)} \tag{15}$$

If absorption is zero ($\alpha = 0$), then,

$$\overline{R} = \frac{(n-1)^2}{(n^2+1)} \tag{16}$$

1.1.2   Absorption and emission of light by impurity atoms

The emission of light from a material originates from two types of mechanisms: thermal emission and luminescence. While all the atoms composing the solid participate in the light emission in the thermal process, in the luminescence process a very small number of atoms (impurities in most cases or crystal defects) are excited and take part in the emission of light. The impurity atom or defect and its surrounding atoms form a luminescent or an emitting center. In most phosphors, the luminescence center is formed by intentionally incorporated impurity atoms called activators.

This section treats the absorption and emission of light by these impurity atoms or local defects.

1.1.2.1   Classical harmonic oscillator model of optical centers

The absorption and emission of light by an atom can be described in the most simplified scheme by a linear harmonic oscillator, as shown in Figure 1, composed of a positive charge ($+e$) fixed at $z = 0$ and an electron bound and oscillating around it along the z-axis. The electric dipole moment of the oscillator with a characteristic angular frequency ω_0 is given by:

$$M = ez = M_o \exp(i\omega_o t) \tag{17}$$

and its energy, the sum of the kinetic and potential energies, is $\left(m_e\omega_0^2/2e^2\right)M_o^2$, where m_e is the mass of the electron. Such a vibrating electric dipole transfers energy to electromagnetic radiation at an average rate of $\left(\omega_0^4/12\pi\varepsilon_0 c^3\right)M_o^2$ per second, and therefore has a total energy decay rate given by:

$$A_0 = \frac{e^2\omega_o^2}{6\pi\varepsilon_0 m_e c^3} \tag{18}$$

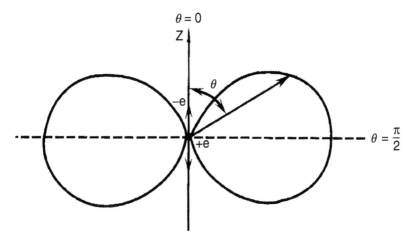

Figure 1   Electromagnetic radiation from an electric dipole oscillator. The length of the arrow gives the intensity of the radiation to the direction.

When the change of the energy of this oscillator is expressed as an exponential function e^{-t/τ_0}, its time constant τ_0 is equal to A_0^{-1}, which is the radiative lifetime of the oscillator, i.e., the time it takes for the oscillator to lose its energy to e^{-1} of the initial energy. From Eq. 8, the radiative lifetime of an oscillator with a 600-nm ($\omega_0 = 3 \times 10^{15}$ s^{-1}) wavelength is $\tau_0 \approx 10^{-8}$ s. The intensity of the emission from an electric dipole oscillator depends on the direction of the propagation, as shown in Figure 1.

A more detailed analysis of absorption and emission processes of light by an atom will be discussed using quantum mechanics in the following subsection.

1.1.2.2   Electronic transition in an atom

In quantum mechanics, the energy of the electrons localized in an atom or a molecule have discrete values as shown in Figure 2. The absorption and emission of light by an

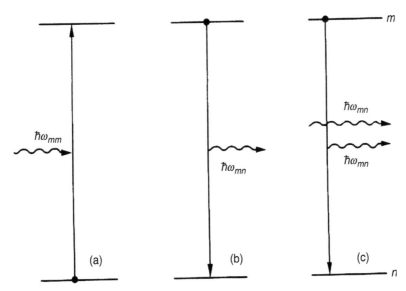

Figure 2   Absorption (a), spontaneous emission (b), and induced emission (c) of a photon by a two-level system.

atom, therefore, is not a gradual and continuous process as discussed in the above section using a classical dipole oscillator, but is an instantaneous transition between two discrete energy levels (states), m and n in Figure 2, and should be treated statistically.

The energy of the photon absorbed or emitted at the transition $m \leftrightarrow n$ is:

$$\hbar\omega_{mn} = E_m - E_n, \qquad (E_m > E_n) \tag{19}$$

where E_n and E_m are the energies of the initial and final states of the transition, respectively, and $\omega_{mn} (= 2\pi\nu_{mn})$ is the angular frequency of light.

There are two possible emission processes, as shown in Figure 2; one is called spontaneous emission (b), and the other is stimulated emission (c). The stimulated emission is induced by an incident photon, as is the case with the absorption process (a). Laser action is based on this type of emission process.

The intensity of the absorption and emission of photons can be enumerated by a transition probability per atom per second. The probability for an atom in a radiation field of energy density $\rho(\omega_{mn})$ to absorb a photon, making the transition from n to m, is given by

$$W_{mn} = B_{n \to m} \rho(\omega_{mn}) \tag{20}$$

where $B_{n \to m}$ is the transition probability or Einstein's B-coefficient of optical absorption, and $\rho(\omega)$ is equal to $I(\omega)/c$ in which $I(\omega)$ is the light intensity, i.e., the energy per second per unit area perpendicular to the direction of light.

On the other hand, the probability of the emission of light is the sum of the spontaneous emission probability $A_{m \to n}$ (Einstein's A-coefficient) and the stimulated emission probability $B_{m \to n} \rho(\omega_{mn})$. The stimulated emission probability coefficient $B_{m \to n}$ is equal to $B_{n \to m}$.

The equilibrium of optical absorption and emission between the atoms in the states m and n is expressed by the following equation.

$$N_n B_{n \to m} \rho(\omega_{mn}) = N_m \left\{ A_{m \to n} + B_{m \to n}(\omega_{mn}) \rho(\omega_{mn}) \right\}, \tag{21}$$

where N_m and N_n are the number of atoms in the states m and n, respectively. Taking into account the Boltzmann distribution of the system and Plank's equation of radiation in thermodynamic equilibrium, the following equation is obtained from Eq. 21 for the spontaneous mission probability.

$$A_{m \to n} = \frac{\hbar\omega_{mn}^3}{\pi^2 c^3} B_{m \to n} \tag{22}$$

Therefore, the probabilities of optical absorption, and the spontaneous and induced emissions between m and n are related to one another.

1.1.2.3 *Electric dipole transition probability*

In a quantum mechanical treatment, optical transitions of an atom are induced by perturbing the energy of the system by $\Sigma_i(-er_i) \cdot E$, in which r_i is the position vector of the electron from the atom center and, therefore, $\Sigma_i(-er_i)$ is the electric dipole moment of the atom (see Eq. 17). In this electric dipole approximation, the transition probability of optical absorption is given by:

$$W_{mn} = \frac{\pi}{3\varepsilon_0 c\hbar^2} I(\omega_{mn}) |M_{mn}|^2 \tag{23}$$

Here, the dipole moment, M_{mn} is defined by:

$$M_{mn} = \int \psi_m^* \left(\sum_i e\mathbf{r}_i \right) \psi_n d\tau \qquad (24)$$

where ψ_m and ψ_n are the wavefunctions of the states m and n, respectively. The direction of this dipole moment determines the polarization of the light absorbed or emitted. In Eq. 23, however, it is assumed that the optical center is isotropic and then $|(M_{mn})_z|^2 = |M_{mn}|^2/3$ for light polarized in the z-direction.

Equating the right-hand side of Eq. 23 to that of Eq. 20, the absorption transition probability coefficient $B_{n\rightarrow m}$ and then, from Eq. 22, the spontaneous emission probability coefficient $A_{m\rightarrow n}$ can be obtained as follows:

$$B_{n\rightarrow m} = \frac{\pi}{3\varepsilon_0 \hbar^2} |M_{mn}|^2$$

$$A_{m\rightarrow n} = \frac{\omega_{mn}^3}{3\pi\varepsilon_0 \hbar c^3} |M_{mn}|^2 \qquad (25)$$

1.1.2.4 Intensity of light emission and absorption

The intensity of light is generally defined as the energy transmitted per second through a unit area perpendicular to the direction of light. The spontaneous emission intensity of an atom is proportional to the energy of the emitted photon, multiplied by the transition probability per second given by Eq. 25.

$$I(\omega_{mn}) \propto \hbar\omega_{mn} A_{m\rightarrow n} = \frac{\omega_{mn}^4}{3\pi\varepsilon_0 c^3} |M_{mn}|^2 \qquad (26)$$

Likewise, the amount of light with intensity $I_0(\omega_{mn})$ to be absorbed by an atom per second is equal to the photon energy ω_{mn} multiplied by the absorption probability coefficient and the energy density I_0/c.

It is more convenient, however, to use a radiative lifetime and absorption cross-section to express the ability of an atom to make an optical transition than to use the amount of light energy absorbed or emitted by the transition.

The radiative lifetime τ_{mn} is defined as the inverse of the spontaneous emission probability $A_{m\rightarrow n}$.

$$\tau_{mn}^{-1} = A_{m\rightarrow n} \qquad (27)$$

If there are several terminal states of the transition and the relaxation is controlled only by spontaneous emission processes, the decay rate of the emitting level is determined by the sum of the transition probabilities to all final states:

$$A_m = \sum_n A_{m\rightarrow n} \qquad (28)$$

and the number of the excited atoms decreases exponentially, $\propto \exp(-t/\tau)$, with time a constant $\tau = A_m^{-1}$, called the natural lifetime. In general, however, the real lifetime of the excited state m is controlled not only by radiative processes, but also by nonradiative ones (see 1.7).

The absorption cross-section σ represents the probability of an atom to absorb a photon incident on a unit area. (If there are N absorptive atoms per unit volume, the absorption coefficient α in Eq. 10 is equal to σN. Therefore, since the intensity of the light with a photon per second per unit area is $I_0 = \omega_{mn}$ in Eq. 23, the absorption cross-section is given by:

$$\sigma_{nm} = \frac{\pi \omega_{mn}}{3\varepsilon_0 c \hbar} \left| M_{mn} \right|^2 \tag{29}$$

1.1.2.5   Oscillator strength

The oscillator strength of an optical center is often used in order to represent the strength of light absorption and emission of the center. It is defined by the following equation as a dimensionless quantity.

$$f_{mn} = \frac{2m_e \omega_{mn}}{\hbar e^2} \left| \left(M_{mn} \right)_z \right|^2 = \frac{2m_e \omega_{mn}}{3\hbar e^2} \left| M_{mn} \right|^2 \tag{30}$$

The third term of this equation is given by assuming that the transition is isotropic, as it is the case with Eq. 24.

The radiative lifetime and absorption cross-section are expressed by using the oscillator strength as:

$$\tau_{mn}^{-1} = A_{m \to n} = \frac{e^2 \omega_{mn}^2}{2\pi \varepsilon_0 m c^3} f_{mn} \tag{31}$$

$$\sigma_{nm} = \frac{\pi e^2}{2\varepsilon_0 m c} f_{mn} \tag{32}$$

Now one can estimate the oscillator strength of a harmonic oscillator with the electric dipole moment $M = -er$ in a quantum mechanical manner. The result is that only one electric dipole transition between the ground state ($n = 0$) and the first excited state ($m = 1$) is allowed, and the oscillator strength of this transition is $f_{10} = 1$. Therefore, the summation of all the oscillator strengths of the transition from the state $n = 0$ is also $\Sigma_m f_{m0} = 1$ ($m \neq 0$). This relation is true for any one electron system; for N-electron systems, the following f-sum rule should be satisfied; that is,

$$\sum_{m \neq n} f_{mn} = N \tag{33}$$

At the beginning of this section, the emission rate of a linear harmonic oscillator was classically obtained as A_0 in Eq. 18. Then, the total transition probability given by Eq. 32 with $f = 1$ in a quantum mechanical scheme coincides with the emission rate of the classical linear oscillator A_0, multiplied by a factor of 3, corresponding to the three degrees of freedom of the motion of the electron in the present system.

1.1.2.6 *Impurity atoms in crystals*

Since the electric field acting on an impurity atom or optical center in a crystal is different from that in vacuum due to the effect of the polarization of the surrounding atoms, and the light velocity is reduced to c/n (see Eq. 7), the radiative lifetime and the absorption cross-section are changed from those in vacuum. In a cubic crystal, for example, Eqs. 31 and 32 are changed, by the internal local field, to:

$$\tau_{mn}^{-1} = \frac{n\left(n^2 + 2\right)^2}{9} \cdot \frac{e^2 \omega_{mn}^2}{2\pi\varepsilon_0 mc^3} f_{nm} \tag{34}$$

$$\sigma_{nm} = \frac{\left(n^2 + 2\right)^2}{9n} \cdot \frac{\pi e^2}{2\pi\varepsilon_0 mc} f_{nm} \tag{35}$$

1.1.2.7 *Forbidden transition*

In the case that the electric dipole moment of a transition M_{mn} of Eq. 25 becomes zero, the probability of the electric dipole (E1) transition in Eq. 25 and 26 is also zero. Since the electric dipole transition generally has the largest transition probability, this situation is usually expressed by the term forbidden transition. Since the electric dipole moment operator in the integral of Eq. 24 is an odd function (odd parity), the electric dipole moment is zero if the initial and final states of the transition have the same parity; that is, both of the wavefunctions of these states are either an even or odd function, and the transition is said to be parity forbidden. Likewise, since the electric dipole moment operator in the integral of Eq. 24 has no spin operator, transitions between initial and final states with different spin multiplicities are spin forbidden.

In Eq. 24 for the dipole moment, the effects of the higher-order perturbations are neglected. If the neglected terms are included, the transition moment is written as follows:

$$\left|M_{mn}\right|^2 = \left|(er)_{mn}\right|^2 + \left|\left(\frac{e}{2mc} r \times p\right)_{mn}\right|^2 + \frac{3\pi\omega_{mn}^2}{40c^2}\left|(er \cdot r)_{mn}\right|^2 \tag{36}$$

where the first term on the right-hand side is the contribution of the electric dipole (E1) term previously given in Eq. 24; the second term, in which p denotes the momentum of an electron, is that of magnetic dipole (M1); and the third term is that of an electric quadrupole transition (E2). Provided that $(r)_{mn}$ is about the radius of a hydrogen atom (0.5 Å) and ω_{mn} is 10^{15} rad/s for visible light, radiative lifetimes estimated from Eq. 26 and 36 are $\sim 10^{-8}$ s for E1, $\sim 10^{-3}$ s for M1, and $\sim 10^{-1}$ s for E2.

E1-transitions are forbidden (parity forbidden) for *f-f* and *d-d* transitions of free rare-earth ions and transition-metal ions because the electron configurations, and hence the parities of the initial and final states, are the same. In crystals, however, the E1 transition is partially allowed by the odd component of the crystal field, and this partially allowed or forced E1 transition has the radiative lifetime of $\sim 10^{-3}$ s. (See 2.2).

1.1.2.8 *Selection rule*

The selection rule governing whether a dipole transition is allowed between the states m and n is determined by the transition matrix elements $(er)_{mn}$ and $(r \times p)_{mn}$ in Eq. 36. However, a group theoretical inspection of the symmetries of the wavefunctions of these states and the operators er and $r \times p$ enables the determination of the selection rules without calculating the matrix elements.

When an atom is free or in a spherical symmetry field, its electronic states are denoted by a set of the quantum numbers S, L, and J in the LS-coupling scheme. Here, S, L, and J denote the quantum number of the spin, orbital, and total angular momentum, respectively, and ΔS, for example, denotes the difference in S between the states m and n. Then the selection rules for E1 and M1 transitions in the LS-coupling scheme are given by:

$$\Delta S = 0, \quad \Delta L = 0 \quad \text{or} \quad \pm 1 \tag{37}$$

$$\Delta J = 0 \quad \text{or} \quad \pm 1 \quad (J = 0 \to J = 0, \quad \text{not allowed}) \tag{38}$$

If the spin-orbit interaction is too large to use the **LS**-coupling scheme, the **JJ**-coupling scheme might be used, in which many (**S**, **L**)-terms are mixed into a **J**-state. In the **JJ**-coupling scheme, therefore, the ΔS and ΔL selection rules in Eqs. 37 ad 38 are less strict, and only the ΔJ selection rule applies.

While the E1 transitions between the states with the same parity are forbidden, as in the case of the *f-f* transitions of free rare-earth ions, they become partially allowed for ions in crystals due to the effects of crystal fields of odd parity. The selection rule for the partially allowed E1 *f-f* transition is $|\Delta J| \leq 6$ ($J = 0 - 0, 1, 3, 5$ are forbidden). M1 transitions are always parity allowed because of the even parity of the magnetic dipole operator $r \times p$ in Eq. 36.

Reference

1. *Practical Applications of Phosphors*, Yen, W.M., Shionoya, S., and Yamamoto, H., Eds., CRC Press, Boca Raton, 2006.

chapter one — section two

Fundamentals of luminescence

Shigetoshi Nara and Sumiaki Ibuki

Contents

1.2   Electronic states and optical transition of solid crystals

1.2.1   Outline of band theory

First, a brief description of crystal properties is given. As is well known, a crystal consists of a periodic configuration of atoms, which is called a *crystal lattice*. There are many different kinds of crystal lattices and they are classified, in general, according to their symmetries, which specify invariant properties for translational and rotational operations. Figure 3 shows a few, typical examples of crystal structures, i.e., a rock-salt (belonging to one of the cubic groups) structure, a zinc-blende (also a cubic group) structure, and a wurtzeite (a hexagonal group) structure, respectively.

Second, consider the electronic states in these crystals. In an isolated state, each atom has electrons that exist in discrete electronic energy levels, and the states of these bound electrons are characterized by atomic wavefunctions. Their discrete energy levels, however, will have finite spectral width in the condensed state because of the overlaps between electronic wavefunctions belonging to different atoms. This is because electrons can become itinerant between atoms, until finally they fall into delocalized electronic states called *electronic energy bands*, which also obey the symmetries of crystals. In these energy bands, the states with lower energies are occupied by electrons originating from bound electrons of atoms and are called *valence bands*. The energy bands having higher energies are not occupied by electrons and are called *conduction bands*. Usually, in materials having crystal symmetries such as rock-salt, zinc-blende, or wurtzeite structures, there is no electronic state between the top of the valence band (the highest state of occupied bands) and the bottom of the conduction band (the lowest state of unoccupied bands); this region is called the *bandgap*. The reason why unoccupied states are called

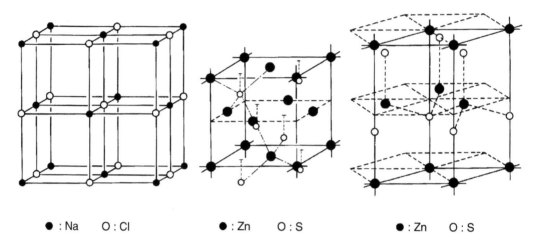

● : Na   O : Cl   ● : Zn   O : S   ● : Zn   O : S

Figure 3   The configuration of the atoms in three important kinds of crystal structures. (a) rock-salt type, (b) zinc-blende type, and (c) wurtzeite type, respectively.

conduction bands is due to the fact that an electron in a conduction band is almost freely mobile if it is excited from a valence band by some method: for example, by absorption of light quanta. In contrast, electrons in valence bands cannot be mobile because of a fundamental property of electrons; as *fermions*, only two electrons (spin up and down) can occupy an electronic state. Thus, it is necessary for electrons in the valence band to have empty states in order for them to move freely when an electric field is applied. After an electron is excited to the conduction band, a hole that remains in the valence band behaves as if it were a mobile particle with a positive charge. This hypothetical particle is called a *positive hole*. The schematic description of these excitations are shown in Figure 4. As noted above, bandgaps are strongly related to the optical properties and the electric conductivity of crystals.

A method to evaluate these electronic band structures in a quantitative way using quantum mechanics is briefly described. The motion of electrons under the influence of electric fields generated by atoms that take some definite space configuration specified by the symmetry of the crystal lattice, can be described by the following Schrödinger equation.

$$-\frac{\hbar^2}{2m}\nabla^2\psi(\mathbf{r}) + V(\mathbf{r})\psi(\mathbf{r}) = E\psi(\mathbf{r}) \qquad (39)$$

where $V(\mathbf{r})$ is an effective potential applied to each electron and has the property of:

$$V(\mathbf{r}+\mathbf{R}_n) = V(\mathbf{r}) \qquad (40)$$

due to the translational symmetry of a given crystal lattice. $\mathbf{R}_n$ is a lattice vector indicating the nth position of atoms in the lattice. In the Fourier representation, the potential $V(\mathbf{r})$ can be written as:

$$V(\mathbf{r}) = \sum_n V_n e^{iG_n \cdot \mathbf{r}} \qquad (41)$$

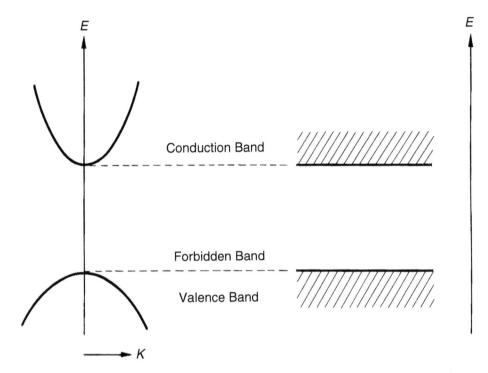

Figure 4 The typical band dispersion near the minimum band gap in a semiconductor or an insulator with a direct bandgap in the Brillouin zone.

where $\mathbf{G}_n$ is a reciprocal lattice vector. (See any elementary book of solid-state physics for the definition of $\mathbf{G}_n$)

It is difficult to solve Eq. 39 in general, but with the help of the translational and rotational symmetries inherent in the equation, it is possible to predict a general functional form of solutions. The solution was first found by Bloch and is called *Bloch's theorem*. The solution $\psi(\mathbf{r})$ should be of the form:

$$\psi(\mathbf{r}) = e^{i\mathbf{k}\cdot\mathbf{r}}u_k(\mathbf{r}) \tag{42}$$

and is called a *Bloch function*. $\mathbf{k}$ is the wave vector and $u_k(\mathbf{r})$ is the periodic function of lattice translations, such as:

$$u_k(\mathbf{r} + \mathbf{R}_n) = u_k(\mathbf{r}) \tag{43}$$

As one can see in Eq. 40, $u_k(\mathbf{r})$ can also be expanded in a Fourier series as:

$$u_k(\mathbf{r}) = \sum_n C_n(\mathbf{k})e^{iG_n\cdot\mathbf{r}} \tag{44}$$

where $C_n(\mathbf{k})$ is a Fourier coefficient. The form of the solution represented by Eq. 42 shows that the wave vectors $\mathbf{k}$ are well-defined quantum numbers of the electronic states in a given crystal. Putting Eq. 44 into Eq. 42 and using Eq. 41, one can rewrite Eq. 39 in the following form:

$$\left\{ \frac{\hbar^2}{2m} (\mathbf{k} + \mathbf{G}_l)^2 - E \right\} C_l + \sum_n C_n V_{l-n} = 0 \tag{45}$$

where E eigenvalues determined by:

$$\left| \left\{ \frac{\hbar^2}{2m} (\mathbf{k} + \mathbf{G}_l)^2 - E \right\} \delta_{G_l G_n} + V_{G_l - G_n} \right| = 0 \tag{46}$$

Henceforth, the **k**-dependence of the Fourier components $C_n(\mathbf{k})$ are neglected. These formulas are in the form of infinite dimensional determinant equations. For finite dimensions by considering amplitudes of $V_{G_l - G_n}$ in a given crystal, one can solve Eq. 46 approximately. Then the energy eigenvalues $E(\mathbf{k})$ (energy band) may be obtained as a function of wave vector **k** and the Fourier coefficients C_n.

In order to obtain qualitative interpretation of energy band and properties of a wavefunction, one can start with the 0th order approximation of Eq. 46 by taking

$$C_0 = 1, \qquad C_n = 0 \quad (n \neq 0) \tag{47}$$

in Eq. 44 or 45; this is equivalent to taking $V_n = 0$ for all n (a vanishing or constant crystal potential model). Then, Eq. 46 gives:

$$E = \frac{\hbar^2}{2m} \mathbf{k}^2 = E_0(\mathbf{k}) \tag{48}$$

This corresponds to the free electron model.

As the next approximation, consider the case that the nonvanishing components of V_n are only for $n = 0, 1$. Eq. 46 becomes:

$$\begin{vmatrix} \frac{\hbar^2}{2m} \mathbf{k}^2 - E & V_{G_1} \\ V_{-G_1} & \frac{\hbar^2}{2m} (\mathbf{k} + \mathbf{G}_1)^2 - E \end{vmatrix} = 0 \tag{49}$$

This means that, in **k**-space, the two free electrons having $E(\mathbf{k})$ and $E(\mathbf{k} + \mathbf{G})$ are in independent states in the absence of the *crystal potential* even when $\|\mathbf{k}\| = \|\mathbf{k} + \mathbf{G}\|$; this energy degeneracy is lifted under the existence of nonvanishing $V_{\mathbf{G}}$. In the above case, the eigenvalue equation can be solved easily and the solution gives

$$E = \frac{1}{2} \{ E(\mathbf{k}) + E(\mathbf{k} + \mathbf{G}) \} \pm \sqrt{ \left[\frac{E(\mathbf{k}) - E(\mathbf{k} + \mathbf{G})}{2} \right]^2 + V_{\mathbf{G}}^2 } \tag{50}$$

Figure 5 shows the global profile of E as a function of **k** in one dimension. One can see the existence of energy gap at the wave vector that satisfies:

$$\mathbf{k}^2 = (\mathbf{k} + \mathbf{G}_1)^2 \tag{51}$$

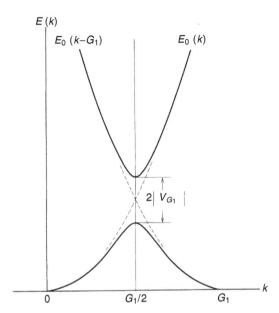

Figure 5 The emergence of a bandgap resulting from the interference between two plane waves satisfying the Bragg condition, in a one-dimensional model.

This is called the *Bragg condition*. In the three-dimensional case, the wave vectors that satisfy Eq. 51 form closed polyhedrons in **k** space and are called the 1st, 2nd, or 3rd, ..., *n*th Brillouin zone.

As stated so far, the electronic energy band structure is determined by the symmetry and Fourier amplitudes of the crystal potential $V(\mathbf{r})$. Thus, one needs to take a more realistic model of them to get a more accurate description of the electronic properties. There are now many procedures that allow for the calculation of the energy band and to get the wavefunction of electrons in crystals. Two representative methods, the *Pseudopotential method* and the *LCAO method* (Linar Combination of Atomic Orbital Method), which are frequently applied to outer-shell valence electrons in semiconductors, are briefly introduced here.

First, consider the pseudopotential method. Eq. 46 is the fundamental equation to get band structures of electrons in crystals, but the size of the determinant equation will become very large if one wishes to solve the equation with sufficient accuracy, because, in general, the Fourier components V_{G_n} do not decrease slowly due to the Coulomb potential of each atom. This corresponds to the fact that the wave functions of valence electrons are free-electron like (plane-wave like) in the intermediate region between atoms and give rapid oscillations (atomic like) near the ion cores.

Therefore, to avoid this difficulty, one can take an effective potential in which the Coulomb potential is canceled by the rapid oscillations of wavefunctions. The rapid oscillation of wavefunctions originates from the orthogonalization between atomic-like properties of wavefunctions near ion cores. It means that one introduces new wavefunctions and a weak effective potential instead of plane waves and a Coulombic potential to represent the electronic states. This effective potential gives a small number of reciprocal wave vectors (**G**) that can reproduce band structures with a corresponding small number of Fourier components. This potential is called the *pseudopotential*. The pseudopotential method necessarily results in some arbitrariness with respect to the choice of these effective potentials, depending on the selection of effective wavefunctions. It is even possible to parametrize a small number of components in V_{G_n} and to determine them empirically.

For example, taking several V_{G_n} values in high symmetry points in the Brillouin zone and, after adjusting them so as to reproduce the bandgaps obtained with experimental measurements, one calculates the band dispersion $E(\mathbf{k})$ over the entire region.

In contrast, the LCAO method approximates the Bloch states of valence electrons by using a linear combination of bound atomic wavefunctions. For example,

$$\psi_k(\mathbf{r}) = \sum_n e^{i\mathbf{k}\cdot\mathbf{R}_n}\phi(\mathbf{r}-\mathbf{R}_n) \tag{52}$$

satisfies the Bloch condition stated in Eq. 42, where $\phi(\mathbf{r})$ is one of the bound atomic wavefunctions. In order to show a simple example, assume a one-dimensional crystal consisting of atoms having one electron per atom bound in the s-orbital. The Hamiltonian of this crystal can be written as:

$$H = -\frac{\hbar^2}{2m}\nabla^2 + V(\mathbf{r}) = H_0 + \delta V(\mathbf{r}) \tag{53}$$

where H_0 is the Hamiltonian of each free atom, and $\delta V(\mathbf{r})$ is the term that represents the effect of periodic potential in the crystal. Using Eq. 53 and the wavefunctions expressed in Eq. 52, the expectation value obtained by multiplying with $\phi^*(\mathbf{r})$ yields:

$$E(\mathbf{k}) = E_0 + \frac{E_1 + \Sigma_{n\neq 0}e^{i\mathbf{k}\cdot\mathbf{R}_n}S_1(\mathbf{R}_n)}{1 + \Sigma_{n\neq 0}e^{i\mathbf{k}\cdot\mathbf{R}_n}S_0(\mathbf{R}_n)} \tag{54}$$

where E_0 is the energy level of s-orbital satisfying $H_0\phi(\mathbf{r}) = E_0\phi(\mathbf{r})$, and E_1 is the energy shift of E_0 due to δV given by $\int\phi^*(\mathbf{r})\delta V(\mathbf{r})\phi(\mathbf{r})d\mathbf{r}$. $S_0(\mathbf{R}_n)$ is called the overlap integral and is defined by:

$$S_0(\mathbf{R}_n) = \int \phi^*(\mathbf{r})\phi(\mathbf{r}-\mathbf{R}_n)d\mathbf{r} \tag{55}$$

Similarly, $S_1(\mathbf{R}_n)$ is defined as:

$$S_1(\mathbf{R}_n) = \int \phi^*(\mathbf{r})\delta V(\mathbf{r})\phi(\mathbf{r}-\mathbf{R}_n)d\mathbf{r} \tag{56}$$

Typically speaking, these quantities are regarded as parameters, and they are fitted so as to best reproduce experimentally observed results. As a matter of fact, other orbitals such as p-, d-orbitals etc. can also be used in LCAO. It is even possible to combine this method with that of pseudopotentials. As an example, Figure 6 reveals two band structure calculations due to Chadi[1]; one is for Si and the other is for GaAs.

In Figure 6, energy = 0 in the ordinate corresponds to the top of the valence band. In both Si and GaAs, it is located at the Γ point ($\mathbf{k}$ = (000) point). The bottom of the conduction band is also located at the Γ point in GaAs, while in Si it is located near the X point ($\mathbf{k}$ = (100) point).

It is difficult and rare that the energy bands can be calculated accurately all through the Brillouin zone with use of a small number of parameters determined at high symmetry points. In that sense, it is quite convenient if one has a simple perturbational method to

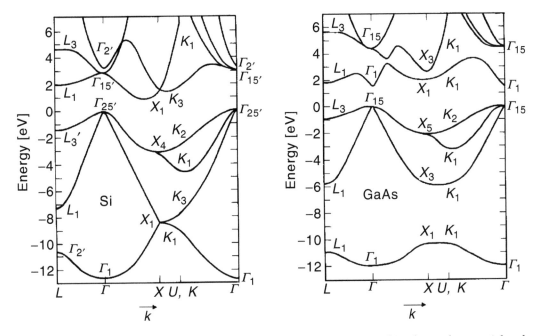

Figure 6 Calculated band structures of (a) Si and (b) GaAs using a combined pseudopotential and LCAO method. (From Chadi, D.J., Phys. Rev., B16, 3572, 1977. With permission.)

calculate band structures approximately at or near specific points in the Brillouin zone (e.g., the top of the valence band or a conduction band minimum). In particular, such procedures are quite useful when the bands are degenerate at some point in the Brillouin zone of interest.

Now, assume that the Bloch function is known at $\mathbf{k} = \mathbf{k}_0$ and is expressed as $\psi_{n\mathbf{k}_0}(\mathbf{r})$. Define a new wavefunction as:

$$\eta_{n\mathbf{k}}(\mathbf{r}) = e^{i\mathbf{k}\cdot\mathbf{r}}\psi_{n\mathbf{k}_0}(\mathbf{r}) \tag{57}$$

and expand the Bloch function in terms of $\eta_{n\mathbf{k}}(\mathbf{r})$ as:

$$\zeta_{n\mathbf{k}} = \sum_{n'} C_{n'}\eta_{n'\mathbf{k}}(\mathbf{r}) \tag{58}$$

Introducing these wavefunctions into Eq. 39 obtains the energy dispersion $E(\mathbf{k}_0 + \mathbf{k})$ in the vicinity of $\mathbf{k}_0$. In particular, near the high symmetry points of the Brillouin zone, the energy dispersion takes the following form:

$$E_n(\mathbf{k}_0 + \mathbf{k}) = E_n(\mathbf{k}_0) + \sum_{ij} \frac{\hbar^2}{2}\left(\frac{1}{m^*}\right)_{ij} k_i k_j \tag{59}$$

where $(1/m^*)_{ij}$ is called the *effective mass tensor*. From Eq. 59, the effective mass tensor is given as:

$$\left(\frac{1}{m^*}\right)_{ij} = \frac{1}{\hbar^2} \cdot \frac{\partial^2 E}{\partial k_i \partial k_j} \qquad (i, j = x, y, z) \tag{60}$$

For the isotropic case, Eq. 60 gives the scalar effective mass m^* as:

$$\frac{1}{m^*} = \frac{1}{\hbar^2} \cdot \frac{d^2 E}{dk^2} \tag{61}$$

Eq. 61 indicates that m^* is proportional to the inverse of curvature near the extremal points of the dispersion relation, E vs. $\mathbf{k}$. Furthermore, Figure 5 illustrates the two typical cases that occur near the bandgap, that is, a positive effective mass at the bottom of the conduction band and a negative effective mass at the top of the valence band, depending on the sign of d^2E/dk^2 at each extremal point. Hence, under an applied electric field $\mathbf{E}$, the specific charge e/m^* of an electron becomes negative, while it becomes positive for a hole. This is the reason why a hole looks like a particle with a positive charge.

In the actual calculation of physical properties, the following quantity is also important:

$$N(E)dE = \frac{1}{3\pi\hbar^2}\left(2m^*\right)^{3/2} E^{1/2} dE \tag{62}$$

This is called the *density of states* and represents the number of states between E and $E + dE$. We assume in Eq. 62 that space is isotropic and m^* can be used.

The band structures of semiconductors have been intensively investigated experimentally using optical absorption and/or reflection spectra. As shown in Figure 7, in many compound semiconductors (most of III-V and II-VI combination in the periodic table), conduction bands consist mainly of s-orbitals of the cation, and valence bands consist principally of p-orbitals of the anion. Many compound semiconductors have a direct bandgap, which means that the conduction band minimum and the valence band maximum are both at the Γ point ($\mathbf{k} = 0$). It should be noted that the states just near the maximum of the valence band in zinc-blende type semiconductors consist of two orbitals, namely Γ_8 which is twofold degenerate and Γ_7 without degeneracy; these originate from the *spin-orbit interaction*. It is known that the twofold degeneracy of Γ_8 is lifted in the $\mathbf{k} \neq 0$ region corresponding to a light and a heavy hole, respectively. On the other hand, in wurzite-type crystals, the valence band top is split by both the spin-orbit interaction and the crystalline field effect; the band maximum then consists of three orbitals: Γ_9, Γ_9, and Γ_7 without degeneracy. In GaP, the conduction band minimum is at the X point ($\mathbf{k} = [100]$), and this compound has an indirect bandgap, as described in the next section.

1.2.2 *Fundamental absorption, direct transition, and indirect transition*

When solid crystals are irradiated by light, various optical phenomena occur: for example, transmission, reflection, and absorption. In particular, absorption is the annihilation of light (photon) resulting from the creation of an electronic excitation or lattice excitation in crystals. Once electrons obtain energy from light, the electrons are excited to higher states. In such quantum mechanical phenomena, one can only calculate the probability of excitation. The probability depends on the distribution of microscopic energy levels of

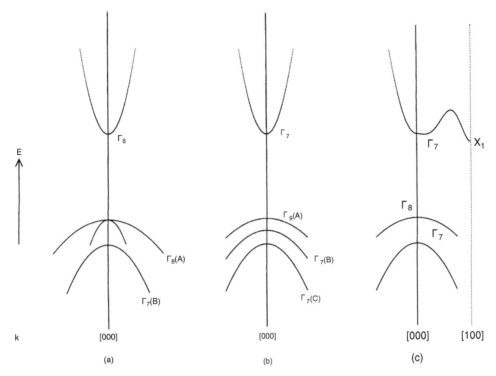

Figure 7 The typical band dispersion near Γ-point (**k** = 0) for II-VI or III-V semiconductor compounds. (a) a direct type in zinc-blende structure; (b) a direct type in wurzeite structure; and (c) an indirect type in zinc-blende structure (GaP).

electrons in that system. The excited electrons will come back to their initial states after they release the excitation energy in the form of light emission or through lattice vibrations.

Absorption of light by electrons from valence bands to conduction bands results in the fundamental absorption of the crystal. Crystals are transparent when the energy of the incident light is below the energy gaps of crystals; excitation of electrons to the conduction band becomes possible at a light energy equal to, or larger than the bandgap. The intensity of absorption can be calculated using the absorption coefficient $\alpha(h\nu)$ given by the following formula:

$$\alpha(h\nu) = A \sum p_{if} n_i n_f \qquad (63)$$

where n_i and n_f are the number density of electronic states in an initial state (occupied by electron) and in a final state (unoccupied by electron), respectively, and p_{if} is the transition probability between them.

In the calculation of Eq. 63, quantum mechanics requires that two conditions are satisfied. The first is *energy conservation* and the second is *momentum conservation*. The former means that the energy difference between the initial state and the final state should be equal to the energy of the incident photon, and the latter means that the momentum difference between the two states should be equal to the momentum of the incident light. It is quite important to note that the momentum of light is three or four orders of magnitude smaller than that of the electrons. These conditions can be written as $\left(\hbar^2/2m^*\right)k_f^2 = \left(\hbar^2/2m^*\right)k_i^2 + h\nu$ (energy conservation); $\hbar k_f = \hbar(k_i + q)$ (momentum

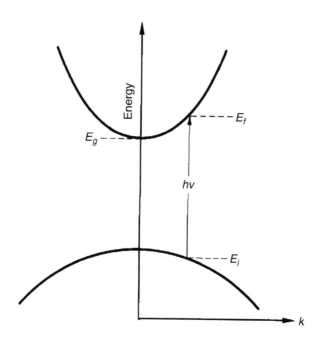

Figure 8 The optical absorption due to a direct transition from a valence band state to a conduction band state.

conservation); and $v = cq$ if one assumes a free-electron-like dispersion for band structure $E(\mathbf{k})$, where k_f and k_i are the final and initial wave vectors, respectively, c is the light velocity, and q is the photon momentum. One can neglect the momentum of absorbed photons compared to those of electrons or lattice vibrations. It results in optical transitions occurring almost vertically on the energy dispersion curve in the Brillouin zone. This rule is called the *momentum selection rule* or **k**-*selection rule*.

As shown in Figure 8, consider first the case that the minimum bandgap occurs at the top of valence band and at the bottom of conduction band; in such a case, the electrons of the valence band are excited to the conduction band with the same momentum. This case is called a *direct transition*, and the materials having this type of band structure are called *direct gap materials*. The absorption coefficient, Eq. 63, is written as:

$$\alpha(h\nu) = A^*\left(h\nu - E_g\right)^{1/2} \tag{64}$$

with the use of Eqs. 63 and 64. A^* is a constant related to the effective masses of electrons and holes. Thus, one can experimentally measure the bandgap E_g, because the absorption coefficient increases steeply from the edge of the bandgap. In actual measurements, the absorption increases exponentially because of the existence of impurities near E_g. In some materials, it can occur that the transition at $\mathbf{k} = 0$ is forbidden by some selection rule; the transition probability is then proportional to $(h\nu - E_g)$ in the $\mathbf{k} \neq 0$ region and the absorption coefficient becomes:

$$\alpha(h\nu) = A'\left(h\nu - E_g\right)^{3/2} \tag{65}$$

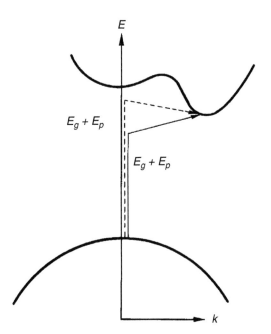

Figure 9 The optical absorption due to an indirect transition from a valence band state to a conduction band state. The momentum of electron changes due to a simultaneous absorption or emission of a phonon.

In contrast to the direct transition, in the case shown in Figure 9, both the energy and the momentum of electrons are changed in the process; excitation of this type is called an *indirect transition*. This transition corresponds to cases in which the minimum bandgap occurs between two states with different **k**-values in the Brillouin zone. In this case, conservation of momentum cannot be provided by the *photon*, and the transition necessarily must be associated with the excitation or absorption of *phonons* (lattice vibrations). This leads to a decrease in transition probability due to a higher-order stochastic process. The materials having such band structure are called *indirect gap materials*. An expression for the absorption coefficient accompanied by *phonon absorption* is:

$$\alpha(h\nu) = A\left(h\nu - E_g + E_p\right)^2 \left(\exp\left(\frac{E_p}{k_B T}\right) - 1\right)^{-1} \tag{66}$$

while the coefficient accompanied by *phonon emission* is:

$$\alpha(h\nu) = A\left(h\nu - E_g - E_p\right)^2 \left(1 - \exp\left(\frac{-E_p}{k_B T}\right)\right)^{-1} \tag{67}$$

where, in both formulas, E_p is the phonon energy.

In closing this section, the light emission process is briefly discussed. The intensity of light emission R can be written as:

$$R = B \sum p_{ul} n_u n_l \tag{68}$$

where n_u is the number density of electrons existing in upper energy states and n_l is the number density of *empty states* with lower energy. The large difference from absorption is in the fact that, usually speaking, at a given temperature electrons are found only in the vicinity of conduction band minimum and light emission is observed only from these electrons. Then, Eq. 68 can be written as:

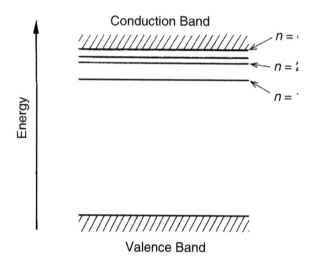

Figure 10   Energy levels of a free exciton.

$$L = B'\left(h\nu - E_g + E_p\right)^{1/2} \exp\left(-\frac{h\nu - E_g}{k_B T}\right) \tag{69}$$

confirming that emission is only observed in the vicinity of E_g. In the case of indirect transitions, light emission occurs from electronic transitions accompanied by phonon emission (cold band); light emission at higher energy corresponding to phonon absorption (hot band) has a relatively small probability since it requires the presence of thermal phonons. Hot-band emission vanishes completely at low temperatures.

1.2.3   Exciton

Although all electrons in crystals are specified by the energy band states they occupy, a characteristic excited state called the *exciton*, which is not derived from the band theory, exists in almost all semiconductors or ionic crystals. Consider the case where one electron is excited in the conduction band and a hole is left in the valence band. An attractive Coulomb potential exists between them and can result in a bound state analogous to a hydrogen atom. This configuration is called an *exciton*. The binding energy of an exciton is calculated, by analogy, to a hydrogen atom as:

$$G_{ex} = -\frac{m_r^* e^4}{32\pi^2 \hbar^2 \epsilon^2} \cdot \frac{1}{n^2} \tag{70}$$

where n (= 1, 2, 3, …) is a quantum number specifying the states, ϵ is the dielectric constant of crystals, and m_r^* is the reduced mass of an exciton.

An exciton can move freely through the crystal. The energy levels of the free exciton are shown in Figure 10. The state corresponding to the limit of $n \to \infty$ is the minimum of the conduction band, as shown in the figure. The energy of the lowest exciton state obtained by putting $n = 1$ is:

$$E_{ex}(n = 1) = E_g - |G_{ex}| \tag{71}$$

Two or three kinds of excitons can be generated, depending on the splitting of the valence band, as was shown in Figure 7. They are named, from the top of the valence band, as A- and B-excitons in zinc-blende type crystals; and A, B, and C-excitons in wurzite-type crystals. There are two kinds of A-excitons in zinc-blende materials originating from the existence of a light and heavy hole, as has already been noted.

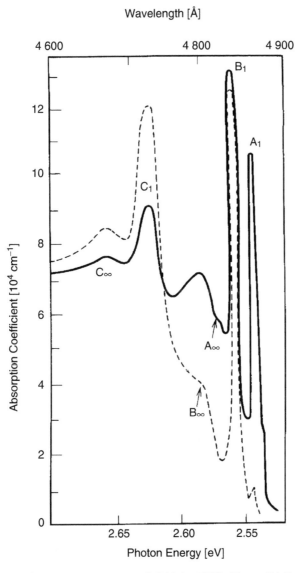

Figure 11 The exciton absorption spectrum of CdS (at 77K). The solid line and the broken line correspond to the cases that the polarization vector of incident light are parallel and perpendicular to the *c*-axis of the crystal, respectively. (From Mitsuhashi, H. and Fujishiro, Y., personal communication. With permission.)

Excitons create several sharp absorption lines in the energy region just below E_g. Figure 11 shows the absorption spectra of excitons in CdS.[2] One can easily recognize the absorption peaks due to A-, B-, and C-excitons with $n = 1$, and the beginning of the inter-band absorption transition corresponding to $n \rightarrow \infty$ (E_g). The order of magnitude of absorption coefficient reaches 10^5 cm^{-1} beyond E_g, as seen from the figure. As noted previously, the absorption coefficient in the neighborhood of E_g in a material with indirect transition, like GaP, is three to four orders of magnitude smaller than the case of direct transition.

An exciton in the $n = 1$ state of a direct-gap material can be annihilated by the recombination of the electron-hole pair; this produces a sharp emission line. The emission from the states corresponding to the larger n states is usually very weak because such states relax rapidly to the $n = 1$ state and emission generally occurs from there.

With intense excitation, excitons of very high concentrations can be produced; exci-tonic molecules (also called biexcitons) analogous to hydrogen molecules are formed from two single excitons by means of covalent binding. The exciton concentration necessary for the formation of excitonic molecules is usually of the order of magnitude of about 10^{16} cm^{-3}. The energy of the excitonic molecule is given by:

$$E_m = 2E_{ex} - G_m \tag{72}$$

where G_m is the binding energy of the molecule. The ratio of G_m to G_{ex} depends on the ratio of electron effective mass to hole effective mass, and lies in the range of 0.03 to 0.3. An excitonic molecule emits a photon of energy $E_{ex} - G_m$, leaving a single exciton behind.

If the exciton concentration is further increased by more intense excitation, the exciton system undergoes the insulator-metal transition, the so-called Mott transition, because the Coulomb force between the electron and hole in an exciton is screened by other electrons and holes. This results in the appearance of the high-density electron-hole plasma state. This state emits light with broad-band spectra.

References

1. Chadi, D.J., *Phys. Rev.*, B16, 3572, 1977.
2. Mitsuhashi, H. and Fujishiro, Y., unpublished data.

Fundamentals of luminescence

Hajime Yamamoto

Contents

1.3   Luminescence of a localized center

1.3.1   Classification of localized centers

When considering optical absorption or emission within a single ion or a group of ions in a solid, it is appropriate to treat an optical transition with a localized model rather than the band model described in Section 1.2. Actually, most phosphors have localized luminescent centers and contain a far larger variety of ions than delocalized centers. The principal localized centers can be classified by their electronic transitions as follows (below, an arrow to the right indicates optical absorption and to the left, emission):

1. $1s \rightleftarrows 2p$; an example is an F center.
2. $ns^2 \rightleftarrows nsnp$. This group includes Tl^+-type ions; i.e., Ga^+, In^+, Tl^+, Ge^{2+}, Sn^{2+}, Pb^{2+}, Sb^{3+}, Bi^{3+}, Cu^-, Ag^-, Au^-, etc.
3. $3d^{10} \rightleftarrows 3d^9 4s$. Examples are Ag^+, Cu^+, and Au^+. Acceptors in IIb-VIb compounds are not included in this group.
4. $3d^n \rightleftarrows 3d^n, 4d^n \rightleftarrows 4d^n$. The first and second row transition-metal ions form this group.
5. $4f^n \rightleftarrows 4f^n, 5f^n \rightleftarrows 5f^n$; rare-earth and actinide ions.

6. $4f^n \rightleftharpoons 4f^{n-1}5d$. Examples are Ce^{3+}, Pr^{3+}, Sm^{2+}, Eu^{2+}, Tm^{2+}, and Yb^{2+}. Only absorption transitions are observed for Tb^{3+}.
7. A charge-transfer transition or a transition between an anion p electron and an empty cation orbital. Examples are intramolecular transitions in complexes such as VO_4^{3-}, WO_4^{2-}, and MoO_4^{2-}. More specifically, typical examples are a transition from the $2p$ orbital of O^{2-} to the $3d$ orbital of V^{5+} in VO_4^{3-}, and transitions from $O^{2-}(2p)$ or $S^{2-}(3p)$ to $Yb^{3+}(4f)$. Transitions from anion p orbitals to Eu^{3+} or transition metal ions are observed only as absorption processes.
8. $\pi \rightleftharpoons \pi^*$ and $n \rightleftharpoons \pi^*$. Organic molecules having π electrons make up this group. The notation n indicates a nonbonding electron of a heteroatom in an organic molecule.

1.3.2   *Configurational coordinate model*[1-5]

1.3.2.1   *Description by a classical model*

The configurational coordinate model is often used to explain optical properties, particularly the effect of lattice vibrations, of a localized center. In this model, a luminescent ion and the ions at its nearest neighbor sites are selected for simplicity. In most cases, one can regard these ions as an isolated molecule by neglecting the effects of other distant ions. In this way, the huge number of actual vibrational modes of the lattice can be approximated by a small number or a combination of specific normal coordinates. These normal coordinates are called the *configurational coordinates*. The *configurational coordinate model* explains optical properties of a localized center on the basis of potential curves, each of which represents the total energy of the molecule in its ground or excited state as a function of the configurational coordinate (Figure 12). Here, the total energy means the sum of the electron energy and ion energy.

To understand how the configurational coordinate model is built, one is first reminded of the adiabatic potential of a diatomic molecule, in which the variable on the abscissa is simply the interatomic distance. In contrast, the adiabatic potential of a polyatomic molecule requires a multidimensional space, but it is approximated by a single configurational coordinate in the one-dimensional configurational coordinate model. In this model, the totally symmetric vibrational mode or the "breathing mode" is usually employed. Such a simple model can explain a number of facts qualitatively, such as:

1. Stokes' law; i.e., the fact that the energy of absorption is higher than that of emission in most cases. The energy difference between the two is called the Stokes' shift.
2. The widths of absorption or emission bands and their temperature dependence.
3. Thermal quenching of luminescence. It must be remarked, however, that the one-dimensional model gives only a qualitative explanation of thermal quenching. A quantitatively valid explanation can be obtained only by a multidimensional model.[6]

Following the path of the optical transition illustrated in Figure 12, presume that the bonding force between the luminescent ion and a nearest-neighbor ion is expressed by Hooke's law. The deviation from the equilibrium position of the ions is taken as the configurational coordinate denoted as Q. The total energy of the ground state, U_g, and that of the excited state, U_e, are given by the following relations.

$$U_g = K_g \frac{Q^2}{2} \qquad \text{(73a)}$$

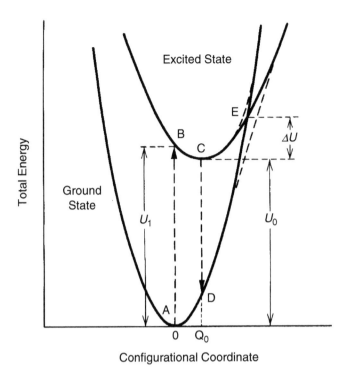

Figure 12 A schematic illustration of a configurational coordinate model. The two curves are modified by repulsion near the intersection (broken lines). The vertical broken lines A ⇌ B and C ⇌ D indicate the absorption and emission of light, respectively.

$$U_e = K_e \frac{(Q - Q_0)^2}{2} + U_0 \tag{73b}$$

where K_g and K_e are the force constants of the chemical bond, Q_0 is the interatomic distance at the equilibrium of the ground state, and U_0 is the total energy at $Q = Q_0$.

The spatial distribution of an electron orbital is different between the ground and excited states, giving rise to a difference in the electron wavefunction overlap with neighboring ions. This difference further induces a change in the equilibrium position and the force constant of the ground and excited states, and is the origin of the Stokes' shift. In the excited state, the orbital is more spread out, so that the energy of such an electron orbital depends less on the configuration coordinate; in other words, the potential curve has less curvature.

In Figure 12, optical absorption and emission processes are indicated by vertical broken arrows. As this illustration shows, the nucleus of an emitting ion stays approximately at the same position throughout the optical processes. This is called the Franck-Condon principle. This approximation is quite reasonable since an atomic nucleus is heavier than an electron by 10^3 to 10^5 times. At 0K, the optical absorption proceeds from the equilibrium position of the ground state, as indicated by the arrow A → B. The probability for an excited electron to lose energy by generating lattice vibration is 10^{12} to 10^{13} s^{-1}, while the probability for light emission is at most 10^9 s^{-1}. Consequently, state B relaxes to the equilibrium position C before it emits luminescence. This is followed by the emission process C → D and the relaxation process D → A, completing the cycle. At finite temperature, the electron state oscillates around the equilibrium position along the configurational coordinate curve up to the thermal energy of kT. The amplitude of this oscillation causes the spectral width of the absorption transition.

When two configurational coordinate curves intersect with each other as shown in Figure 12, an electron in the excited state can cross the intersection E assisted by thermal energy and can reach the ground state nonradiatively. In other words, one can assume a nonradiative relaxation process with the activation energy ΔU, and with the transition probability per unit time N given by:

$$N = s \exp \frac{-\Delta U}{kT} \tag{74}$$

where s is a product of the transition probability between the ground and excited states and a frequency, with which the excited state reaches the intersection E. This quantity s can be treated as a constant, since it is only weakly dependent on temperature. It is called the *frequency factor* and is typically of the order of 10^{13} s^{-1}.

By employing Eq. 74 and letting W be the luminescence probability, the luminescence efficiency η can be expressed as:

$$\eta = \frac{W}{W + N} = \left[1 + \frac{s}{W} \exp \frac{-\Delta U}{kT} \right]^{-1} \tag{75}$$

If the equilibrium position of the excited state C is located outside the configurational coordinate curve of the ground state, the excited state intersects the ground state in relaxing from B to C, leading to a nonradiative process.

It can be shown by quantum mechanics that the configurational coordinate curves can actually intersect each other only when the two states belong to different irreducible representations. Otherwise, the two curves behave in a repulsive way to each other, giving rise to an energy gap at the expected intersection of the potentials. It is, however, possible for either state to cross over with high probability, because the wavefunctions of the two states are admixed near the intersection. In contrast to the above case, the intersection of two configurational coordinate curves is generally allowed in a multidimensional model.

1.3.2.2   *Quantum mechanical description*

The classical description discussed above cannot satisfactorily explain observed phenomena, e.g., spectral shapes and nonradiative transition probabilities. It is thus necessary to discuss the configurational coordinate model based on quantum mechanics.

Suppose that the energy state of a localized center involved in luminescence processes is described by a wavefunction Ψ. It is a function of both electronic coordinates $\mathbf{r}$ and nuclear coordinates $\mathbf{R}$, but can be separated into the electronic part and the nuclear part by the *adiabatic approximation*:

$$\Psi_{nk}(\mathbf{r}, \mathbf{R}) = \psi_k(\mathbf{r}, \mathbf{R}) \chi_{nk}(\mathbf{R}) \tag{76}$$

where n and k are the quantum numbers indicating the energy states of the electron and the nucleus, respectively. For the nuclear wavefunction $\chi_{nk}(\mathbf{R})$, the time-independent Schrödinger equation can be written as follows:

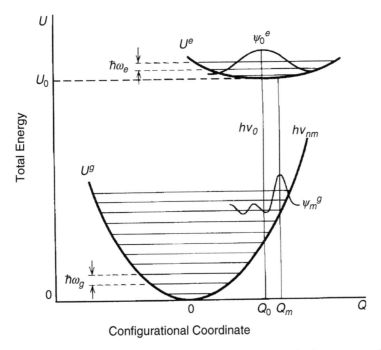

Figure 13 Discrete energy levels due to lattice vibration, each with the energy of $\hbar\omega$ and the wavefunctions ψ_0^e and ψ_m^g of harmonic oscillators representing the two states. The notation ν_0 means the frequency at the emission peak. A luminescent transition can occur at ν_{nm}.

$$\left\{-\sum_\alpha \left(\hbar^2/2M_\alpha\right)\Delta \mathbf{R}_\alpha + U_k(\mathbf{R})\right\}\chi_{nk}(\mathbf{R}) = E_{nk}\chi_{nk}(\mathbf{R}) \tag{77}$$

with α being the nuclear number, M_α the mass of the αth nucleus, $\Delta\mathbf{R}_\alpha$ the Laplacian of $\mathbf{R}_\alpha$, and E_{nk} the total energy of the localized center. The energy term $U_k(\mathbf{R})$ is composed of two parts: the energy of the electrons and the energy of the electrostatic interaction between the nuclei around the localized center. Considering Eq. 77, one finds that $U_k(\mathbf{R})$ plays the role of the potential energy of the nuclear wavefunction χ_{nk}. (Recall that the electron energy also depends on $\mathbf{R}$.) Thus, $U_k(\mathbf{R})$ is an adiabatic potential and it forms the configurational coordinate curve when one takes the coordinate Q as $\mathbf{R}$. When $U_k(\mathbf{R})$ is expanded in a Taylor series up to second order around the equilibrium position of the ground state, the potentials are expressed by Eq. 73. For a harmonic oscillation, the second term is the first nonvanishing term, while the first term is non-zero only when the equilibrium position is displaced from the original position. In the latter case, the first term is related to the Jahn-Teller effect. Sometimes, the fourth term in the expansion may also be present, signaling anharmonic effects. In the following, consider for simplicity only a single coordinate or a two-dimensional model.

Consider a harmonic oscillator in a potential shown by Eq. 73. This oscillator gives discrete energy levels inside the configurational coordinate curves, as illustrated in Figure 13.

$$E_m = \left(m + 1/2\right)\hbar\omega \tag{78}$$

where ω is the proper angular frequency of the harmonic oscillator.

The electric dipole transition probability, W_{nm}, between the two vibrational states n and m is given by:

$$W_{nm} = \left| \iint \psi_e \chi_{en}^* er\psi_g \chi_{gm} drdQ \right|^2 = \left| \int \chi_{en}^* \chi_{gm} M_{eg}(Q) dQ \right|^2 \tag{79}$$

Here,

$$M_{eg}(Q) \equiv \int \psi_e^*(r,Q) er\psi_g(r,Q) dr \tag{80}$$

When the transition is allowed, M_{eg} can be placed outside the integral, because it depends weakly on Q. This is called the Condon approximation and it makes Eq. 79 easier to understand as:

$$W_{nm} = \left| M_{eg}(Q) \right|^2 \cdot \left| \int \chi_{en}^* \chi_{gm} dQ \right|^2 \tag{81}$$

The wavefunction of a harmonic oscillator has the shape illustrated in Figure 13. For m (or n) = 0, it has a Gaussian shape; while for m (or n) ≠ 0, it has maximum amplitude at both ends and oscillates m times with a smaller amplitude between the maxima. As a consequence, the integral $\left| \int \chi_{en}^* \chi_{gm} dQ \right|$ takes the largest value along a vertical direction on the configurational coordinate model. This explains the Franck-Condon principle in terms of the shapes of wavefunctions. One can also state that this is the condition for which $W_{nm} \propto \left| \int \chi_{en}^* \chi_{gm} dQ \right|^2$ holds. The square of the overlap integral $\left| \int \chi_{en}^* \chi_{gm} dQ \right|^2$ is an important quantity that determines the strength of the optical transition and is often called the Franck-Condon factor.

1.3.3   Spectral shapes

As described above, the shape of an optical absorption or emission spectrum is decided by the Franck-Condon factor and also by the electronic population in the vibrational levels at thermal equilibrium. For the special case where both ground and excited states have the same angular frequency ω, the absorption probability can be calculated with harmonic oscillator wavefunctions in a relatively simple form:

$$W_{nm} = e^{-S} \left[\frac{m!}{n!} \right] S^{n-m} \left[L_m^{n-m}(S) \right]^2 \tag{82}$$

Here $L_\beta^\alpha(z)$ are Laguerre's polynomial functions. The quantity S can be expressed as shown below, with K being the force constant of a harmonic oscillator and Q_0 the coordinate of the equilibrium position of the excited state.

$$S = \frac{1}{2} \frac{K}{\hbar\omega} (Q - Q_0)^2 \tag{83}$$

As can be seen in Figure 14, S is the number of emitted phonons accompanying the optical transition. It is commonly used as a measure of electron-phonon interaction and is called the *Huang-Rhys-Pekar factor*. At 0K or $m = 0$, the transition probability is given by the simple relation:

$$W_{no} = S^n \frac{e^{-S}}{n!} \tag{84}$$

A plot of W_{n0} against n gives an absorption spectrum consisting of many sharp lines. This result is for a very special case, but it is a convenient tool to demonstrate how a spectrum varies as a function of the intensity of electron-phonon interaction or the displacement of the equilibrium position in the excited state. The results calculated for $S = 20$ and 2.0 are shown in Figures 14 (a) and (b),[7] respectively. The peak is located at $n \cong S$. For $S \cong 0$ or weak electron-phonon interaction, the spectrum consists only of a single line at $n = 0$. This line (a zero-phonon line) becomes prominent when S is relatively small. For luminescence, transitions accompanied by phonon emission show up on the low-energy side of the zero-phonon line in contrast to absorption shown in Figure 14(b). If the energy of the phonon, $\hbar\omega$, is equal both for the ground and excited states, the absorption and emission spectra form a mirror image about the zero-phonon line. Typical examples of this case are the spectra of YPO:$_4$Ce^{3+} shown in Figure 15,[8] and that of ZnTe:O shown in Figure 16.[9]

Examples of other S values are described. For the A emission of KCl:Tl$^+$ having a very broad band width, S for the ground state is found to be 67, while for the corresponding A absorption band, S of the excited state is about 41.[10] Meanwhile, in Al$_2$O$_3$:Cr^{3+} (ruby), $S = 3$ for the narrow $^4A_2 \rightarrow {}^4T_2$ absorption band, and $S \approx 10^{-1}$ for the sharp R lines ($^4A_2 \leftrightarrow {}^4T_2$) were reported.[11] A very small value similar to that of R lines is expected for sharp lines due to $4f^n$ intraconfigurational transitions. The spectra of YPO$_4$:Ce^{3+} in Figure 15, which is due to $4f \leftrightarrow 5d$ transition, show $S \approx 1$.[8]

The above discussion has treated the ideal case of a transition between a pair of vibrational levels (gm) and (en) resulting in a single line. The fact is, however, that each line has a finite width even at 0K as a result of zero-point vibration.

Next, consider a spectral shape at finite temperature T. In this case, many vibrational levels at thermal equilibrium can act as the initial state, each level contributing to the transition with a probability proportional to its population density. The total transition probability is the sum of such weighted probabilities from these vibrational levels. At sufficiently high temperature, one can treat the final state classically and assume the wavefunction of the final state is a δ-function and the population density of the vibrational levels obeys a Boltzmann distribution. By this approximation, the absorption spectrum has a Gaussian shape given by:

$$W(\hbar\omega) = \frac{1}{\sqrt{2\pi}\sigma_a} \exp\left[\frac{-(\hbar\omega - U_1)^2}{2\sigma_a^2}\right] \tag{85}$$

Here,

$$U_1 \equiv U_0 + \frac{K_e}{2Q_0^2} \tag{86}$$

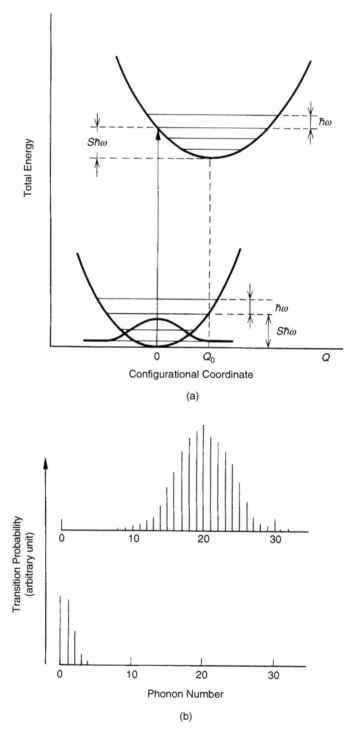

(a)

(b)

Figure 14 (b) shows the spectral shape calculated for the configurational coordinate model, in which the vibrational frequency is identical in the ground and excited states shown in (a). The upper figure in (b) shows a result for $S = 20$, while the lower figure is for $S = 2.0$. The ordinate shows the number of phonons n accompanying the optical transition. The transition for $n = 0$ is the zero-phonon line.

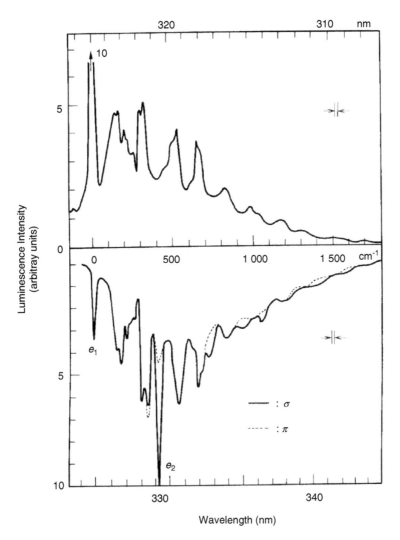

Figure 15 Optical spectra of $5d \rightleftarrows 4f(^2F_{3/2})$ transition of Ce^{3+} doped in a YPO_4 single crystal. The upper figure is an excitation spectrum, with the lower luminescence spectrum at 4.2K. The two spectra are positioned symmetrically on both sides of the zero-phonon line at 325.0 nm. Vibronic lines are observed for both spectra. The notations π and σ indicate that the polarization of luminescence is parallel or perpendicular to the crystal c-axis, respectively.

$$\{\sigma_a(T)\}^2 \equiv S_e \frac{(\hbar\omega_e)^3}{\hbar\omega_g} \coth \frac{\hbar\omega_g}{2kT} \tag{87}$$

$$\approx 2S_e \cdot kT \cdot \frac{(\hbar\omega_e)^3}{(\hbar\omega_g)^2} \tag{88}$$

where $\hbar\omega$ is the energy of an absorbed phonon, and S_e denotes S of the excited state. The coefficient on the right-hand side of Eq. 85 is a normalization factor defined to give $\int W(\hbar\omega)d\omega = 1$. By defining w as the spectral width, which satisfies the condition $W(U_1 + w) = W(U_1)/e$, one finds:

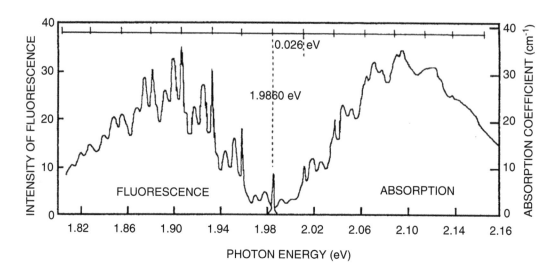

Figure 16 Absorption and luminescence spectra of ZnTe:O at 20K. (From Merz, J.L., *Phys. Rev.*, 176, 961, 1968. With permission.)

$$w = \sqrt{2\sigma_a} \qquad\qquad\qquad (89)$$

At sufficiently high temperature, the spectral width w is proportional to $\sqrt{T}$ and the peak height is inversely proportional to $\sqrt{T}$. The relations for the luminescence process are found simply by exchanging the suffixes e and g of the above equations.

In experiments, a Gaussian shape is most commonly observed. It appears, however, only when certain conditions are satisfied, as is evident from the above discussion. In fact, more complicated spectral shapes are also observed. A well-known example is the structured band shape of a transition observed for Tl+-type ions in alkali halides.[6] It has been shown that this shape is induced by the Jahn-Teller effect and can be described by a configurational coordinate model based on six vibrational modes around a Tl+-type ion. Another example is the asymmetric luminescence band of $Zn_2SiO_4{:}Mn^{2+}$. To explain this shape, a configurational coordinate model with a small difference between the excited- and ground-state potential minima ($S = 1.2$) has been proposed.[12]

In summarizing the discussion of the spectral shape based on the configurational coordinate model, one can review the experimental results on luminescence bandwidths. In Figure 17,[13] the halfwidth of the luminescence band of typical activators in phosphors is plotted against the peak wavelength.[13] The activators are classified by the type of optical transition described in the Section 1.3.1. When the $d \rightleftarrows d$ (Mn^{2+}), $f \rightleftarrows d$ (Eu^{2+}), and $s^2 \rightleftarrows sp$ transitions (Sn^{2+}, Pb^{2+}, and Sb^{3+}) are sequentially compared, one finds that the halfwidth increases in the same order. This is apparently because the overlap of the electron wavefunctions between the excited and ground states increases in the above order. The difference in the wavefunction overlap increases the shift of the equilibrium position of the excited state, Q_0, and consequently the Stokes' shift and the halfwidth increase as well.

Weak electron-phonon interactions give line spectra. The line width in this case results from factors other than those involved in the configurational coordinate model. Such factors are briefly reviewed below.

1.3.3.1 Line broadening by time-dependent perturbation

The most fundamental origin of the line width is the energy fluctuation of the initial and final states of an optical transition caused by the uncertainty principle. With τ being the

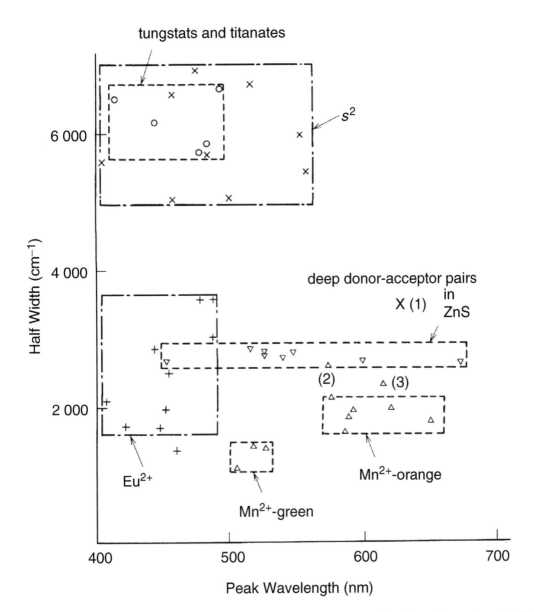

Figure 17 A plot of peak wavelength and half-width of various phosphors. The points (1)-(3) indicate the following materials. The luminescence of (2) and (3) originates from Mn^{2+} principally. (1) $(Sr,Mg)_3(PO_4)_2:Sn^{2+}$; (2) $Sr_5(PO_4)_3F:Sb^{3+},Mn^{2+}$ (3) $CaSiO_3:Pb^{2+}, Mn^{2+}$. (From Narita, K., Tech. Digest Phosphor Res. Soc. 196th Meeting, 1983 (in Japanese). With permission.)

harmonic mean of the lifetimes of the initial and final states, the spectral line width is given by $\hbar/\tau$ and the spectral shape takes a Lorentzian form:

$$I(\nu) = \frac{1}{\pi} \cdot \frac{1/\nu_L}{1+(\nu-\nu_0)^2/\nu_L^2} \qquad (90)$$

where $\nu_L \equiv (\tau_i^{-1} + \tau_f^{-1})/4\pi c$, ν is the frequency of light, ν_0 the frequency at the line center, and τ_i and τ_f are the lifetimes of the initial and final states, respectively.

In addition to the spectral width given by Eq. 89, there are other kinds of time-dependent perturbation contributing to the width. They are absorption and emission of a photon, which makes "the natural width," and absorption and emission of phonons. The fluorescent lifetime of a transition-metal ion or a rare-earth ion is of the order of 10^{-6} s at the shortest, which corresponds to 10^{-6} cm^{-1} in spectral width. This is much sharper than the actually observed widths of about 10 cm^{-1}; the latter arise from other sources, as discussed below.

At high temperatures, a significant contribution to the width is the Raman scattering of phonons. This process does not have any effect on the lifetime, but does make a Lorentzian contribution to the width. The spectral width due to the Raman scattering of phonons, ΔE, depends strongly on temperature, as can be seen below:

$$\Delta E = \alpha \left(\frac{T}{T_D}\right)^7 \int_0^{X_0} \frac{x^6 e^x}{\left(e^x - 1\right)^2} dx, \quad X_0 = \frac{\hbar \omega_a}{KT} \tag{91}$$

where T_D is Debye temperature and α is a constant that includes the scattering probability of phonons.

1.3.3.2   Line broadening by time-independent perturbation

When the crystal field around a fluorescent ion has statistical distribution, it produces a Gaussian spectral shape.

$$I(\nu) = \frac{1}{\sqrt{2\pi\sigma}} \exp\left\{-\frac{(\nu - \nu_0)}{2\sigma^2}\right\} \tag{92}$$

with σ being the standard deviation.

Line broadening by an inhomogeneously distributed crystal field is called *inhomogeneous broadening*, while the processes described in Section 1.3.3.1 result in *homogeneous broadening*.

1.3.4   Nonradiative transitions

The classical theory describes a *nonradiative transition* as a process in which an excited state relaxes to the ground state by crossing over the intersection of the configurational coordinate curve through thermal excitation or other means (refer to Section 1.3.2). It is often observed, however, that the experimentally determined activation energy of a non-radiative process depends upon temperature.

This problem has a quantum mechanical explanation: that is, an optical transition accompanied by absorption or emission of $m - n$ phonons can take place when an nth vibrational level of the excited state and an mth vibrational level of the ground state are located at the same energy. The probability of such a transition is also proportional to a product of the Franck-Condon coefficient and thermal distribution of population in the ground state, giving the required temperature-dependent probability. When the phonon energy is the same both at the ground and excited states, as shown in Figure 14, the nonradiative relaxation probability is given by:

$$N_p = N_{eg} \cdot \exp\left\{-S(2\langle n \rangle + 1)\right\} \sum_{j=0}^{\infty} \frac{\left(S\langle n \rangle\right)^j \left\{S(1 + \langle n \rangle)\right\}^{p+j}}{j!(p+j)!} \tag{93}$$

Here, let $p \equiv m - n$, and $\langle n \rangle$ denotes the mean number of the vibrational quanta n at temperature T expressed by $\langle n \rangle = \{\exp(\hbar\omega/kT) - 1\}^{-1}$. The notation N_{eg} implies the overlap integral of the electron wavefunctions.

The temperature dependence of N_p is implicitly included in $\langle n \rangle$. Obviously, Eq. 93 does not have a form characterized by a single activation energy. If written in a form such as $N_p \propto \exp(-E_p/kT)$, one obtains:

$$E_p = \left(\langle n \rangle_p - \langle n \rangle\right)\hbar\omega \tag{94}$$

where $\langle n \rangle_p \hbar\omega$ is the mean energy of the excited state subject to the nonradiative process. The energy E_p increases with temperature and one obtains $E_p < \Delta U$ at sufficiently low temperature.

If $S < 1/4$ or if electron-phonon interaction is small enough, Eq. 93 can be simplified by leaving only the term for $j = 0$.

$$N_p = N_{eg} \cdot \exp\{-S(1 + 2\langle n \rangle)\}\{-S(1 + \langle n \rangle)\}^p / p! \tag{95}$$

In a material that shows line spectra, such as rare-earth ions, the dominating nonradiative relaxation process is due to multiphonon emission. If E_{gap} is the energy separation between two levels, the nonradiative relaxation probability between these levels is given by an equation derived by Kiel:[25]

$$N_p = A_K \epsilon^p \left(1 + \langle n \rangle\right)^p \tag{96}$$

$$p\hbar\omega = E_{gap} \tag{97}$$

where A_K is a rate constant and ϵ is a coupling constant.

Eq. 95 can be transformed to the same form as Eq. 96 using the conditions $S \approx 0$, $\exp\{-S(1 + 2\langle n \rangle)\} \approx 1$, $S^p/p! \approx \epsilon^p$ and $A_K = N_{eg}$, although Eq. 95 was derived independently of the configurational coordinate model. If two configurational coordinate curves have the same curvature and the same equilibrium position, the curves will never cross and there is no relaxation process by thermal activation between the two in the framework of the classical theory. However, thermal quenching of luminescence can be explained for such a case by taking phonon-emission relaxation into account, as predicted by Kiel's equation.

References

1. Klick, C.C. and Schulman, J.H., *Solid State Physics*, Vol. 5, Seitz, F. and Turnbull, D., Eds., Academic Press, 1957, pp. 97-116.
2. Curie, D., *Luminescence in Crystals*, Methuen & Co., 1963, pp. 31-68.
3. Maeda, K., *Luminescence*, Maki Shoten, 1963, pp. 6-10 and 37-48 (in Japanese).
4. DiBartolo, B., *Optical Interactions in Solids*, John Wiley & Sons, 1968, pp. 420-427.
5. Kamimura, A., Sugano, S., and Tanabe, Y., *Ligand Field Theory and Its Applications*, First Edition, Shokabo, 1969, pp. 269-321 (in Japanese).
6. Fukuda, A., *Bussei*, 4, 13, 1969 (in Japanese).
7. Keil, T., *Phys. Rev.*, 140, A601, 1965.

8. Nakazawa, E. and Shionoya, S., *J. Phys. Soc. Jpn.*, 36, 504, 1974.
9. Merz, J.L., *Phys. Rev.*, 176, 961, 1968.
10. Williams, F.E., *J. Chem. Phys.*, 19, 457, 1951.
11. Fonger, W.H. and Struck, C.W., *Phys. Rev.*, B111, 3251, 1975.
12. Klick, C.C. and Schulman, J.H., J. Opt. Soc. Am., 42, 910, 1952.
13. Narita, K., *Tech. Digest Phosphor Res. Soc. 196th Meeting*, 1983 (in Japanese).
14. Struck, C.W. and Fonger, W.H., *J. Luminesc.*, B111, 3251, 1975.
15. Kiel, A., *Third Int. Conf. Quantum Electronics*, Paris, Grivet, P. and Bloembergen, N., Eds., Columbia University Press, p. 765, 1964.

Fundamentals of luminescence

Sumiaki Ibuki

Contents

1.4   Impurities and luminescence in semiconductors

1.4.1   Impurities in semiconductors

As is well known, when semiconductors are doped with impurities, the lattices of the semiconductors are distorted and the energy level structures of the semiconductors are also affected. For example, when in Si an As atom (Group V) is substituted for a Si atom (Group IV), one electron in the outermost electronic orbit in the N shell of the As atom is easily released and moves freely in the Si lattice, because the number of electrons in the N shell of As (5) is one more than that in the M shell of Si (4). Thus, impurities that supply electrons to be freed easily are called *donors*.

On the contrary, when a Ga atom (Group III) is substituted for a Si atom, one electron is attracted from a Si atom, forming a hole that moves freely in the Si lattice; this is because the number of electrons in the N shell of Ga (3) is one less than that in the M shell of Si (4). Thus, impurities that supply free holes easily are called *acceptors*. In compound semiconductors, it is easily understood in a similar way what kinds of impurities play the role of donors and acceptors.

Usually, in compound semiconductors such as ZnS and GaAs, the stoichiometry does not hold strictly. Therefore, when more positive ions exit, negative ion vacancies are created and work as donors. Similarly, when more negative ions exit, positive ion vacancies work as acceptors.

In a donor, one excess electron orbits around the positively charged nucleus, as in a hydrogen atom. This electron moves around in a semiconductor crystal (which usually has a large dielectric constant) so that the Coulomb interaction between the nucleus and

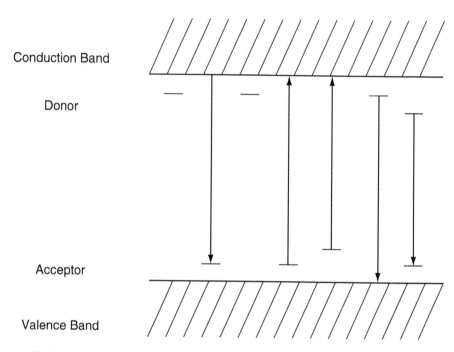

Figure 18 Shallow impurity levels in a semiconductor.

the electron is weakened. The radius of the electron orbit becomes large under these conditions and the electron is greatly affected by the periodic potential of the crystal. For example, when the effective mass of the electron is 0.5 and the dielectric constant of the crystal is 20, the Bohr radius of the electron becomes 40 times larger than that in the hydrogen atom. Therefore, the excess electron of the donor can be released from its binding to the nucleus by an excitation of small energy. This means that the donor level is located very close to the bottom of the conduction band, as shown in Figure 18. Similarly, the acceptor level is located very close to the top of the valence bond. Impurity levels with small ionization energies are called *shallow* impurity levels. Other impurity levels can also be located at *deep* positions in the forbidden land.

Light absorption takes place between the valence band and impurity levels, or between impurity levels and the conduction band. When a large quantity of impurities exists, the band shape can be observed in absorption.

Luminescence takes place through these impurity levels with wavelengths longer than the bandgap wavelength. When the dopant impurity is changed, the luminescent wavelength and efficiency also change. It is usually found that in *n*-type semiconductors, luminescence between the conduction band and acceptor levels is strong; whereas in *p*-type semiconductors, luminescence between donor levels and the valence band is strong.

1.4.2   *Luminescence of excitons bound to impurities*

The number of impurities included in semiconductors is of the order of magnitude of 10^{14} to 10^{16} cm^{-3}, even in so-called pure semiconductors. Therefore, excitons moving in a crystal are generally captured by these impurities and *bound exciton* states are created. Luminescence from such bound excitons is, in ordinary crystals, stronger than that from *free excitons*. Excitons bound to donors or acceptors create H_2 molecule-type complexes. Those bound to ionized donors or acceptors create H_2^+ molecular ion-type complexes. Binding energies of excitons in these complexes depend on the effective mass ratio of electron to hole, and

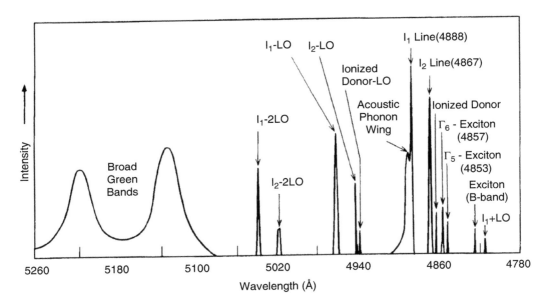

Figure 19 Luminescence spectrum near the band gap of CdS (1.2K). (From Litton, C.W., Reynolds, D.C., Collins, T.C., and Park, Y.S., *Phys. Rev. Lett.*, 25, 1619, 1970. With permission.)

are about 0.1 to 0.3 of ionization energies of the donor or acceptor impurities. The radiative recombination of bound excitons takes place efficiently with energy less than that of free excitons. The halfwidths of luminescence lines are very narrow.

As an example, a luminescence spectrum of CdS, a II-VI compound of the *direct transition type*, near the band edge is shown in Figure 19.[1] In the figure, the I_1, I_2, and I_3 lines correspond to the luminescence of excitons bound to neutral acceptors, neutral donors, and ionized donors, respectively. They were identified by measurements of their Zeeman effect. The binding energies of these bound excitons are 19, 8, and 5 meV, respectively. The halfwidths of the luminescence line are very narrow, about 2–3 cm^{-1}, and are much less than those of the free exciton lines shown as Γ_5 and Γ_6 excitons. In II-VI compounds like CdS, excitons couple strongly with the longitudinal optical (LO) phonons that generate a polarized electric field. As a result, exciton luminescence lines accompanied by simultaneous emission of one, two, or more LO phonons are observed strongly, as shown in the figure.

The oscillator strength of the I_2 bound exciton was obtained from the area of the absorption spectrum and found to be very large, about 9.[2] The oscillator strength of the free exciton is 3×10^{-3}, so that of the bound exciton is enhanced by ~10^3. This enhancement effect is called the giant oscillator strength effect. From a theoretical point of view, the ratio of the oscillator strength of the bound exciton to the free exciton is given by the ratio of the volume in which the bound exciton moves around, to that of the unit cell. In CdS, this ratio is ~10^3, so that the very large value observed for the I_2 bound exciton is reasonable. This value gives a calculated lifetime of 0.4 ns for I_2. The lifetimes of excitons are determined from luminescence decay measurement.[3] For the I_2 bound exciton, a value of 0.5 ± 0.1 ns was obtained, which agrees well with the calculated value. This also indicates that the luminescence quantum efficiency of the I_2 bound exciton is close to 1.

In the case of *indirect transition-type* semiconductors, on the other hand, the luminescence efficiency of bound excitons is very low. A typical example is the case of S donors in GaP. The luminescence quantum efficiency has been estimated to be $1/(700 \pm 200)$.[4] The reason for the low efficiency is ascribed to the Auger effect. The state in which an exciton

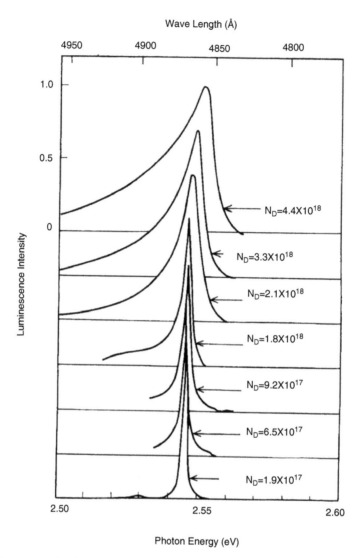

Figure 20 Changes of luminescence spectra of excitons bound to Cl donors in CdS (1.8K) with the Cl concentration. N_D:Cl donor concentration (cm^{-3}). (From Kukimoto, H., Shionoya, S., Toyotomi, S., and Morigaki, K., J. Phys. Soc. Japan, 28, 110, 1970. With permission.)

is bound to a neutral donor includes two electrons and one hole, so when one electron and one hole recombine, the recombination energy does not result in light emission, but is instead transferred to the remaining electron to raise it into the conduction band.

Next, the effect of high concentrations of impurities on the bound exciton luminescence is discussed. As an example, consider the case of an I_2 bound exciton in CdS:Cl as shown in Figure 20.[5] With increasing Cl donor concentration N_D, the spectral width broadens. Beyond $N_D \sim 2 \times 10^{18}$ cm^{-3}, the emission peak shifts toward the high-energy side with further increases of N_D. Simultaneously, the spectral width broadens more and the shape becomes asymmetric, having long tails toward the low-energy side.

These facts can be interpreted theoretically.[6] At higher N_D, an exciton bound to a donor collides with other donors. Donor electrons can thus be virtually excited and can exert the screening effect on the bound excitons through changes of the dielectric constant. This brings about the high-energy shift of the emission peak. The asymmetry of the spectral shape with long tails is interpreted as being caused by the Stark effect due to ionized impurities, i.e., compensated donors and acceptors.

1.4.3 Luminescence of isoelectronic traps

In semiconductor crystals, if an isoelectronic element, (i.e., an element belonging to the same column in the periodic table as a constituent element) is substituted for a constituent element, either a free electron or a hole in the semiconductor is attracted to the isoelectronic element. This is because of the differences in electronegativity between the isoelectronic element and the mother element. Such isoelectronic elements are called *isoelectronic traps*.

When an electron is trapped in an isoelectonic trap, a hole is attracted to the top trap by the Coulomb force, and an exciton bound to an isoelectronic trap is created. This state produces luminescence that is quite different from that due to an exciton bound to a donor or acceptor. In such cases, an electron or hole is attracted to the donor or acceptor by a long-range Coulomb force. On the other hand, the isoelectric trap attracts an electron or a hole by the short-range type force that comes from the difference in the electronegativity. Therefore, the wavefunctions of the electron or hole trapped at the isoelectronic trap is very much localized in real space and, instead, is greatly extended in **k**-space. This plays an important role in the case of indirect transition-type semiconductors.

Figure 21 shows the wavefunction of the electron bound to an N isoelectronic trap in GaP.[7] The bottom of the conduction band of GaP is located at the X point in **k**-space, and the electron has a relatively large amplitude, even at the Γ point. Therefore, the electron can recombine with a hole at the Γ point with a high probability for conditions applicable to direct transitions. The emission spectrum is shown in Figure 22.[8] The recombination probability is 100 times larger than that of an exciton bound to a neutral S donor, for which only the indirect transition is possible. Moreover, in the GaP:N system, there is no third particle (electron or hole), so the Auger nonradiative recombination does not occur, and the recombination probability is actually close to 1. When the concentration of N traps is high, luminescence of an exciton strongly bound to a pair of N traps closely located to each other is also observed at a slightly longer wavelength. Other isoelectric traps in GaP, Zn-O, and Cd-O centers, in which two elements are located in the nearest neighbor sites, are known. These centers also produce efficient luminescence, as do isoelectric traps in direct transition-type semiconductors, of which CdS:Te[9] and ZnTe:O[10] have been identified.

1.4.4 Luminescence of donor-acceptor pairs

When the wavefunction of an electron trapped at a donor overlaps to some extent with the wavefunction of a hole located at an acceptor, both particles can recombine radiatively. The luminescence thus produced has some interesting characteristics because the electron and the hole in this pair are located in lattice sites apart from each other. As explained below, the luminescence wavelength and probability will depend on the electron-hole distance in a pair.

As shown in Figure 23, at the start of luminescence, the electron is located at the donor D and the hole at the acceptor A. The energy of this initial stage is expressed, taking the origin of the energy axis to be the acceptor level A, as $E_i = E_g - (E_D + E_A)$, where E_g, E_D, and E_A are the bandgap energy, ionization energy of a neutral donor, and that of a neutral acceptor, respectively. After the recombination, a positive effective charge is left in the donor and a negative effective charge in the acceptor. The final state is determined by the Coulomb interaction between them, giving the final state energy to be $E_f = -e^2/4\pi\varepsilon r$, where ε is the static dielectric constant of the crystal, and r is the

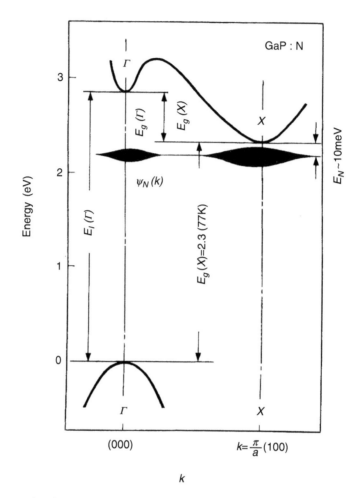

Figure 21 Energy level and wavefunction of N isoelectronic trap in GaP in the **k**-space. (From Holonyak, N., Campbell, J.C., Lee, M.H., et al., *J. Appl. Phys.*, 44, 5517, 1973. With permission.)

distance between the donor and acceptor in the pair. Therefore, the recombination energy E_r is given by:

$$E_r = E_i - E_f$$
$$= E_g - \left(E_D + E_A\right) + e^2/4\pi\varepsilon r \tag{98}$$

In this formula, r takes discrete values. For smaller r values, each D-A pair emission line should be separated, so that a series of sharp emission lines should be observed. For larger r values, on the other hand, intervals among each emission line are small, so that they will not be resolved and a broad emission band will be observed.

The transition probability should be proportional to the square of the overlap of the electron and hole wavefunctions. Usually, the wavefunction of a donor electron is more widely spread than that of an acceptor hole. The electron wavefunction of a hydrogen-like donor is assumed to decrease exponentially with r. Therefore, the transition probability W(r) is expressed as:

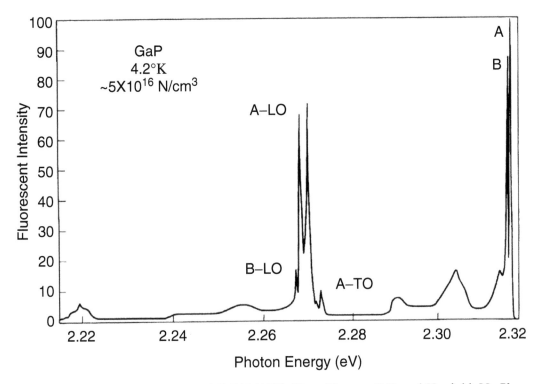

Figure 22 Luminescence spectrum of GaP:N (4.2K). (From Thomas, D.G. and Hopfield, J.J., *Phys. Rev.*, 150, 680, 1966. With permission.)

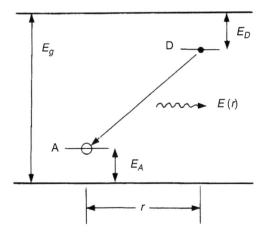

Figure 23 Energy levels of a donor-acceptor pair.

$$W(r) = W_0 \exp(-2r/r_B), \tag{99}$$

where r_B is the Bohr radius of the donor electron and W_0 is a constant related to the D-A pairs.

As a typical example of D-A pair luminescence, a spectrum of S donor and Si acceptor pairs in GaP is shown in Figure 24.[11] Both S and Si substitute for P. The P site, in other words the site of one of the two elements constituting GaP, composes a face-centered cubic lattice. In this lattice, r is given by $\{(1/2)m\}^{1/2}a$, where m is the shell number and a is the lattice constant. For the shell numbers m = 1, 2, ... 12, 13, 15, 16, ..., there exists atoms;

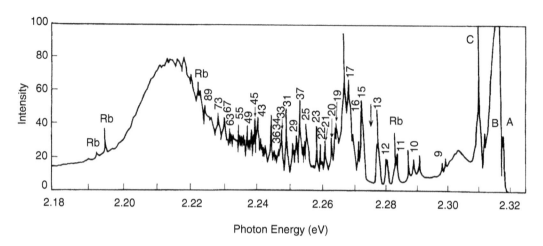

Figure 24 Luminescence spectrum of D-A pairs in GaP:Si,S (1.6K). (From Thomas, D.G., Gershenzon, M., and Trumbore, F.A., *Phys. Rev.*, 133, A269, 1964. With permission.)

but for m = 14, 30, 46, …, atoms do not exist. Assuming that the position of each emission line is given by Eq. 98 with r given in this way and $E_D + E_A = 0.14$ eV ($E_g = 2.35$ eV), it is possible to determine the shell number for each line, as shown in Figure 24. As expected, lines for m = 14, 30, … do not appear as seen in the figure. Agreement between experiment and theory is surprisingly good.

As understood from Eqs. 98 and 99, the smaller the r value is, the shorter the luminescence wavelength emitted and the higher the transition probability becomes; in other words, the shorter the decay time. Therefore, if one observes a time-resolved emission spectrum for a broad band composed of many unresolved pair lines, the emission peak of the broad band should shift to longer wavelengths with the lapse of time. The broad band peaking at 2.21 eV in Figure 24 is the ensemble of many unresolved pair lines. Figure 25[12] shows time-resolved luminescence spectra of this band. It is clearly seen that the peak shifts to longer wavelengths with time, as expected. Similar time shifts in D-A pair luminescence have been observed in II-VI compounds such as ZnSe and CdS. (See 2.7.)

1.4.5   Deep levels

As the final stage of this section, luminescence and related phenomena caused by deep levels in semiconductors are discussed. Certain defects and impurities create deep localized levels with large ionization energies. In these deep levels, electron-lattice interactions are generally strong, so that the nonradiative recombination takes place via these levels, thus lowering the luminescence efficiencies of emitting centers. Further, these deep centers sometimes move and multiply by themselves in crystals, and cause the deterioration of luminescence devices because of the local heating by multiphonon emission.

Changes of the states of deep levels caused by photoexcitation are studied from measurements of conductivity, capacitance, and magnetic properties. In this way, the structure, density, position of energy levels, and capture and release probabilities for carriers have been determine for various deep levels. Calculations of binding energies of deep levels using wavefunctions of the conduction and valence bands have also been performed. In this way, binding energies of O in GaP and GaAs and those of Ga and As vacancies in GaAs are obtained. Calculations are further made for complex defects including O, for example, a complex of O and Si or Ge vacancy, and atoms occupying antisites.[13]

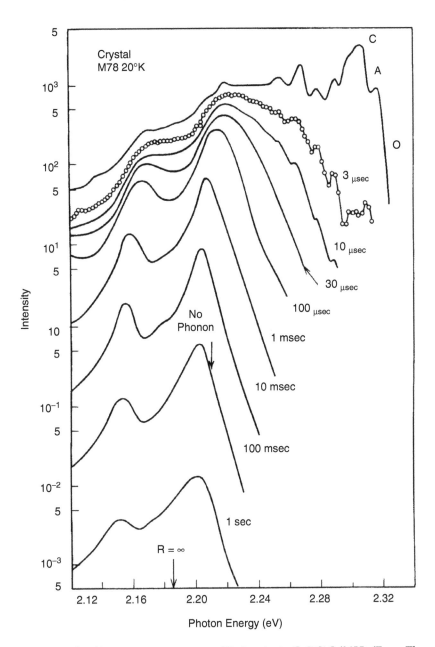

Figure 25 Time-resolved luminescence spectra of D-A pairs in GaP:Si,S (20K). (From Thomas, D.G., Hopfield, J.J., and Augustniak, W.M., *Phys. Rev.*, 140, A202, 1965. With permission.)

Transition metals incorporated in semiconductors usually create deep levels and exhibit luminescence. Since electron-lattice interactions are strong, broad-band spectra with relatively weak zero-phonon lines are usually observed. Figure 26 shows luminescence spectra of Cr^{3+} in $GaAs$[14] as an example. Coupling with phonons results in the phonon sidebands shown in Figure 26.

As for the nonradiative recombination through defects, not only the Auger recombination process but also many phonon emission process are observed. The transition probabilities of the latter increase when related levels are deep and crystal temperatures are high. In certain cases, the energy level of a localized trap is shallow before trapping

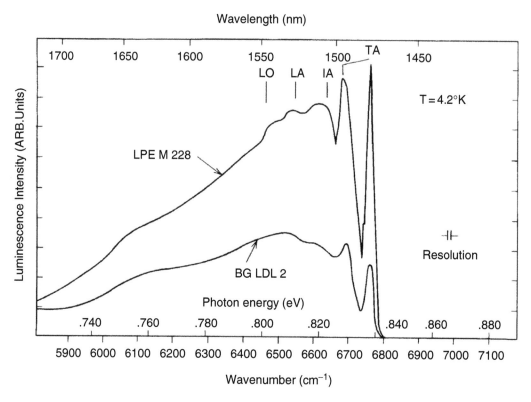

Figure 26 Luminescence spectrum of Cr^{3+} in GaAs (4.2K). (From Stocker, H.J. and Schmidt, M., *J. Appl. Phys.*, 47, 2450, 1976. With permission.)

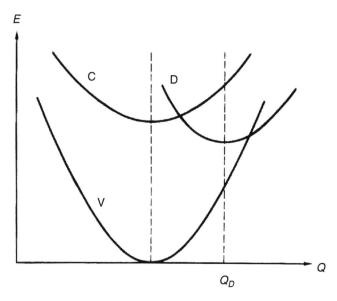

Figure 27 Configurational coordinate model of deep defect level. (C: conduction band, V: valence band, D: deep defect.)

an electron; but after trapping, lattice relaxation and the rearrangement of surrounding atoms take place and the energy level is made deep, as shown by the configurational coordinate model (1.3.2) in Figure 27. In this state, the difference between the optical activation energy and thermal activation energy is large, and nonradiative recombination through the emission of many phonons occurs with high probability.[15]

References

1. Litton, C.W., Reynolds, D.C., Collins, T.C., and Park, Y.S., *Phys. Rev. Lett.*, 25, 1619, 1970.
2. Thomas, D.G. and Hopfield, J.J., *Phys. Rev.*, 175, 1021, 1968.
3. Henry, C.H. and Nassau, K., *Phys. Rev.*, B1, 1628, 1970.
4. Nelson, D.F., Cuthbert, J.D., Dean, P.J., and Thomas, D.G., *Phys. Rev. Lett.*, 17, 1262, 1966.
5. Kukimoto, H., Shionoya, S., Toyotomi, S., and Morigaki, K., *J. Phys. Soc. Jpn.*, 28, 110, 1970.
6. Hanamura, E., *J. Phys. Soc. Jpn.*, 28, 120, 1970.
7. Holonyak, Jr., N., Campbell, J.C., Lee, M.H., Verdeyen, J.T., Johnson, W.L., Craford, M.G., and Finn, D., *J. Appl. Phys.*, 5517, 1973.
8. Thomas, D.G. and Hopfield, J.J., *Phys. Rev.*, 150, 680, 1966.
9. Aften, A.C. and Haaustra, J.H., *Phys. Lett.*, 11, 97, 1964.
10. Merz, J.L., *Phys. Rev.*, 176, 961, 1968.
11. Thomas, D.G., Gershenzon, M., and Trumbore, F.A., *Phys. Rev.*, 133, A269, 1964.
12. Thomas, D.G., Hopfield, J.J., and Augustyniak, W.M., *Phys. Rev.*, 140, A202, 1965.
13. Alt, H.Ch., *Materials Science Forum*, 143-147, 283, 1994.
14. Stocker, H.J. and Schmidt, M., *J. Appl. Phys.*, 47, 2450, 1976.
15. Kukimoto, H., *Solid State Phys.*, 17, 79, 1982 (in Japanese).

chapter one — section five

Fundamentals of luminescence

Chihaya Adachi and Tetsuo Tsutsui

Contents

1.5   Luminescence of organic compounds

1.5.1   Origin of luminescence in organic compounds

The luminescence of organic compounds is essentially based on localized π-electron systems within individual organic molecules[1]. This is in clear contrast to inorganic phosphors where luminescence is determined by their lattice structures, and thus their luminescence is altered or disappears altogether when the crystals melt or decompose. In organic luminescent compounds, in contrast, it is the π-electron systems of individual molecules that are responsible for luminescence. Therefore, even when organic crystals melt into amorphous aggregates, luminescence still persists. Further, when molecules are in vapor phase or in solution, they basically demonstrate similar luminescence spectrum as in solid films.

Luminescence from organic compounds can be classified into two categories: luminescence from electronically excited singlet (S_1) or triplet (T_1) states. Emission from singlet excited states, called "fluorescence," is commonly observed in conventional organic compounds. Emission from triplet excited states, called "phosphorescence," is rarely observed in conventional organic compounds at ambient temperatures due to the small radiative decay rate of phosphorescence.

Electronically excited states of organic compounds are easily produced not only via photoexcitation but also by other excitation methods (such as chemical reactions, electrochemical reactions, mechanical forces, heat, and electric charge recombination) capable of producing electronically excited states in organic molecules, as depicted in Figure 28.

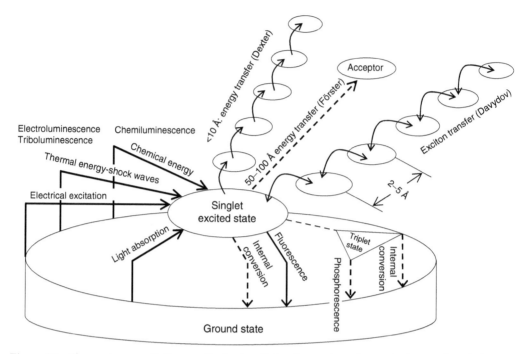

Figure 28 The various excitation methods, i.e., light absorption, thermal, chemical and charged-particle, and decay processes, i.e., photoluminescence, thermal deactivation, and energy transfer and migration, in organic molecules.

It should be emphasized that any kind of luminescence in organic compounds is due to well-defined electronically neutral singlet or triplet excited states in the organic molecules, even though luminescence can be produced by a variety of excitation methods having different names like photoluminescence, chemiluminescence, electrochemiluminescence, triboluminescence, thermoluminescence, and electroluminescence. In addition to radiative decay, the excited molecules also decay nonradiatively through thermal deactivation and energy transfer and migration.

1.5.2 Electronically excited states of organic molecules and their photoluminescence

Electronic transitions in organic molecules are described by the molecular orbitals of σ-electrons and π-electrons. Each molecular orbital can accept two electrons with antiparallel spins according to Pauli's exclusion principle, and both σ and π-electrons participate in chemical bonding. Here, π-electrons demonstrate a variety of photo- and electronic activities compared with σ-electrons, since σ-electrons become located at deeper energy levels compared with those of π-electrons (Figure 29). The ground state is characterized by the π-electrons in the highest occupied molecular orbital (HOMO). In order to produce an electronically excited state, a molecule must absorb energy equal to or greater than the energy difference between the HOMO and the lowest unoccupied molecular orbital (LUMO) levels

$$\Delta E = E_{LUMO} - E_{HOMO}. \tag{100}$$

With absorption of energy, an electron is promoted from HOMO to LUMO, and this constitutes an electronic transition from the ground state (S_0) to an electronically excited state (S_1). Here, the energy level diagrams (Jablonski diagram) for molecular orbitals for the ground

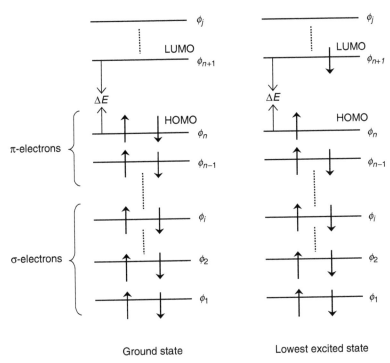

Figure 29 Energy level diagrams of molecular orbitals and electron configurations for ground and singlet excited states.

and excited states are commonly used for the description of electronic transitions in organic molecules (Figure 30). In Figure 30, the transition of an electron from HOMO to LUMO is expressed in terms of its spin states. The electronic transitions are expressed in terms of the difference in energy between the ground and excited states in the electronic-state diagram. The spin multiplicity of the states is implicitly indicated by the notations of S (singlet) or T (triplet). The ground state, S_0, and lowest singlet and triplet states, S_1 and T_1, are composed of multiple vibrational states, due to vibronic and rotation energy levels of the molecules.

When an energy larger than the HOMO–LUMO energy difference is absorbed by a molecule, either higher vibronic states within the S_1 states or higher singlet excited states S_2 and S_3 are produced. The higher vibronic states of S_1 relax to the lowest vibronic state of S_1 within a timescale of ~10^{-12} s. The higher energy singlet states such as S_2 and S_3 relax to the S_1 state via nonradiative, internal conversion (IC) processes. Triplet excited states are usually produced via an intersystem crossing (ISC) process from $S_1 \rightarrow T_1$, since the transition probability of direct excitation from S_0 into T_1 is very small. Also, the higher energy triplet states such as T_2 and T_3 relax to the T_1 state via nonradiative processes. Thus, radiative transitions take place as an electronic transition from the lowest excited states of S_1 or T_1 to the ground state S_0. The radiative transition from $S_1 \rightarrow S_0$ is classified as a spin-allowed transition and therefore the timescale of the transition is of the order of ~10^{-9} s. On the other hand, the timescale of the $T_1 \rightarrow S_0$ transition is much longer, ranging from micro- to milliseconds because the process is intrinsically spin-forbidden. The emission spectra of organic molecules often exhibit a vibronic structure because the ground state also contains vibronic and rotational fine structures.

Figure 31 shows schematically the relation between absorption and emission spectra. An emission spectrum looks like the mirror image of the electronic absorption spectrum of a molecule due to the presence of vibrational levels in each energy level. The emission wavelength for the radiative transition from the lowest S_1 state to the lowest S_0 state, the 0–0

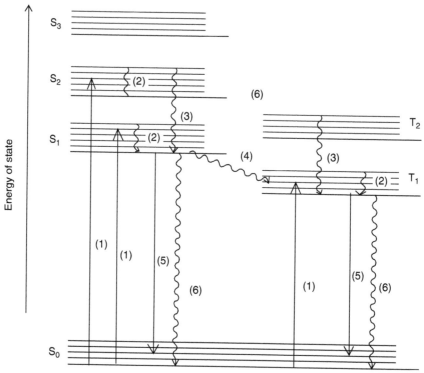

Figure 30 The Jablonski diagram, which explains photophysical processes in molecular systems: (1) light absorption, (2) vibrational relaxation, (3) internal conversion (IC), (4) intersystem crossing (ISC), (5) radiative transition, and (6) nonradiative transition.

emission, corresponds to the wavelength for the electronic transition from the lowest S_0 state to the lowest S_1 state, the 0–0 absorption. However, in the actual absorption and emission spectra, the peaks of the 0–0 transitions do not coincide with each other because of Stokes' shift.

1.5.3 Fluorescence of organic molecules in a solid state

Fluorescence in organic solids is essentially the same as that from the individual molecules of the solid. This is because molecular orbitals assumed for isolated molecules are only weakly perturbed in the solid state by the presence of weak van der Waals interactions among the molecules. However, one has to note that drastic changes of fluorescence can appear in solids because of the formation of intermolecular complexes and due to energy migration and transfer among the molecules.

Intermolecular complexes that are formed in their excited states, called excimers and exciplexes, give characteristic emissions at wavelengths different from those of the component molecules. An excimer is an excited-state complex formed between two same molecules. Aromatic hydrocarbons such as anthracene and perylene, for example, show a characteristic broad, featureless excimer fluorescence. On the other hand, exciplexes are excited-state complexes made of two different molecules. Further, charge–transfer interactions between donor and acceptor molecules sometimes cause the formation of CT-complexes, which have broad and redshifted weak fluorescence spectra. Such intermolecular complexes formed in their ground states also show changes in their absorption spectra. The absorption spectra of molecular aggregates with parallel arrangements are shifted significantly to shorter wavelengths compared with those of isolated component molecules, and their fluorescence spectra are also different. Molecular aggregates called

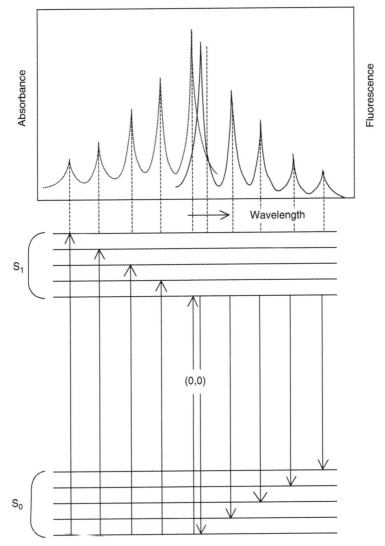

Figure 31 Explanation for the relationship between absorption and emission spectra based on the energy state diagram.

J-aggregates in which molecules are arranged in a head-to-tail fashion, in particular, show very sharp, redshifted absorption and fluorescence.

Very rapid and efficient migration or transfer of excitation energy among molecules occurs in a solid state, and these mechanisms induce drastic changes in fluorescence spectra. Anthracene crystals doped with a trace amount of tetracene, for example, never give characteristic anthracene emission, but yield the tetracene emission spectrum instead, although the absorption spectrum still looks the same as that of anthracene. The excited states produced in anthracene molecules by photoexcitation are efficiently transferred from anthracene to tetracene molecules. Therefore, the effects of trace amounts of impurities incorporated in organic solids should be carefully considered. In some cases, impurities act as effective exciton-quenching centers, reducing the intrinsically high fluorescence yields of organic solids. Here, we make a distinction between energy migration and energy transfer. Energy migration implies coherent energy transfer between like molecules, where energy is transferred as an exciton. Therefore, it is the dominant mechanism in single crystals where identical molecules are regularly aligned. On the other hand, energy transfer involving dipole–dipole interaction via the near-field electromagnetic interaction, i.e., the Förster mechanism, dominantly occurs in donor–acceptor molecular combinations where the emission spectrum of the donor and the absorption spectrum of the acceptor overlap. Therefore, the Förster mechanism occurs between singlet–singlet energy transfers where spin conservation is maintained. Further, the Dexter mechanism involving direct electron exchange between adjacent molecules dominates in the case of triplet–triplet energy transfer.

1.5.4 Quantum yield of fluorescence

The fluorescence "quantum yield" is given as the ratio of the emitted and absorbed photons and is normally determined experimentally through careful photoluminescence measurements using an integrated sphere. A more general definition of quantum yield of fluorescence is the ratio of the radiative transition rate, k_r, and total (radiative and non-radiative) transition rate, $k_r + k_{nr}$, from a singlet excited state to the ground state.

$$\phi_f = \frac{k_r}{k_r + k_{nr}} \tag{101}$$

Here, nonradiative processes include the direct radiation-less transition from S_1 to S_0 and the ISC from S_1 to T_1. In a solid state, energy migration and energy transfer processes also need to be included. The experimentally determined quantum efficiency values of isolated molecules are not always useful for the evaluation of the efficiency of organic solids, because molecular aggregation occurs. For example, coumarins are laser dyes with high quantum yields of ~90% in dilute solutions, but they yield weak fluorescence in bulk solid states. Deactivation processes intrinsic to solid states, called concentration-quenching processes, occur in many compounds, although some aromatic hydrocarbons such as anthracene and pyrene show high quantum yields both as single molecules and crystals. The quantum yields of fluorescence are strongly dependent on temperature because thermal nonradiative decay processes are highly dependent on temperature. At temperatures lower than the liquid nitrogen temperature, almost all rotational and vibrational motions of pendant groups are frozen and quantum yields tend to be high.

1.5.5 Organic fluorescent and phosphorescence compounds with high quantum yields

Organic molecules with well-developed π-conjugated systems usually show intense electronic absorption in the ultraviolet to visible wavelength regions due to a transition from the bonding π^* orbital into the antibonding π^* orbital (i.e., a transition from S_0 to S_1 states). The absorption maxima shift to longer wavelengths as the length of the π-conjugated systems increases. Incorporation of heteroatoms such as nitrogen, oxygen, and sulfur within the π-conjugated systems usually causes a redshift in the absorption peaks. Attachment of electron-donating groups such as $-NH_2$, $-CH_3$, and $-OCH_3$ or electron-accepting groups such as $-CN$ and $-NO_2$ also causes redshifts in the absorption spectra. Organic compounds with strong absorptions in the near ultraviolet to visible regions usually exhibit visible fluorescence. The highest bonding π orbital and the lowest antibonding π^* orbital mainly govern the quantum yields of fluorescence, as well as do the emission peak wavelengths. However, it should be noted that nonbonding π orbitals play a role in determining quantum yields in some cases.[2] Organic fluorescent molecules can be classified into the following categories:[3]

1. Aromatic hydrocarbons and their derivatives. This category includes polyphenyl hydrocarbons, hydrocarbons with fused aromatic nuclei, and hydrocarbons with arylethylene and arylacetylene groups

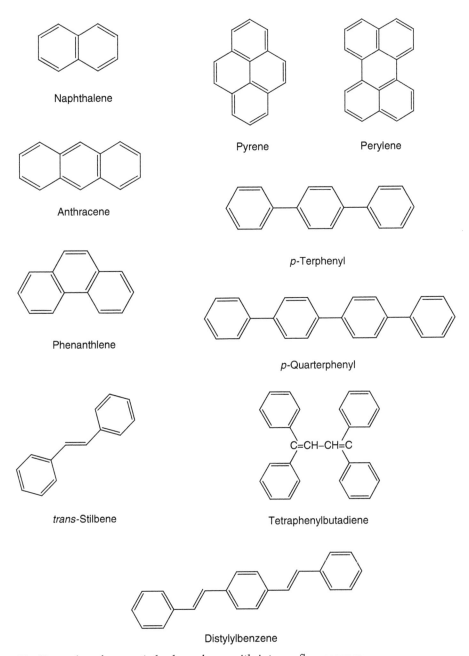

Figure 32 Examples of aromatic hydrocarbons with intense fluorescence.

2. Compounds with five- and six-membered heterocycles
3. Compounds with carbonyl groups
4. Complexes of metals with organic ligands

Aromatic hydrocarbons form one of the most important groups among fluorescent compounds, which emit light in the violet to blue regions, and their quantum yields are fairly high even in crystalline states as well as in solutions[4]. Introduction of substituents can shift the fluorescence toward longer wavelengths. Naphthalene, anthracene, phenan-thlene, pyrene, perylene, *p*-terphenyl, and *p*-quarterphenyl are typical examples of this class of compounds (Figure 32). Some polyphenyl hydrocarbons and hydrocarbons with fused aromatic nuclei show poor solubility in conventional organic solvents, which may limit their practical applications. Polyphenyl hydrocarbons with arylethylene and arylacetylene groups, such as *trans*-stilbenes, tetraphenylbutadiene, and distylylbenzenes, show very similar optical characteristics with polyphenyl hydrocarbons. Syntheses of these molecules are relatively easy, and therefore, these compounds are useful in various applications.

One of the most popular classes of fluorescent compounds comprises derivatives of five- and six-membered heterocycles. Intense fluorescence is observed, simply when heterocycles are incorporated in developed π-conjugated systems. Almost all the compounds of this group show this property, both in solution and in the crystalline state, with spectral ranges spanning the violet to the red. 2,5-Diphynyloxazole, 2-pheny-5-(4-biphenyl)-1,3,4-oxadiazole, and 1,3,5-triphenyl-2-pyrazoline, shown in Figure 32, are typical fluorescent compounds with five-membered heterocycles.

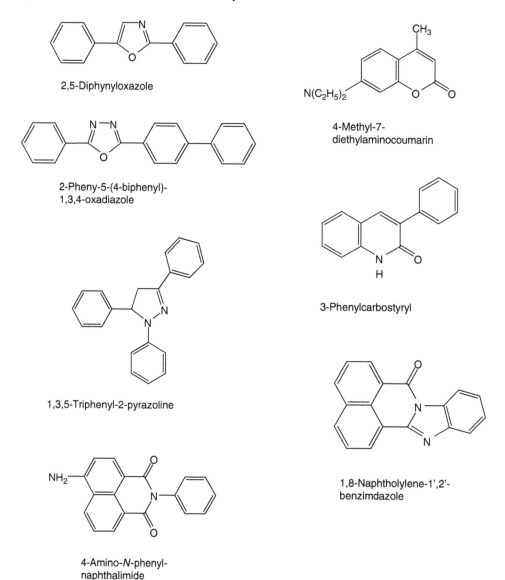

2,5-Diphynyloxazole

4-Methyl-7-
diethylaminocoumarin

2-Pheny-5-(4-biphenyl)-
1,3,4-oxadiazole

3-Phenylcarbostyryl

1,3,5-Triphenyl-2-pyrazoline

1,8-Naphtholylene-1',2'-
benzimdazole

4-Amino-*N*-phenyl-
naphthalimide

Figure 33 Examples of fluorescence molecules containing heteroatoms.

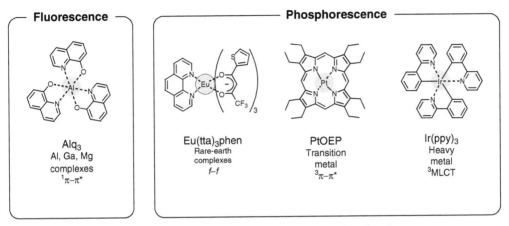

Figure 34 Examples of metal complexes demonstrating intense photoluminescence.

It is more difficult to describe the fluorescence behavior of aromatic compounds containing carbonyl groups, because a delicate balance exists between the π–π^* transition of π-conjugated systems, and the π–π^* transitions related to the carbonyl groups govern the transition from singlet excited states to the ground states. One finds a variety of useful compounds with intense fluorescence among carbonyl-containing molecules, for example, coumarins, carbostyryls, naphthalimides, and naphtholylene benzimdazoles. Typical carbonyl-containing compounds are depicted in Figure 33.

Metal complexes with organic ligands are another important class of fluorescent compounds (Figure 34).[5] Some of these compounds exhibit rather broad fluorescence spectra similar to those of organic ligands. One such compound is (8-hydroxyquinolino)aluminum, which has been used for organic thin-film electroluminescent devices. Metal complexes, such as some europium and terbium complexes, that exhibit narrowband luminescence specific to incorporated metal ions are also known. Further, heavy metal complexes such as Ir, Pt, Ru, and Au complexes were recently revealed to have intense phosphorescence based on the transition of the metal-to-ligand charge transfer (MLCT) complex triplet state. Since the MLCT has a nature of mixing singlet and triplet excited states, radiative decays are allowed, leading to intense phosphorescence. In particular, Ir complexes have been widely developed in recent years, and iridium tris(phenypyridine), in particular, shows almost 100% phosphorescence efficiency.[6]

References

1. Pope, M. and Swenberg, C.E., *Electronic Processes in Organic Crystals*, Oxford University Press, New York, 1982, chap. 1.
2. Krasovitskii, B.M. and Bolotin, B.M., *Organic Luminescent Materials*, VCH Publishers, New York, 1988, chap. 7.
3. Krasovitskii, B.M. and Bolotin, B.M., *Organic Luminescent Materials*, VCH Publishers, New York, 1988, chap. 18.
4. Becker, R.S., *Theory and Interpretation of Fluorescence and Phosphorescence*, John Wiley & Sons, New York, 1969, chap. 13.
5. Yersin, H., Transition Metal and Rare Earth Compounds: Excited States, Transitions, Interactions I (Topics in Current Chemistry), Springer Verlag.
6. C. Adachi, Baldo, M.A., and Forrest, S.R., Nearly 100% internal quantum efficiency in an organic light-emitting device, *J. Appl. Phys.*, 90, 5048, 2001.

chapter one — section six

Fundamentals of luminescence

Yasuaki Masumoto

1.6 Luminescence of low-dimensional systems

Low-dimensional systems discussed in this section are two-dimensional (2D) systems, one-dimensional (1D) systems, and zero-dimensional (0D) systems. 2D systems include layered materials and quantum wells; 1D systems include linear chain-like materials and quantum wires; and 0D systems include small microcrystallites and quantum dots. Optical properties of low-dimensional systems are substantially different from those of three-dimensional (3D) systems. The most remarkable modification comes from different distributions of energy levels and densities of states originating from the spatial confinement of electrons and holes.

Different distributions of energy levels in low-dimensional systems arise from the quantum confinement of electrons and holes. The simplest model for 2D systems is that of a particle in a box with an infinitely deep well potential, as is shown in Figure 35. The wavefunctions and energy levels in the well are known from basic quantum mechanics and are described by:

$$\psi_n(z) = \left(2/L_z\right)^{1/2} \cos\left(n\pi z/L_z\right) \tag{102}$$

$$E_n = \frac{\hbar^2}{2m}\left(\frac{\pi n}{L_z}\right)^2, \qquad n = 1,\, 2,\, 3,\, \dots \tag{103}$$

In type-I semiconductor quantum wells, both electrons and holes are confined in the same wells. The energy levels for electrons and holes are described by:

$$E_e = E_g + \frac{\hbar^2}{2m_e^{*}}\left(\frac{\pi n_e}{L_z}\right)^2 + \frac{\hbar^2}{2m_e^{*}}\left(k_x^2 + k_y^2\right) \tag{104}$$

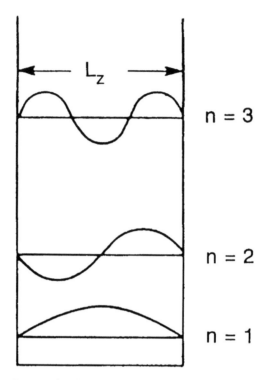

Figure 35 A particle in a box made of infinitely tall potential barriers.

$$E_h + \frac{\hbar^2}{2m_h^*}\left(\frac{\pi n_h}{L_z}\right)^2 + \frac{\hbar^2}{2m_h^*}\left(k_x^2 + k_y^2\right) \tag{105}$$

where m_e^* and m_h^* are the effective masses of electron and hole, respectively.

If electric dipole transitions are allowed from the valence band to the conduction band, the optical transition occurs from the state described by n_h, k_x, and k_y to the state described by n_e, k_x, and k_y. Therefore, the optical transition takes place at an energy:

$$E = E_e + E_h = E_g + \frac{\hbar^2}{2\mu}\left(\frac{\pi n}{L_z}\right)^2 + \frac{\hbar^2}{2\mu}\left(k_x^2 + k_y^2\right) \tag{106}$$

where μ is the reduced mass given by $\mu^{-1} = m_e^{*-1} + m_h^{*-1}$

It is well known that the joint density of states ρ_{3D} for the 3D for an allowed and direct transition in semiconductors is represented by:

$$\rho_{3D}(E) = \frac{1}{2\pi^2}\left(\frac{2\mu}{\hbar^2}\right)^{3/2}\left(E - E_g\right)^{1/2} \tag{107}$$

Here, E_g is the bandgap energy and μ is the reduced mass as above. The joint densities of states for 2D, 1D, and 0D systems are given, respectively, by the expressions

$$\rho_{2D}(E) = \frac{\mu}{\pi\hbar^2}\sum_n \theta\left(E - E_n - E_g\right) \tag{108}$$

(1)

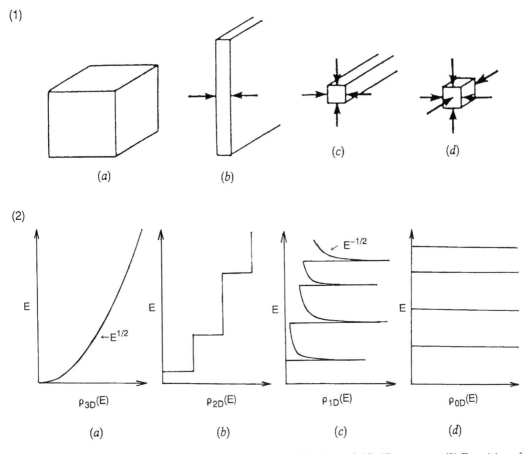

(2)

Figure 36 (1) Schematic illustrations of (a) 3D, (b) 2D, (c) 1D, and (d) 0D systems. (2) Densities of states for (a) 3D, (b) 2D, (c) 1D, and (d) 0D systems are shown.

$$\rho_{1D}(E) = \frac{(2\mu)^{1/2}}{\pi\hbar} \sum_{m,n} \frac{1}{\left(E - E_m - E_n - E_g\right)^{1/2}} \tag{109}$$

$$\rho_{0D}(E) = 2 \sum_{l,m,n} \delta\left(E - E_l - E_m - E_n - E_g\right) \tag{110}$$

where θ is a step function and δ is a delta function. The sum of quantum confinement energies of electrons and holes are represented by E_l, E_m, and E_n, where E_l, E_m, and E_n refer to the three directions of spatial confinement. Figure 36 shows schematically the joint densities of states for 3D, 2D, 1D, and 0D systems given by Eqs. 107-110.

The optical absorption spectrum $\alpha(E)$ is proportional to the joint density of states, if the energy dependence of the optical matrix element and the other slowly varying energy dependence are neglected. As a result, the absorption spectral shapes of 3D, 2D, 1D, and 0D systems are well described by the joint density of states as shown in Figure 36.

Although the exciton effect has been neglected thus far, it dominates the absorption spectrum around the bandgap. The exciton is a composite of an electron and a hole due to the Coulomb attraction. As in the hydrogen atom, the Coulomb attraction forms bound levels of the exciton. The lowest-energy bound state is characterized by the effective

Rydberg energy Ry^* and the effective Bohr radius a_B^*. The lowest exciton state is lower than the unbound continuum state by Ry^*, and its radius is given by a_B^*. In the 3D case, the effective Rydberg energy and the effective Bohr radius are given by:

$$Ry^* = \frac{\mu}{\varepsilon^2 m_0} Ry \tag{111}$$

$$a_B^* = \frac{\varepsilon m_0}{\mu} a_B \tag{112}$$

where ε is the dielectric constant, m_0 is the electron mass, and $Ry = 13.6$ eV and $a_B = 52.9$ pm are the Rydberg energy and Bohr radius of the hydrogen atom, respectively. The exciton energy levels are described by:

$$E_n = E_g - \frac{Ry^*}{n^2}, \qquad (n = 1,\ 2,\ 3,\ ...) \tag{113}$$

and the absorption spectrum is modified as shown in Figure 37.[1]

   In the 2D case, the binding energy of the lowest-energy exciton is enhanced to be $4Ry^*$, because the exciton energy levels are described by:

$$E_{n,m} = E_g + E_n - \frac{Ry^*}{\left(m + \dfrac{1}{2}\right)^2}, \qquad (m = 0,\ 1,\ 2,\ ...) \tag{114}$$

where n is the quantum number for electrons and holes, and m is the quantum number for their relative motion. The wavefunction of a 2D exciton shrinks in the 2D plane and its radius becomes $(\sqrt{3/4})\, a_B^*$. This means that the overlap between the electron wavefunction and hole wavefunction is enhanced compared with the 3D case. As a result, the oscillator strength of a 2D exciton is larger than that of a 3D exciton. The oscillator strength of a 2D n^{th} exciton per unit layer f_n^{2D} is written as:

$$f_n^{2D} = \frac{n^3}{\left(n + \dfrac{1}{2}\right)^3}\, a_B^* f_n^{3D} \tag{115}$$

where f_n^{3D} is the oscillator strength of the nth 3D exciton. The enhancement of the exciton binding energy and the oscillator strength lead to the stability of the 2D exciton at room temperature. Figure 38 shows an example of the observation of a 2D exciton in the optical absorption spectrum of GaAs quantum wells at room temperature.[2]

   The binding energy and the oscillator strength of an exciton increase with a decrease in size and dimension.[3] Figure 39 shows the increase of the binding energy of 2D, 1D, and 0D excitons with the decrease in size and dimensionality. Here, the 1D exciton is confined in a square parallelepiped and the 0D exciton is confined in a cube, where the side-length of the square or the cube is L. Since the radiative lifetime is inversely proportional to the oscillator strength, it decreases with a decrease in size and dimensionality. Shortening of the radiative lifetime of the exciton with decreasing size is observed in GaAs quantum wells.[4]

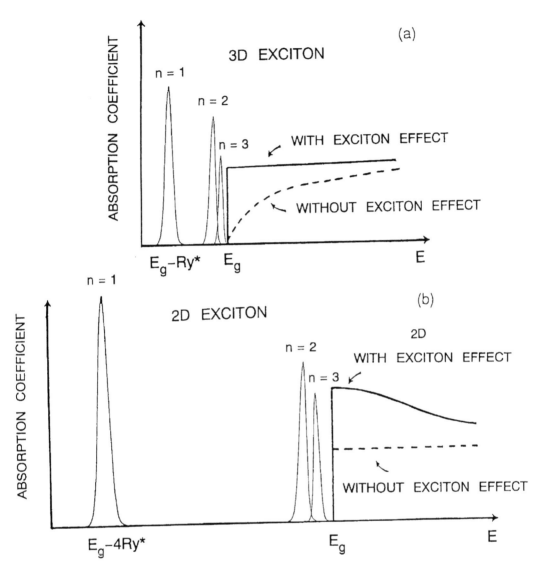

Figure 37   Absorption spectra for (a) 3D or (b) 2D excitons. Above the energy gap E_g, the absorption coefficient is enhanced from its value by the Sommerfeld factor, as a result of the Coulomb interaction between electrons and holds. (From Weisbuch, C. and Vinter, B., *Quantum Semiconductor Structures*, Academic Press, Boston, 1991. With permission.)

Discussion thus far has focused on the optical properties of low-dimensional systems, with special emphasis on semiconductor quantum wells. However, in phosphors, the more important low-dimensional systems are small semiconductor microcrystallites and quantum dots. Many kinds of nanometer-size microcrystallites made by various means behave as quantum dots. For example, microcrystallites of II-VI and I-VII compounds can be chemically grown in polymers, solutions, and zeolites. Porous Si, made by means of electrochemical etching, is regarded as an ensemble of quantum dots and quantum wires. III-V semiconductors GaAs and GaInAs microcrystallites can be epitaxially grown on the oriented GaAs crystal surface.

The above-mentioned, nanometer-size semiconductor microcrystallites can be regarded as quantum dots in the sense that they show the quantum size effect. That is with a decrease in size, the absorption bands show a blue-shift due to this effect, because

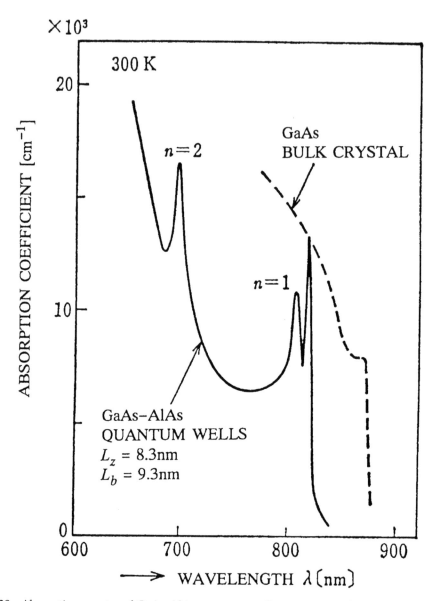

Figure 38 Absorption spectra of GaAs-AlAs quantum well structures and a bulk GaAs crystal at 300K. (From Ishibashi, T., Tarucha, S., and Okamoto, H., *Inst. Phys. Conf. Ser. No. 63*, 1982, chap. 12, 587. With permission.)

the spatial confinement of electrons, holes, and excitons increases the kinetic energy of these particles. Simultaneously, the same spatial confinement increases the Coulomb interaction between electrons and holes. The quantum confinement effect can be classified into three categories[5]: the weak confinement, the intermediate confinement, and the strong confinement regimes, depending on the relative size of the radius of microcrystallites R compared to an electron a_e^*, a hole a_h^*, and an exciton Bohr radius a_B^*, respectively. Here, the microcrystallites are assumed to be spheres, and a_e^* and a_h^* are defined, respectively, by:

$$a_e^* = \frac{\hbar^2 \varepsilon}{m_e^* e^2} \quad \text{and} \quad a_h^* = \frac{\hbar^2 \varepsilon}{m_h^* e^2} \tag{116}$$

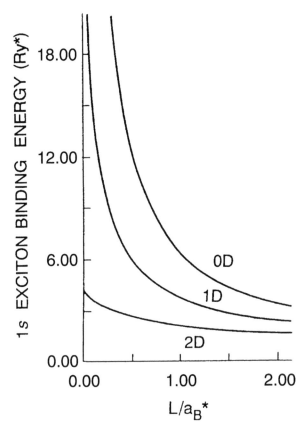

Figure 39 Binding energies of 1s excitons in a plane L thick, a square parallelepiped, and a cube of a side L. (From Matsuura, M. and Kamizato, T., *Surf. Sci.*, 174, 183, 1986. With permission.)

The exciton Bohr radius a_B^* is given by:

$$a_B^* = \frac{\hbar^2 \varepsilon}{\mu e^2} \tag{117}$$

and an inequality $a_e^*, a_h^* < a_B^*$ holds.

- *Strong confinement* ($R \ll a_e^*, a_h^* < a_B^*$). The individual motion of electrons and holes is quantized and the Coulomb interaction energy is much smaller than the quantized kinetic energy. Nanometer-size GaAs, CdS, CdSe, and CdTe microcrystallites are good examples of the strong confinement regime. The ground-state energy is:

$$E(R) = E_g + \frac{\hbar^2 \pi^2}{2\mu R^2} - \frac{1.786 e^2}{\varepsilon R} - 0.248 Ry^* \tag{118}$$

where the second term is the kinetic energy of electrons and holes, the third term is the Coulomb energy, and the last term is the correlation energy.

- *Intermediate confinement* ($a_h^* < R < a_e^*$). In this case, the electron motion is quantized, while the hole is bound to the electron by their Coulombic attraction. Many II-VI microcrystallites belong to the intermediate confinement regime.

- *Weak confinement* (R >> $a_B^* > a_e^*$, a_h^*). In this case, the center-of-mass motion of excitons is quantized. Nanometer-size CuCl microcrystallites are typical examples of the weak confinement regime; the ground-state energy is written as:

$$E = E_g - Ry^* + \frac{\hbar^2 \pi^2}{2MR^2} \tag{119}$$

where M = $m_e^* + m_h^*$ is the translational mass of the exciton.

Typical experimental data for three categories are shown in Figure 40.[6] CdS, CuBr, and CuCl microcrystallites belong to strong, intermediate, and weak confinement regimes, respectively. With a decrease in microcrystallite size, the continuous band changes into a series of discrete levels in CdS, although the levels are broadened because of the size distribution. In the case of CuCl, the exciton absorption bands show blue-shifts with a decrease in size.

The luminescence of semiconductor microcrystallites not only depends on the microcrystallites themselves, but also on their surfaces and their surroundings since the surface:volume ratio in these systems is large. The luminescence spectrum then depends on the preparation conditions of microcrystallites. Thus it is that some samples show donor-acceptor pair recombination, but other samples do not; in others, the edge luminescence at low temperature consists of exciton and bound exciton luminescence. The exciton luminescence spectrum of many samples shows Stokes shift from the absorption spectrum, indicating the presence of localized excitons. Typical examples of the luminescence spectra of CdSe microcrystallites and CuCl microcrystallites are shown in Figures 41 and 42.[7,8]

Impurities or defects in insulating crystals often dominate their luminescence spectra; this is also the case with semiconductor microcrystallites, but additional effects occur in the latter. Nanometer-size semiconductor microcrystallites can be composed of as few as 10^3–10^6 atoms; if the concentration of centers is less than ppm, considerable amounts of the microcrystallites are free from impurities or defects. This conjecture is verified in AgBr microcrystallites, which are indirect transition materials.[9] Figure 43 shows luminescence spectra of AgBr microcrystallites with average radii of 11.9, 9.4, 6.8, and 4.2 nm. The higher-energy band observed at 2.7 eV is the indirect exciton luminescence, and the lower-energy band observed at 2.5 eV is the bound exciton luminescence of iodine impurities. In contrast to AgBr bulk crystals, the indirect exciton luminescence is strong compared with the bound exciton luminescence at iodine impurities. The ratio of the indirect exciton luminescence to the bound exciton luminescence at iodine impurities increases with the decrease in size of AgBr microcrystallites. This increase in ratio shows that the number of impurity-free microcrystallites increases with the decrease in size. Simultaneously, the decay of the indirect exciton luminescence approaches single exponential decay approximating the radiative lifetime of the free indirect exciton. The blue-shift of the indirect exciton luminescence shown in Figure 43 is due to the exciton quantum confinement effect, as discussed previously.

Nanometer-size semiconductor microcrystallites can be used as a laser medium.[10] Figure 44 shows the lasing spectrum of CuCl microcrystallites. When the microcrystallites embedded in a NaCl crystal are placed in a cavity and excited by a nitrogen laser, lasing occurs at a certain threshold. The emission spectrum below the threshold, shown in Figure 44, arises from excitonic molecule (biexciton) luminescence. Above the threshold, the broad excitonic molecule emission band is converted to a sharp emission spectrum having a maximum peak at 391.4 nm. In this case, the lasing spectrum is composed of a few longitudinal modes of the laser cavity consisting of mirrors separated by 0.07 nm. The optical gain of the CuCl microcrystallites compared with that in a bulk CuCl sample is found to be much larger. The high optical gain of CuCl microcrystallites comes from the spatial confinement of excitons, resulting in the enhanced formation efficiency of excitonic molecules.

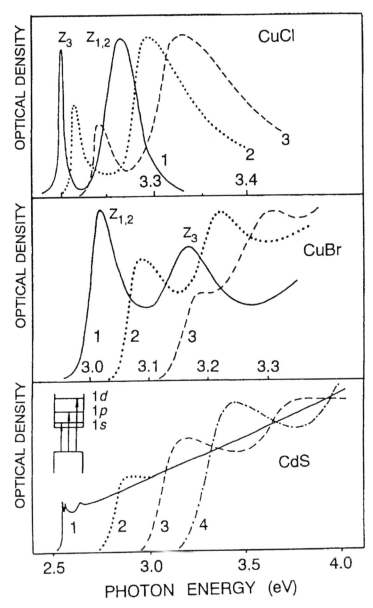

Figure 40 Absorption spectra at 4.2K for CuCl, CuBr, and CdS microcrystallites. For CuCl, the average radius R = 31 nm (1), 2.9 nm (2), and 2.0 nm (3); for CuBr, R = 24 nm (1), 3.6 nm (2), and 2.3 nm (3); for CdS, R = 33 nm (1), 2.3 nm (2), 1.5 nm (3), and 1.2 nm (4). (From Ekimov, A.I., *Phyica Scripta T*, 39, 217, 1991. With permission.)

After the initial report of visible photoluminescence from porous Si,[11] much effort has been devoted to clarify the mechanism of the photoluminescence. Figure 45 shows a typical example of a luminescence spectrum from porous Si. As the first approximation, porous Si made by electrochemical etching of Si wafers can be treated as an ensemble of quantum dots and quantum wires. However, real porous Si is a much more complicated system, consisting of amorphous Si, SiO_2, Si-oxygen-hydrogen compounds, Si microcrystallites, and Si wires. This complexity obscures the quantum size effect with other effects, and the physical origin of the visible luminescence from porous Si remains a puzzle. Depending on the sample preparation method, porous Si shows red, green, or blue luminescence.

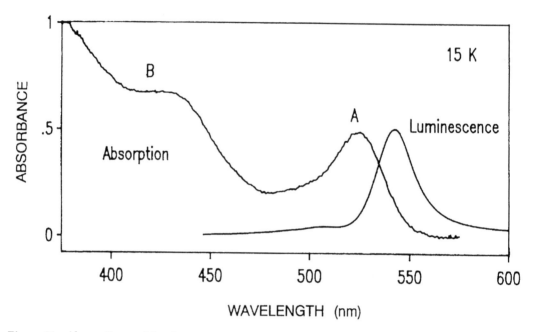

Figure 41 Absorption and luminescence spectra of wurtzite CdSe microcrystallites (R = 1.6 nm). (From Bawendi, M.G., Wilson, W.L., Rothberg, L., et al., *Phys. Rev. Lett.*, 65, 1623, 1990. With permission.)

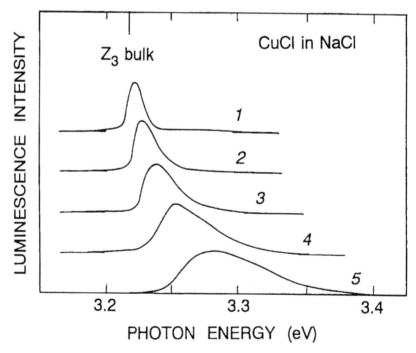

Figure 42 Exciton luminescence spectra of CuCl microcrystallites at 77K. The average radius R is 5.7 nm (1), 4.9 nm (2), 3.4 nm (3), 2.7 nm (4), and 2.2 nm (5). The energy of Z_3 exciton for bulk CuCl crystals at 77K is indicated by a vertical bar. (From Itoh, T., Iwabuchi, Y., and Kataoka, M., *Phys. Stat. Solidi (b)*, 145, 567, 1988. With permission.)

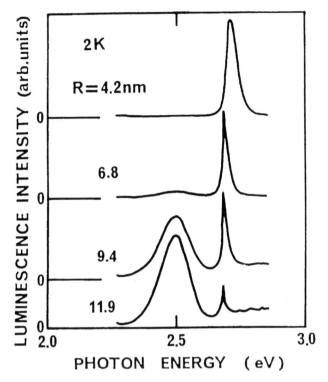

Figure 43 Photoluminescence spectra of AgBr microcrystallites at 2K. The average radius *R* of microcrystallites is 11.9, 9.4, 6.8, and 4.2 nm. The luminescence spectra are normalized by their respective peak intensities. The 2.7-eV band is indirect exciton luminescence, and the 2.5-eV band is bound exciton luminescence at iodine impurities. (From Masumoto, Y., Kawamura, T., Ohzeki, T., and Urabe, S., *Phys. Rev.*, B446, 1827, 1992. With permission.)

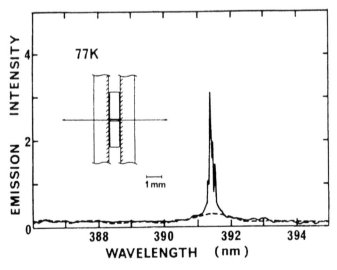

Figure 44 Emission spectra of the laser device made of CuCl microcrystallites at 77K below and above the lasing threshold. The threshold I_{th} is about 2.1 MW cm^{-2}. The solid line shows the spectrum under the excitation of 1.08 I_{th}. The dashed line shows the spectrum under the excitation of 0.86 I_{th}. (From Matsumoto, Y., Kawamura, T., and Era, K., *Appl. Phys. Lett.*, 62, 225, 1993. With permission.)

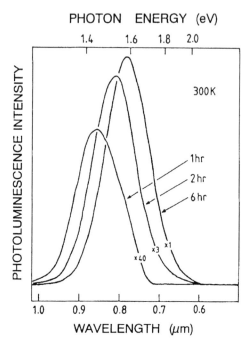

Figure 45   Room-temperature photoluminescence from the porous Si. Anodization times are indicated. (From Canham, L.T., *Appl. Phys. Lett.*, 57, 1046, 1990. With permission.)

The photoluminescence quantum efficiency of porous Si exhibiting red luminescence is as high as 35%, but its electroluminescence quantum efficiency is 0.2% Light-emitting diodes made of porous Si have also been demonstrated but the quantum efficiency is too low for practical application. If the electroluminescence quantum efficiency is improved substantially, porous Si will be used in light-emitting devices because Si is the dominant material in the semiconductor industry.

Note: An updated discussion on the size affect on radiative properties alluded in Reference 10 below appears elsewhere.[13]

References

1. Weisbuch, C. and Vinter, B., *Quantum Semiconductor Structures*, Academic Press, Boston, 1991.
2. Ishibashi, T., Tarucha, S., and Okamoto, H., Int. Symp. GaAs and Related Compounds, Oiso, 1981, Inst. Phys. Conf. Ser. No. 63, 1982, chap. 12, 587.
3. Matsuura, M. and Kamizato, T., *Surf. Sci.*, 174, 183, 1986; Masumoto, Y. and Matsuura, M., *Solid State Phys. (Kotai Butsuri)*, 21, 493, 1986 (in Japanese).
4. Feldmann, J., Peter, G., Göbel, E.O., Dawson, P., Moore, K., Foxon, C., and Elliott, R.J., *Phys. Rev. Lett.*, 59, 2337, 1987.
5. Yoffe, A.D., *Adv. Phys.*, 42, 173, 1993.
6. Ekimov, A.I., *Phyica Scripta T*, 39, 217, 1991.
7. Bawendi, M.G., Wilson, W.L., Rothberg, L., Carroll, P.J., Jedju, T.M., Steigerwald, M.L., and Brus, L.E., *Phys. Rev. Lett.*, 65, 1623, 1990.
8. Itoh, T., Iwabuchi, Y., and Kataoka, M., *Phys. Stat. Solidi (b)*, 145, 567, 1988.
9. Matsumoto, Y., Kawamura, T., Ohzeki, T., and Urabe, S., *Phys. Rev.*, B446, 1827, 1992.
10. Masumoto, Y., Kawamura, T., and Era, K., *Appl. Phys. Lett.*, 62, 225, 1993.
11. Canham, L.T., *Appl. Phys. Lett.*, 57, 1046, 1990.
12. *Properties of Porous Silicon*, Canham, L., ed., The Institute of Electrical Engineers, 2005.
13. *Practical Applications of Phosphors*, Yen, W.M., Shionoya, S., and Yamamoto, H., Eds., CRC Press, Boca Raton, 2006.

Fundamentals of luminescence

Eiichiro Nakazawa

Contents

1.7   Transient characteristics of luminescence

This section focuses on transient luminescent phenomena, that is, time-dependent emission processes such as luminescence after-glow (phosphorescence), thermally stimulated emission (thermal glow), photo (infrared)-stimulated emission, and photoquenching. All of these phenomena are related to a quasistable state in a luminescent center or an electron or hole trap.

1.7.1   Decay of luminescence

Light emission that persists after the cessation of excitation is called *after-glow*. Following the terminology born in the old days, luminescence is divided into fluorescence and phosphorescence according to the duration time of the after-glow. The length of the duration time required to distinguish the two is not clearly defined. In luminescence phenomena in inorganic materials, the after-glow that can be perceived by the human eye, namely that persisting for longer than 0.1 s after cessation of excitation, is usually called phosphorescence. Fluorescence implies light emission during excitation. Therefore, fluorescence is the process in which the emission decay is ruled by the lifetime (<10 ms) of the emitting state of a luminescence center, while the phosphorescence process is ruled by a quasistable state of a center or a trap.

In organic molecules, fluorescence and phosphorescence are distinguished by a quite different definition. The two are distinguished by whether the transition to emit light is allowed or forbidden by spin selection rules. Light emission due to an allowed transition is called *fluorescence*, while that due to a forbidden transition, usually showing a long after-glow, is called *phosphorescence* (see 1.5).

1.7.1.1 Decay of fluorescence

The decay process of the luminescence intensity $I(t)$ after the termination of excitation at $t = 0$ is generally represented by an exponential function of the elapsed time after the excitation.

$$I(t) = I_0 \exp(-t/\tau) \tag{120}$$

where τ is the decay time constant of the emission. It should be noted that the emission decay curve of nonlocalized centers, donor-acceptor pairs for example, is not always represented in the exponential form of Eq. 120. (See 1.4.)

If one denotes the number of excited luminescence centers in a unit volume by n^*, and the radiative and nonradiative transition probabilities by W_R and W_{NR}, respectively, then the rate equation for n^* is:

$$\frac{dn^*}{dt} = -(W_R + W_{NR})n^* \tag{121}$$

and the solution of this equation is:

$$n^*(t) = n_0^* \exp\left[-(W_R + W_{NR})t\right] \tag{122}$$

Here, n_0^* is the value at $t = 0$, that is, at the end-point of excitation or, in other words, at the start point of the after-glow.

Therefore, the lifetime of the center, which corresponds to the elapsed time for n^* to be decreased by the factor of e^{-1} of n_0^*, is $(W_R + W_{NR})^{-1}$. Since the emission intensity is proportional to n^*, the decay time of the after-glow in Eq. 120 is equal to the lifetime of the center:

$$\tau = (W_R + W_{NR})^{-1} \tag{123}$$

and the luminescence efficiency of the center is given by:

$$\eta = \frac{W_R}{W_R + W_{NR}} \tag{124}$$

The radiative transition probability W_R of an emitting state is the summation of the spontaneous emission probability $A_{m \to n}$ from the state m to all the final states n, (see Eq. 29)

$$W_R = \sum_n A_{m \to n} = \sum_n \frac{1}{\tau_{mn}} \tag{125}$$

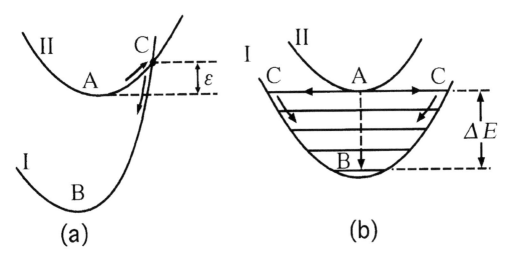

Figure 46 Configurational coordinate models of nonradiative relaxation processes: thermal activation type (a), and multiphonon type (b).

where τ_{mn} is given by Eqs. 32 or 35. The ratio of the transition probability to a particular final state n to W_R, $A_{m \to n}/W_R$, is called the *branching ratio*.

While the nonradiative transition probability W_{NR} is generally ruled by thermal relaxation processes (i.e., the emission of energy into lattice vibrations), it is also increased by the effect of resonant energy transfer between optical centers. (See 1.8.1.)

The thermal relaxation in a luminescence center can be divided into two types of mechanisms as shown by the two configurational coordinate diagrams (a) and (b) in Figure 46. In the first type (a), the center is thermally activated from point A, the point of the lowest energy on the exited state II, to the crossing point C where the electronic states of the excited and ground states are intermixed, and then thermally released from C to B on the ground state I. The energy ε necessary to excite the center from A to C is called the *thermal activation energy*. The probability that the center will make the transition from state II to state I by thermal activation via point C is generally given by:

$$a = s \exp\left(\frac{-\varepsilon}{kT}\right) \tag{126}$$

Therefore, the nonradiative transition probability by thermal activation is given by:

$$W_{NR} = a_{II \to I} = s \exp\left(\frac{-\varepsilon}{kT}\right) \tag{127}$$

where k is the Boltzmann constant and s is the frequency factor. This type of nonradiative transition is strongly dependent on temperature, resulting in thermal quenching, that is, the decrease of emission efficiency and shortening of the emission decay time at high temperature (see Eqs. 123 and 124).

An example of the thermal quenching effect is shown in Figure 47 for $Y_2O_2S:Yb^{3+}$.[1] The emission from the charge-transfer state (CTS) of Yb^{3+} ions at 530 nm is strongly reduced by thermal quenching at high temperature. The $4f$ emission under CTS excitation (313 nm), however, is increased at high temperature due to the increased amount of excitation transfer from the CTS. The Figure also shows that, as expected, the emission from the

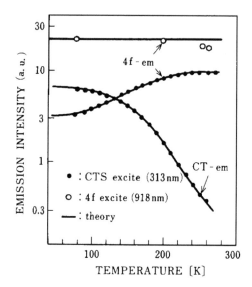

Figure 47 Temperature dependence of two types of the emission in $Y_2O_2S:Yb^{3+}$, from the CTS (charge-transfer state) and from a 4*f*-emitting level of Yb^{3+} ions. (From Nakazawa, E., *J. Luminesc.*, 18/19, 272, 1979. With permission.)

4*f* level at 930 nm is not so thermally quenched under direct 4*f* excitation into the emitting level (918 nm).

The second type of nonradiative transition is a multiphonon process shown in Figure 46b. This type is often observed in the relaxation between the 4*f* excited levels of rare-earth ions, where no cross-point exists between curves I and II in the configuration coordinate diagram because of the similarity of the electronic states. The transition between states I and II occurs at point A, where an energy gap ΔE exists between the states: namely, the transition from the pure electronic state of II to the electron-phonon-coupled state of I with *n* phonons takes place at A, which is followed by the instantaneous transfer to point C and relaxation to B. The nonradiative transition probability is, therefore, dependent on ΔE or *n*, the number of phonons necessary to fill the energy gap, since $\Delta E = n\ \omega_p$, where ω_p is the largest phonon energy. The nonradiative multiphonon transition probability is then given by:[2]

$$W_{NR}(\Delta E) = W_{NR}(0)e^{-\alpha\Delta E} \tag{128}$$

where α depends on the character of the phonon (lattice vibration). Since the process is mainly due to the spontaneous emission of phonons, the temperature dependence of the probability is small. An experimental result[2] showing the applicability of Eq. 128 is shown in Figure 48.

1.7.1.2　Quasistable state and phosphorescence

If one of the excited states of a luminescent center is a quasistable state (i.e., an excited state with very long lifetime), a percentage of the centers will be stabilized in that state during excitation. After excitation has ceased, after-glow is caused by the thermal activation of the state. This situation is illustrated using the configurational coordinate diagram in Figure 49, where state III is a quasistable state and state II is an emitting state with a radiative transition probability W_R. The center, once stabilized at *A'*, transfers from state III to state II by thermal activation via point C. The probability of this activation, $a_{III\to II}$, is given by Eq. 126. Then, if $W_R \gg a_{III\to II}$, the decay time constant of the emission becomes

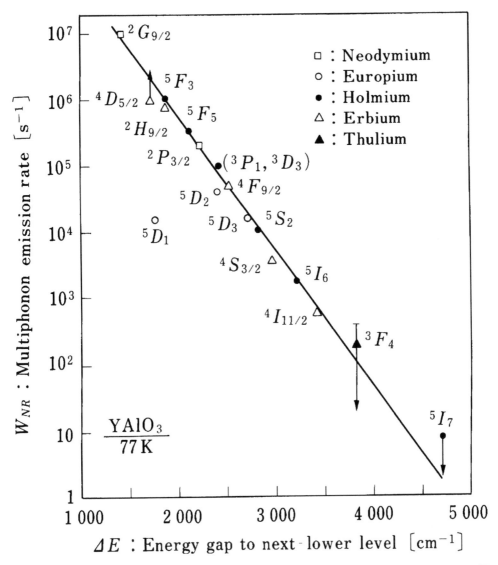

Figure 48 Energy gap law of nonradiative relaxation due to multiphonon processes. (From Weber, M.J., *Phys. Rev.*, B8, 54, 1993. With permission.)

almost equal to $1/a_{III \to II}$, that is, the lifetime of the quasistable state. The decay curve of the after-glow is represented by an exponential function that is similar to Eq. 120, and is strongly temperature dependent. The decay time constant of an emitting center with quasistable states is not usually longer than a second.

1.7.1.3   Traps and phosphorescence

Excited electrons and holes in the conduction and valence bands of a phosphor can often be captured by impurity centers or crystal defects before they are captured by an emitting center. When the probability for the electron (hole) captured by an impurity or defect center to recombine with a hole (electron) or to be reactivated into the conduction band (valence band) is negligibly small, the center or defect is called a *trap*.

The electrons (holes) captured by traps may cause phosphorescence (i.e., long after-glow) when they are thermally reactivated into the conduction band (valence band) and then radiatively recombined at an emitting center. The decay time of phosphorescence

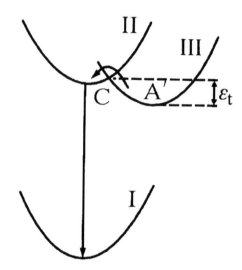

Figure 49 Configurational coordinate model of the luminescence after-glow (phosphorescence) via a quasistable state.

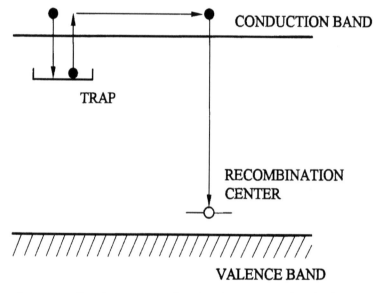

Figure 50 Luminescence after-glow process via a trap in an energy band scheme.

due to traps can be as long as several hours and is often accompanied by photoconductive phenomena.

The decay curve of the after-glow due to traps is not generally represented by a simple exponential function. The form of the curve is dependent on the concentration of the traps and on the electron capture cross-sections of the trap and the emitting center. Furthermore, it also depends on the excitation intensity level.

While several kinds of traps usually exist in practical phosphors, only one kind of electron trap is presumed to exist in the simple model shown in Figure 50. Let N be the trap concentration, and n_c and n_t the number of electrons per unit volume in the conduction band and trap states, respectively. The number of holes denoted by p is equal to $n_c + n_t$. The rate equation representing the decaying processes of the concentration of electrons and holes after the termination of excitation is:

$$\frac{dn_t}{dt} = -an_t + b(N - n_t)n_c$$

$$\frac{dp}{dt} = -rpn_c \tag{129}$$

where a is the probability per second for a trapped electron to be thermally excited into the conduction band and is given by the same form as Eq. 126 with the density of states in the conduction band included in s. The probabilities that a free electron in the conduction band will be captured by a trap or to recombine with a hole are given by b and r, respectively. It is supposed that the number of the electrons n_c in the conduction band in the after-glow process is so small that $p \simeq n_t$ and $dp/dt \simeq dn_t/dt$. Then, the above two equations give:

$$\frac{dn_t}{dt} = \frac{-an_t^2}{n_t + (b/r)(N - n_t)} \tag{130}$$

This equation can be solved analytically for two cases: $b \ll r$ and $b \simeq r$.

First, the case of $b \ll r$, which presumes that the electrons once released from traps are not retrapped in the after-glow process. Eq. 130 then simplifies to:

$$\frac{dn_t}{dt} = -an_t \tag{131}$$

Since the emission intensity is given by $I(t) \propto dp/dt$, and $dp/dt \simeq dn_t/dt$ as mentioned above, then

$$I(t) = I_o \exp(-at) \tag{132}$$

This simple exponential decay of after-glow is the same as the one due to the quasistable state mentioned previously and is called a first-order or monomolecular reaction type in the field of chemical reaction kinetics.

In the case $b \simeq r$, which means that the traps and emitting centers have nearly equal capturing cross-sections, Eq. 130 can be simplified to:

$$\frac{dn_t}{dt} = -\frac{a}{N}n_t^2 \tag{133}$$

and then the number of trapped electrons per unit volume is given by:

$$n_t = \frac{n_{t0}}{1 + (N/an_{t0})t} \tag{134}$$

Approximating $I(t) \propto dn_t/dt$ as before, the decay curve of the after-glow is obtained as:

$$I(t) = \frac{I_o}{(1 + \gamma t)^2}, \quad (\gamma = N/an_t) \tag{135}$$

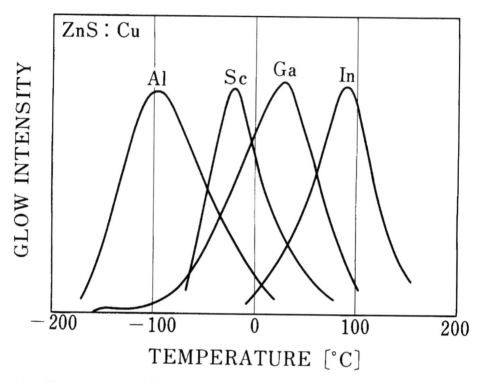

Figure 51 Glow curves of ZnS:Cu phosphors co-activated with Al, Sc, Ga, and In. (From Hoogen-straaten, W., *Philips Res. Rept.*, 13, 515, 1958. With permission.)

This form is called the second-order or bimolecular reaction type, where the decay curve is changed by excitation intensity as well as by temperature.

While the treatise mentioned above is a simple model presuming a single kind of trap, the real phosphors may have several kinds of traps of different depths. In many real cases, therefore, the after-glow decay curve is not represented simply by a monomolecular or bimolecular reaction curve. It often fits into the following equation.

$$I(t) = \frac{I_o}{(1 + \gamma t)^n} \qquad (136)$$

where n is around 0.5–2. If $t \gg 1/\gamma$, this decay curve can be approximated by $I(t) \propto t^{-n}$. The decay time constant of an after-glow is therefore denoted either by the $1/e$ decay time or the 10% decay time.

1.7.2 Thermoluminescence

When a phosphor with deep traps is excited for a while at rather low temperatures and then heated, it shows an increased after-glow called thermally stimulated luminescence due to the recombination of electrons thermally reactivated from the deep traps. This emission is also called thermoluminescence, and the temperature dependence of the emission intensity is called the glow curve, which is a good means to measure the depth (i.e., the activation energy of traps). Figure 51 shows the glow curves of ZnS:Cu phosphors with various co-activators.[3]

The measurement of a glow curve of a phosphor sample proceeds as follows.

1. The sample is cooled to a low temperature (liquid nitrogen is often used as coolant).
2. The sample is excited by UV light until the traps are filled with electrons or holes.
3. The excitation is terminated, and the temperature of the sample is raised at a constant rate, $dT/dt \equiv \beta$, while the intensity is recorded.
4. The temperature dependence of fluorescence is then measured under a constant UV excitation, which is used to calibrate the effect of temperature quenching on the thermoluminescence intensity.

The glow curve thus obtained is analyzed with the following theory. Assume that (1) a single kind of trap exists; (2) the decay of after-glow is of the first-order type given by Eq. 132, and (3) the probability for the trapped electrons to be thermally released into the conduction band is given by Eq. 126. Since the retrapping of the released electrons is neglected in the first-order kinetics, the change in the number of trapped electrons is:

$$\frac{dn_t}{dt} = -n_t s \exp(-\varepsilon/kT) \tag{137}$$

Integrated from a temperature T_0 to T with the relation $dT/dt = \beta$, this equation gives the number of residual electrons in the traps at T as:

$$n_t(T) = n_{t0} \exp\left(-\int_{t_0}^{t} s \exp\left(\frac{-\varepsilon}{kT}\right) \cdot \frac{dT}{\beta}\right) \tag{138}$$

where n_{t_0} is the number of the trapped electrons at the initial temperature T_0. Therefore, the emission intensity at T, approximated by $I \propto dn_t/dt$ as mentioned in the previous section, is given by:

$$I(T) \propto n_{t0} s \exp\left(\frac{-\varepsilon}{kT}\right) \exp\left(-\int_{T_0}^{T} s \exp\left(\frac{-\varepsilon}{kT}\right) \cdot \frac{dT}{\beta}\right) \tag{139}$$

Based on this equation, the following techniques have been proposed for obtaining the trap depth (activation energy ε) from a glow curve.

(a) In the initial rising part of the glow peak on the low-temperature side where the number of trapped electrons is nearly constant, Eq. 139 is approximated by

$$I(T) \propto s \exp\frac{-\varepsilon}{kT} \tag{140}$$

Then the slope of the Arrhenius plot ($1/T$ vs. ln I) of the curve in this region gives the trap depth ε. In fact, however, it is not easy to determine the depth with this method because of the uncertainty in fixing the initial rising portion.

(b) Let the peak position of a glow curve be T_m. Then, the following equation derived from $dI/dT = 0$ should be valid.

$$\frac{\beta\varepsilon}{kT_m^2} = s \exp\left(\frac{-\varepsilon}{kT_m}\right) \tag{141}$$

Table 1

β/s [K]	K [K/eV]	T_n [K]
10^{-4}	833	35
10^{-5}	725	28
10^{-6}	642	22
10^{-7}	577	17
10^{-8}	524	13
10^{-9}	480	10
10^{-10}	441	7
10^{-11}	408	6
10^{-12}	379	6
10^{-13}	353	5
10^{-14}	331	5
10^{-15}	312	4

From Curie, D., Luminescence in Crystals, John Wiley & Sons, 1963, chap. VI. With permission.

If the frequency factor s is obtained in some manner, ε can be estimated by this relation from the observed value of T_m. Note that the temperature rise rate β should be kept constant throughout the measurement for this analysis. Randall and Wilkins[4] performed a numerical calculation based on this theory and obtained the following equation, which approximates the trap depth ε with 1% error.

$$\varepsilon = \frac{T_m - T_o(\beta/s)}{K(\beta/s)} \tag{142}$$

Here, $T_o(\beta/s)$ and $K(\beta/s)$ are the parameters determined by β/s as listed in Table 1.

For ZnS:Cu, $s = 10^9$ s^{-1} is assumed and the following estimations have been proposed for various values of β.[5]

$$\varepsilon|eV| = T_m/500 \qquad (\beta = 1K/s)$$

$$\varepsilon|eV| \approx T_m/400 \qquad (\beta = 0.01K/s)$$

$$\varepsilon|eV| = (T_m - 7)/433 \quad (\beta = 0.06K/s)$$

(c) If the glow curve is measured with two different rise rates, β_1 and β_2, it is apparent that one can obtain the value of ε without assuming the value of s in Eqs. 141 or 142, using the following equation.

$$\frac{\varepsilon}{k}\left(\frac{1}{T_{m2}} - \frac{1}{T_{m1}}\right) = \ln\left(\frac{\beta_1}{\beta_2} \cdot \frac{T_{m2}^2}{T_{m1}^2}\right) \tag{143}$$

Hoogenstraaten[3] extended this method for many raising rates β_i, and, by plotting the curves ($1/T_{mi}$ vs. $\ln(T_{mi}^2/\beta_i)$), obtained the trap depth from the slope ε/k of the straight line connecting the plotted points as shown in Figure 52.[6] A numerical analysis[7] has shown that Hoogenstraaten's method gives the best result among several methods for obtaining trap depths from glow curves.

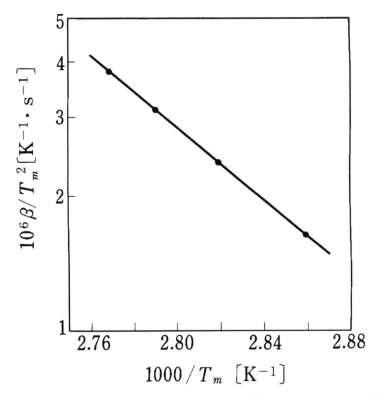

Figure 52 Hoogenstraaten plot showing the dependence of the peak temperature (T_m) of a glow curve on the temperature-raising rate (β). The slope of this line gives the depth (activation energy) of the trap. (From Avouris, P. and Morgan, T.N., *J. Chem. Phys.*, 74, 4347, 1981. With permission.)

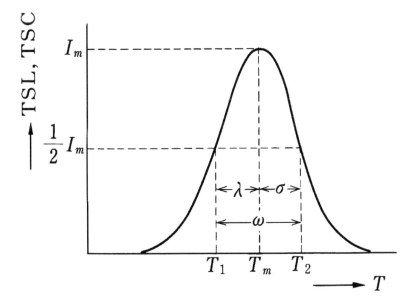

Figure 53 Predicted shape of a glow curve.

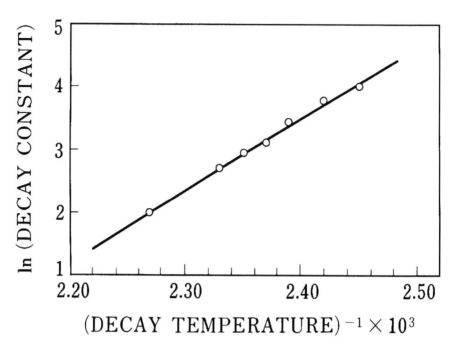

Figure 54 Temperature dependence of the decay time constant of ZnS:Cu. (From Bube, R.H., *Phys. Rev.*, 80, 655, 1950. With permission.)

(d) Many methods for obtaining the depth ε from the width of the peak of a glow curve have been proposed.[5] The results are listed below, where the width for a peak is defined in various ways, as shown in Figure 53.

1. $\varepsilon = kT_m^2 \sigma$
2. $\varepsilon = 2kT_m(1.25T_m/\omega - 1)$
3. $\varepsilon = 1.52\,kT_m^2/\lambda - 3.16\,kT_m$
4. $\varepsilon = (1 + \omega/\lambda)kT_m^2/\sigma$

Note that these methods are usable under certain restricted conditions.[5]

There is a method to obtain trap depth other than the glow-curve method described above. It is to use the temperature dependence of the decay time of after-glow, that is, phosphorescence. As mentioned in relation to Eq. 132, the decay time constant of the exponential after-glow due to the first-order reaction kinetics is equal to the inverse of the thermal detrapping probability, a^{-1}, and its temperature dependence is given by:

$$\tau_{1/e} = s^{-1} \exp\!\left(\frac{\varepsilon}{kT}\right) \tag{144}$$

Therefore, the trap depth ε can be obtained from the measurements of the phosphorescence decay time $\tau_{1/e}$ at several different temperatures (T_i). An example is shown in Figure 54, in which the depth is obtained from the slope of the straight line connecting the Arrhenius plots of the observed values[8] for τ_i and T_i.

A usable method with which trap depths and relative trap densities are obtained more easily and accurately was recently proposed.[9] In this method, the sample is excited periodically under a slowly varying temperature and the after-glow (phosphorescence) intensity is measured at several delay times (t_d) after the termination of excitation in each cycle.

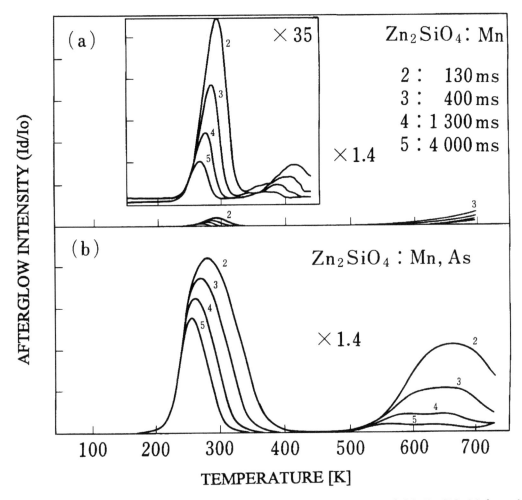

Figure 55 Temperature dependencies of the after-glow intensity of (a) $Zn_2SiO_4:Mn^{2+}$ and (b) $Zn_2SiO_4:Mn^{2+}$, As. The delay time (t_d) is 0.13, 0.4, 1.3, and 4.0 s, respectively, for the curves numbered from 2 through 5 in the figures. (From Nakazawa, E., *Jpn. J. Appl. Phys.*, 23(9), L755, 1984. With permission.)

The temperature dependence of the after-glow intensity at each delay time makes a peak at a certain temperature T_m. From the equation $dI/dT = 0$ and using either Eq. 132 or Eqs. 135 and 136, the following relation is obtained between the peak temperature T_m and the delay time t_d.

$$t_d = s^{-1} \exp\left(\frac{\varepsilon}{kT_m}\right) \qquad (145)$$

Since Eq. 145 is similar to Eq. 144 (i.e., the decay time method), the method used there for obtaining the trap depth ε can be applied hereby, substituting t_d for τ, and T_m for T_i. An example of the measured after-glow intensity curves is shown in Figure 55.

1.7.3 Photostimulation and photoquenching

When a phosphor with deep traps is once excited and then irradiated by infrared (IR) or red light during the decay of its phosphorescence, it sometimes shows photostimulation

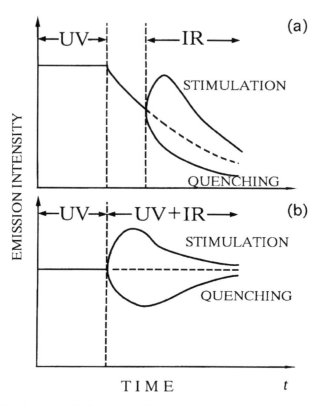

Figure 56 Photostimulation and photoquenching simulation for the case (b) under a constant excitation and (a) in the after-glow process after the termination of excitation.

or photoquenching of luminescence; that is, an increase or decrease of the emission intensity as schematically shown in Figure 56(a). Under a stationary excitation shown in Figure 56(b), the stimulation enhances and the quenching reduces the emission intensity temporarily.

These phenomena are utilized for IR detection and radiographic imaging, in which the intensity of the stimulated emission is used to measure the intensity of IR light or X-rays.[11]

Photostimulation is caused by the radiative recombination of the electrons (holes) released by photoactivation from deep trap levels, as shown in Figure 57(a). On the other hand, photoquenching is caused by the nonradiative recombination of holes (electrons) photoactivated from luminescent centers as shown in Figure 57(b). Figure 58 depicts the configuration coordinate model of photostimulation. The activation energy ε_0 of photostimulation is not generally equal to the thermal activation energy ε_t of trapped electrons discussed before with reference to Figure 49. Since the optical absorption process takes place in a very short time period without changing the configuration of the atoms in the center at that moment, the process is represented by the straight vertical transition in Figure 58 from state III (a trap or the quasi-stable state of emitting centers) to state II (the conduction band or emitting centers). On the other hand, the thermal activation needs energy ε_t to overcome at least the lowest barrier between states II and III in Figure 58; hence, the activation energy ε_t is generally smaller than ε_0. Photostimulation spectra (i.e., excitation spectra for IR stimulation) can be used to measure the optical activation energy ε_0.

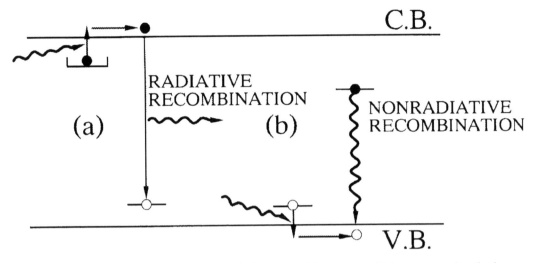

Figure 57 Photostimulation process (a), and photoquenching process (b) in an energy band scheme. C.B. and V.B. indicate the conduction band and the valence band of the host crystal, respectively.

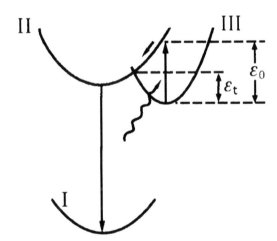

Figure 58 Photostimulation in configurational coordinate models.

References

1. Nakazawa, E., *J. Luminesc.*, 18/19, 272, 1979.
2. Weber, M.J., *Phys. Rev.*, B8, 54, 1973.
3. Hoogenstraaten, W., *Philips Res. Rept.*, 13, 515, 1958.
4. Randall, J.T. and Wilkins, M.H.F., *Proc. Roy. Soc.*, A184, 366, 1945.
5. Curie, D., *Luminescence in Crystals*, John Wiley & Sons, 1963, chap. VI.
6. Avouris, P. and Morgan, T.N., *J. Chem. Phys.*, 74, 4347, 1981.
7. Kivits, P. and Hagebeuk, H.J.L., *J. Luminesc.*, 15, 1, 1977.
8. Bube, R.H., *Phys. Rev.*, 80, 655, 1950.
9. Nakazawa, E., *Jpn. J. Appl. Phys.*, 23(9), L755, 1984.
10. Nakazawa, E., *Oyo Buturi*, 55(2), 145, 1986 (in Japanese).
11. *Practical Applications of Phosphors*, Yen, W.M., Shionoya, S., and Yamamoto, H., Eds., CRC Press, Boca Raton, 2006.

Fundamentals of luminescence

Eiichiro Nakazawa

1.8   Excitation energy transfer and cooperative optical phenomena

1.8.1   Excitation energy transfer

The process of excitation energy transfer from an excited point in a crystal to a luminescent center can be classified into the following two types.

1. Migration of free electrons (holes), electron-hole pairs, or quasi-particles such as excitons and plasmons conveys the excitation energy to luminescent centers. This type of transfer seems to be active especially in such semiconductor-like hosts as ZnS and CdS, which are widely used as cathode-ray tube (CRT) phosphors. In the initial excitation process of CRT phosphors, the local excitation by a high-energy electron produces several hundreds of these particles, and they are dispersed in the crystal by this type of transfer. (See 1.9.)
2. Excitation energy is transferred from an excited center (energy donor) to an unexcited center (energy acceptor) by means of quantum mechanical resonance.[1,2] This type of transfer is practically utilized for the sensitization of luminescence in lamp phosphors, which are mostly oxides or oxoacid salts with less-mobile electrons and holes than in CRT phosphor materials.

In this section, the resonant energy transfer process and related phenomena, such as the sensitization and quenching of luminescence, will be discussed. More recent experimental studies on this topic are referred to in the reference list.[3]

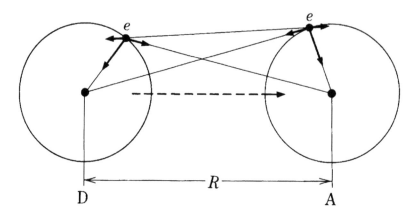

Figure 59 Coulomb interaction in a resonant energy transfer process.

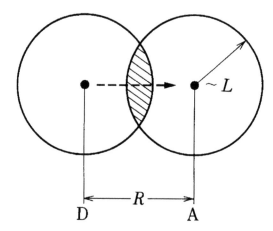

Figure 60 Resonant energy transfer by the exchange interaction, in which the overlapping of the wavefunctions of D and A (shaded region) is necessary.

1.8.1.1 *Theory of resonant energy transfer*

Dexter's theory of resonant energy transfer[2] has elucidated that two optical centers within a certain distance may be in resonance and transfer the excitation energy from one (donor) to the other (acceptor). The close proximity of the centers enables them to be connected by the electrostatic interaction shown in Figure 59 or by the quantum mechanical exchange interaction shown in Figure 60. The energy donor, which is called a *sensitizer* in practical usage, is denoted hereafter by D and the acceptor by A. For resonant energy transfer to take place, it is necessary that the transition energies of D and A be equal.

(a) **Multipolar Interaction.** The mechanisms of resonant energy transfer can be classified into several types based on the character of the transitions in D and A. When both transitions in D and A are of electric dipole character (dipole-dipole interaction), the probability per second of energy transfer from D to A is given by:

$$P_{dd}(R) = \frac{3c^4\hbar^4\sigma_A}{4\pi n^4\tau_D R^6}\int\frac{f_D(E)F_A(E)}{E^4}dE \tag{146}$$

Here, R is the separation between D and A, n is the refractive index of the crystal, σ_A is the absorption cross-section of A, and τ_D is the radiative lifetime of D. Likewise, the transfer probability due to the dipole-quadrupole interaction is:

$$P_{dq}(R) = \frac{135\pi\alpha c^8 \hbar^9}{4n^6 \tau_D \tau_A R^8} \int \frac{f_D(E)F_A(E)}{E^8} dE \qquad (\alpha = 1.266) \qquad (147)$$

In these equations, $f_D(E)$ and $F_A(E)$ represent the shape of the D emission and A absorption spectra, respectively, which are normalized (i.e., $\int f_D(E)dE = 1$ and $\int F_A(E)dE = 1$. The integrals in Eqs. 146 and 147 are, therefore, the overlapping ratios of these two spectra, which is a measure of the resonance condition. The transfer probabilities due to all multipolar interactions—i.e., dipole-dipole (d-d) in Eq. 146, dipole-quadrupole (d-q) in Eq. 147, and quadrupole-quadrupole (q-q), are summarized in Eq. 148 with their R dependence being noticed.

$$P_s(R) = \frac{\alpha_{dd}}{R^6} + \frac{\alpha_{dq}}{R^8} + \frac{\alpha_{qq}}{R^{10}} + \dots = \sum_{s=6,8,10} \frac{\alpha_s}{R^s} \qquad (148)$$

Here, s in the third term is 6, 8, and 10 for (dd), (dq), and (qq), respectively.

If the dipole transition is allowed for both D and A, the magnitudes of α_s are $\alpha_{dd} > \alpha_{dq} > \alpha_{qq}$, and the dipole-dipole interaction has the highest transfer probability. However, if the dipole transition is not completely allowed for D and/or A, as is the case with the f-f transition of rare-earth ions, it is probable that the higher-order interaction, d-q or q-q, may have the larger transfer probability for small distance pairs due to the higher-order exponent of R in Eq. 148.[4,5]

Since the emission intensity and the radiative lifetime of D are decreased by energy transfer, the mechanism of the transfer can be analyzed using the dependence of the transfer probability on the pair distance given by Eq. 148, and hence the dominant mechanism among (dd), (dq), and (qq) can be determined.

When the acceptors are randomly distributed with various distances from a donor D in a crystal, the emission decay curve of D is not an exponential one. It is given by the following equation for the multipolar interactions.[6]

$$\phi(t) = \exp\left[-\frac{t}{\tau_D} - \Gamma\left(1 - \frac{3}{s}\right)\frac{C}{C_0}\left(\frac{t}{\tau_D}\right)^{3/s}\right], \qquad (s = 6, 8, 10) \qquad (149)$$

Here, $\Gamma()$ is the gamma function, and C and C_0 are, respectively, the concentration of A and its critical concentration at which the transfer probability is equal to the radiative probability $(1/\tau_D)$ of D.

Thus, the emission efficiency η and the emission decay time constant τ_m can be estimated using Eq. 149 and the following equations:

$$\frac{\eta}{\eta_0} = \frac{\int_0^\infty \phi(t)dt}{\tau_D} \qquad (150)$$

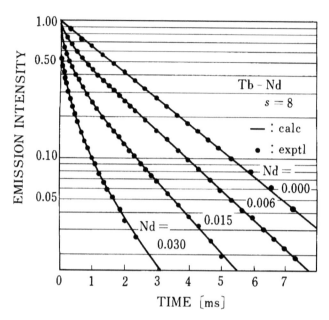

Figure 61 Decay curves of Tb^{3+} emission $(^5D_4 - 7F_j)$ affected by the energy transfer to Nd^{3+} ions in $Ca(PO_3)_2$. (From Nakazawa, E. and Shionoya, S., *J. Chem. Phys.*, 47, 3211, 1967. With permission.)

$$\tau_m = \frac{\int_0^\infty t\phi(t)dt}{\int_0^\infty \phi(t)dt} \tag{151}$$

Figure 61 shows the decay curves of Tb^{3+} emission $(^5D_4 - ^7F_j)$ in $Ca(PO_3)_2$, which are affected, due to energy transfer,[4] by the concentration of the co-activated Nd^{3+} ions. The dependence of the emission intensity and decay time of the donor Tb^{3+} ion on the concentration of the acceptor Nd^{3+} ion in the same system are shown in Figure 62. Theoretical curves in these figures are calculated using Eqs. 150 and 151 with $s = 8$ (quadrupole-dipole interaction).

(b) Exchange Interaction. When an energy donor D and an acceptor A are located so close that their electronic wavefunctions overlap each other as shown in Figure 60, the excitation energy of D could be transferred to A due to a quantum mechanical exchange interaction between the two. If the overlap of wavefunctions varies as $\exp(-R/L)$ with R, the transfer probability due to this interaction becomes:[2]

$$P_{ex}(R) = (2\pi/\hbar)K^2 \exp(-2R/L)\int f_D(E)F_A(E)dE \tag{152}$$

where K^2 is a constant with dimension of energy squared and L is an effective Bohr radius; that is, an average of the radii of D in an excited state and A in the ground state.

The emission decay curve of D, taking into account a randomly distributed A interacting through the exchange mechanism, is given (similar to Eq. 149) by:[6]

$$\phi(t) = \phi_0 \exp\left[-\frac{t}{\tau_D} - \gamma^{-3}\frac{C}{C_0}g\left(\frac{e^{\gamma t}}{\tau_D}\right)\right] \qquad (\gamma = 2R_0/L) \tag{153}$$

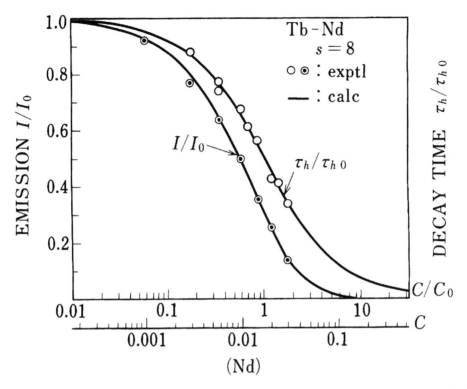

Figure 62 Intensity and decay time of Tb^{3+} emission in the same system as that in Figure 61. Solid curves are theoretical ones for quadrupole-dipole interaction (s = 8). (From Nakazawa, E. and Shionoya, S., *J. Chem. Phys.*, 47, 3211, 1967. With permission.)

The emission efficiency and the averaged emission decay time τ_m can be estimated using Eqs. 150, 151, and 153.

Since the exchange interaction requires the overlapping of the electron clouds of D and A (see Figure 60), the ion can be no further away than the second nearest site in the host crystal.

Note that while Eq. 152 requires the spectral overlap for the resonance condition, it is irrelevant to the spectral intensities. Therefore, if A is located next to D, and if the transitions are not completely electric dipole allowed, the transfer probability by exchange interaction can be larger than for multipolar interactions.

As described later in reference to Figures 64 and 65, the emission of Mn^{2+} in the halophosphate phosphor, the most general lamp phosphor, is excited by the energy transfer from Sb^{3+} due to the exchange interaction since the corresponding transition in the Mn^{2+} ion is a forbidden d-d transition.[7]

Perrin's model[8] treats the emission decay of D under general energy transfer in a simple manner. In this model, it is assumed that the transfer probability is a constant if A exists within some critical distance and is zero outside the range. This model, therefore, is thought to be most applicable to the short-range exchange interaction.

(c) Phonon-Assisted Energy Transfer. Phonon-assisted energy transfer occurs when the resonance condition is not well satisfied between D and A, resulting in the spectral overlap in Eqs. 148 and 149 being small. In this case, the difference δE between the transition energies of D and A is compensated by phonon emission or absorption. The transfer probability[9] is given by:

$$P_{as}(\Delta E) = P_{as}(0)e^{-\beta\Delta E} \tag{154}$$

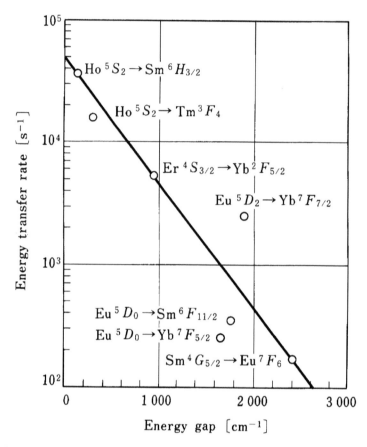

Figure 63 Relation between the nonresonant phonon-assisted energy transfer rate and the energy mismatch δ in Y_2O_3:R^{3+} (R = rare earth ions). (From Yamada, N., Shionoya, S., and Kushida, T., *J. Phys. Soc. Japan*, 32, 1577, 1972. With permission.)

where $P_{as}(0)$ is equal to the resonant transfer probability given by Eqs. 148 or 149, and β is a parameter that depends on the energy and occupation number of participating phonons. The energy gap ΔE is equal to $n\ \omega_p$ with n and ω_p being the number and energy, respectively, of the largest energy phonon in the host. Figure 63 shows the energy transfer rates for various D-A systems of rare-earth ions in Y_2O_3 host[10]; these are in excellent agreement with Eq. 154.

1.8.1.2 Diffusion of excitation

Energy transfer to the same type of ion is called excitation migration or energy migration. While the effect of energy migration among donors (D → D) prior to D → A transfer is neglected in the above discussion, it must be taken into account in the emission decay of sensitized phosphors. The effect of D → D migration on D → A systems is theoretically expressed by the following equation.[11,12]

$$\phi(t) = \exp-\left(\frac{1}{\tau_D} + \frac{1}{\tau_M}\right)t \tag{155}$$

where migration rate is defined as $\tau_M^{-1} = 0.51 \cdot 4\pi N_A \alpha^{1/4} D^{3/4}$, in which D is a diffusion constant, typically 5×10^{-9} $cm^2 s^{-1}$ for the Pr^{3+}-Pr^{3+} pair in $La_{0.8}Pr_{0.2}Cl_3$:Nd and 6×10^{-10} $cm^2 s^{-1}$ for the Eu^{3+}-Eu^{3+} pair in $Eu(PO_3)$ glass.[12,13]

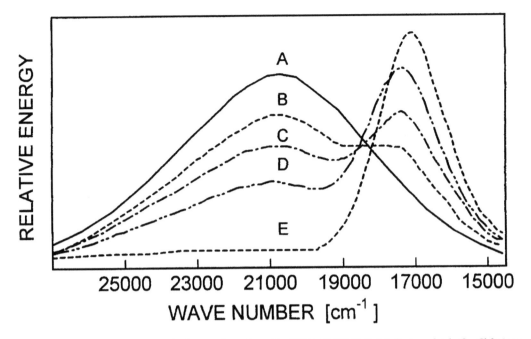

Figure 64 Emission spectra of a lamp phosphor, $Ca_5(PO_4)_3(F,Cl):Sb^{3+}$, Mn^{2+}, in which the Sb^{3+} ions sensitize the emission of Mn^{2+} ions by an energy transfer process. The Sb^{3+} concentration is fixed to be 0.01 mol/mol Ca. The Mn^{2+} concentration is changed, A:0, B:0.005, C:0.010, D:0.020 and E:0.080 mol/mol Ca. (From Batler, K.H. and Jerome, C.W., *J. Electrochem. Soc.*, 97, 265, 1950. With permission.)

1.8.1.3 Sensitization of luminescence

Energy transfer processes are often used in practical phosphors in order to enhance the emission efficiency. The process is called *sensitization of luminescence*, and the energy donor is called a sensitizer. The emission intensity of Mn^{2+}-activated silicate, phosphate, and sulfide phosphors, for example, is sensitized by Pb^{2+}, Sb^{3+}, and Ce^{3+}. In the halo-phosphate phosphor widely used in fluorescent lamps ($3Ca_3(PO_4)_2Ca(F,Cl)_2:Sb^{3+},Mn^{2+}$), Sb^{3+} ions play the role of a sensitizer as well as the role of an activator. As shown in Figure 64,[14] the intensity of the blue component of the emission spectra due to the Sb^{3+} activator of this lamp phosphor decreases with the concentration of Mn^{2+} ions because of the excitation energy transfer from Sb^{3+} to Mn^{2+}. Figure 65 shows the excitation spectra for the blue Sb^{3+} emission of this phosphor and for the red emission component due to Mn^{2+} activator. The similarities between the two spectra are evidence of energy transfer.[15]

The emission of Tb^{3+} is sensitized by Ce^{3+} ions in many oxide and double oxide, green-emitting phosphors.[16,17]

While the energy transfer from a donor to an emitting center causes sensitization of luminescence, the transfer from an emitting center to a nonradiative center causes the quenching of luminescence. A very small amount (~10ppm) of Fe, Co, and Ni in ZnS phosphors, for example, appreciably quenches the original emission as a result of this type of energy transfer (see 2.7.4.1).[18] They are called *killer* or *quencher* ions. In some phosphors, however, these killers are intentionally added for the purpose of reducing the emission decay time, thereby obtaining a fast-decay phosphor at the expense of emission intensity[19] (see Eq. 151).

Two-step or tandem energy transfer from Yb^{3+} donors to Er^{3+} or Tm^{3+} acceptors is used in infrared-to-visible up-conversion phosphors.

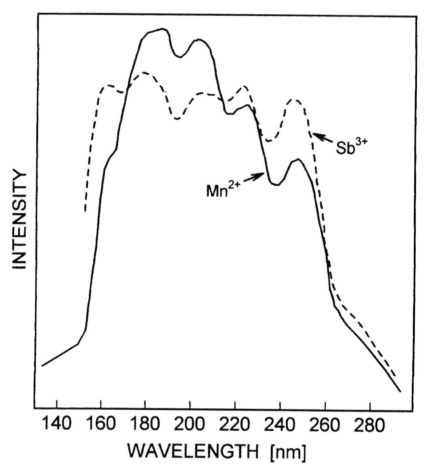

Figure 65 Excitation spectra for the blue Sb^{3+} emission and the red Mn^{2+} emission in the same system as that in Figure 64. (From Johnson, P.D., *J. Electrochem. Soc.*, 108, 159, 1961. With permission.)

1.8.1.4   Concentration quenching of luminescence

If the concentration of an activator is higher than an appropriate value (usually several wt %), the emission of the phosphor is usually lowered, as shown in Figure 66. This effect is called *concentration quenching*. The origin of this effect is thought to be one of the following:

1. Excitation energy is lost from the emitting state due to cross-relaxation (described later) between the activators.
2. Excitation migration due to the resonance between the activators is increased with the concentration (see 1.1.2), so that the energy reaches remote killers or the crystal surface acting as quenching centers.[20,21]
3. The activator ions are paired or coagulated, and are changed to a quenching center.

In some rare-earth activated phosphors, the effect of concentration quenching is so small that even stoichiometric phosphors, in which all (100%) of the host constituent cations are substituted by the activator ions, have been developed. Figure 67 shows the concentration dependence of the emission intensity of the Tb^{3+} activator in $Tb_xLa_{1-x}P_5O_{14}$, a stoichiometric phosphor, in which the emission intensity from the $^5D_4$ emitting level (see the energy level diagram of Tb^{3+} in 2.3) attains the maximum at $x = 1$.[22] This phosphor has the same crystal

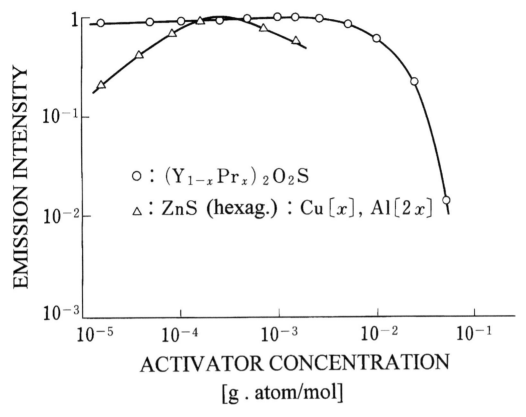

Figure 66 Activator concentration dependence of the cathode-luminescence intensities of $Y_2O_2S:Eu^{3+}$ and ZnS:Cu. (From Kuboniwa, S., Kawai, H., and Hoshina, T., *Jpn. J. Appl. Phys.*, 19, 1647, 1980. With permission.)

structure as NdP_5O_{14}, a typical stoichiometric phosphor, in which each Nd^{3+} ion is isolated by the surrounding PO_4 groups.[23]

When the concentration quenching due to cross-relaxation (relaxation due to resonant energy transfer between the same element atoms or ions [see the insertion in Figure 68]) occurs on a particular level among several emitting levels, the emission color of the phosphor changes with the activator concentration. For example, while the emission color of Tb^{3+}-activated phosphors is blue-white due to mixing of blue emission from the $^5D_3$ emitting level and green emission from the $^5D_4$ level at concentrations below 0.1%, the color changes to green at the higher concentrations. The change is caused by cross-relaxation, as shown in Figure 68,[24] between the $^5D_3$ and $^5D_4$ emitting levels, thereby diminishing the population of Tb^{3+} ions in $^5D_3$ state and increasing the one in the $^5D_4$ state.

1.8.2   Cooperative optical phenomena

In emission and absorption spectra of crystals highly doped with two types of rare-earth ions, labeled A and B, sometimes show weak additional lines other than the inherent spectral lines specific to the A and B ions. These additional lines are due to the cooperative optical processes induced in an AB ion-pair coupled by electrostatic or exchange interactions. The cooperative optical process can be divided into three types as shown in Figure 69. They are: (a) cooperative absorption ($AB + \omega_{A+B} \rightarrow A^*B^*$); (b) Raman lumines-cence ($A^*B \rightarrow AB^* + \omega_{A-B}$); and (c) cooperative luminescence ($A^*B^* \rightarrow AB + \omega_{A+B}$). The observed intensities of all these cooperative spectra are very weak (10^{-5} of the normal f-f

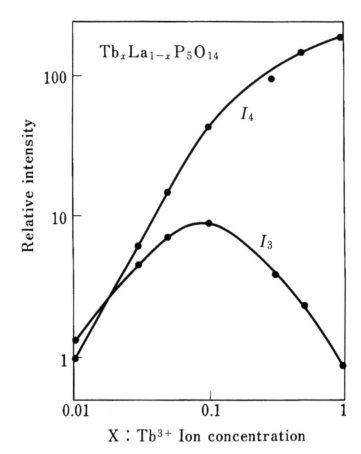

Figure 67 Activator concentration dependence of photoluminescence intensity of a "stoichiometric" phosphor, $Tb_xLa_{1-x}P_5O_{14}$, whose emission reaches the maximum intensity at the complete substitution of host La^{3+} ions by the activator Tb^{3+} ions ($x = 1$). (From Tanaka, S., Nakamura, A., Kobayashi, H., and Sasakura, H.,*Tech. Digest Phosphor Res. Soc., 166th Meeting*, 1977 (in Japanese). With permission.)

transition). Cooperative absorption has been observed for Pr^{3+}-Pr^{3+}, Pr^{3+}-Ce^{3+}, and Pr^{3+}-Ho^{3+} pairs,[25,26] Raman luminescence for Gd^{3+}-Yb^{3+} and Tm^{3+}-Tm^{2+} pairs[27,28], and cooperative luminescence for Yb^{3+}-Yb^{3+} and Pr^{3+}-Pr^{3+} pairs.[29,30]

The cooperative absorption transition $AB + \hbar\omega \rightarrow A^*B^*$ proceeds via an intermediate state A^i or B^i in a manner $AB \rightarrow A^iB^* \rightarrow A^*B^*$.[31] These three states are combined by the multipolar or exchange interaction operator H_{AB}, which is also operative in energy transfer processes described in 1.8.1, and the perturbation P by the radiation field given by $-er{\cdot}E$ for electric dipole transitions as described in 1.1.

Then, the moment of the transition (see 1.1) is given by:

$$M_{AB} = \sum_i \left[\frac{-\langle A^*|P|A^i\rangle\langle A^iB^*|H_{AB}|AB\rangle}{E_A^i + E_B} + \frac{\langle A^*B^*|H_{AB}|A^iB\rangle\langle A^i|P|A\rangle}{E_A^i - E_A - E_B} \right] \quad (156)$$

The cooperative absorption intensity of the Pr^{3+}-Pr^{3+} ion pair in $PrCl_3$ crystals, estimated by this equation, agrees well with that of observed cooperative spectra.[32,33] The estimation indicates that the dq or higher-order multipolar interaction is effective in this pair.[33]

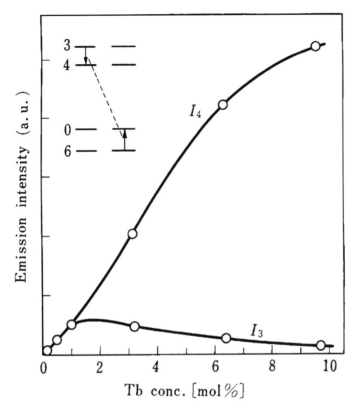

Figure 68 Activator concentration dependence of the emission intensities of the two emitting levels of Tb^{3+}, $I_3(^5D_3)$, and $I_4(^5D_4)$. The relative intensity between the two and therefore the emission color is changed from blue-white to green with the increase of the activator concentration due to cross-relaxation between $^5D_3$-$^5D_4$ and $^7F_6$-$^7F_0$. (From Nakazawa, E. and Shionoya, S., *J. Phys. Soc. Japan*, 28, 1260, 1970. With permission.)

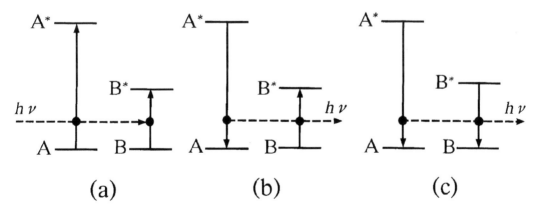

Figure 69 Transition in cooperative optical processes: (a) cooperative absorption, (b) Raman luminescence and (c) cooperative luminescence.

References

1. Foerster, Th., *Ann. Phys.*, 2, 55, 1948.
2. Dexter, D.L., *J. Chem. Phys.*, 21, 836, 1953.
3. Yen, W.M., *Modern Problems in Condensed Matter Science*, Vol. 21, Elsevier, Amsterdam, 1989, pp. 185-249.
4. Nakazawa, E. and Shionoya, S., *J. Chem. Phys.*, 47, 3211, 1967.
5. Kushida, T., *J. Phys. Soc. Japan*, 34, 1313; 1327; 1334; 1983.
6. Inokuti, M. and Hirayama, F., *J. Chem. Phys.*, 43, 1978, 1965.
7. Soules, T.H., Bateman, R.L., Hews, R.A., and Kreidler, E.H., *Phys. Rev.*, B7, 1657, 1973.
8. Perrin, F., *Compt. Rend.*, 178, 1978, 1924.
9. Miyakawa, T. and Dexter, D.L., *Phys. Rev.*, B1, 2961, 1970.
10. Yamada, N., Shionoya, S., and Kushida, T., *J. Phys. Soc. Japan*, 32, 1577, 1972.
11. Yokota, M. and Tanimoto, O., *J. Phys. Soc. Japan*, 22, 779, 1967.
12. Weber, M.J., *Phys. Rev.*, B4, 2934, 1971.
13. Krasutky, N. and Moose, H.W., *Phys. Rev.*, B8, 1010, 1973.
14. Batler, K.H. and Jerome, C.W., *J. Electrochem. Soc.*, 97, 265, 1950.
15. Johnson, P.D., *J. Electrochem. Soc.*, 108, 159, 1961.
16. Shionoya, S. and Nakazawa, E., *Appl. Phys. Lett.*, 6, 118, 1965.
17. Blasse, G. and Bril, A., *J. Chem. Phys.*, 51, 3252, 1969.
18. Tabei, M. and Shionoya, S., *Jpn. J. App. Phys.*, 14, 240, 1975.
19. Suzuki, A., Yamada, H., Uchida, Y., Kohno, H., and Yoshida, M., *Tech. Digest Phosphor Res. Soc. 197th Meeting*, 1983 (in Japanese).
20. Ozawa, L. and Hersh, H.N., *Tech. Digest Phosphor Res. Soc. 155th Meeting*, 1974 (in Japanese).
21. Kuboniwa, S., Kawai, H., and Hoshina, T., *Jpn. J. Appl. Phys.*, 19, 1647, 1980.
22. Tanaka, S., Nakamura, A., Kobayashi, H., and Sasakura, H., *Tech. Digest Phosphor Res. Soc. 166th Meeting*, 1977 (in Japanese).
23. Danielmeyer, H. G., *J. Luminesc.*, 12/13, 179, 1976.
24. Nakazawa, E. and Shionoya, S., *J. Phys. Soc. Japan*, 28, 1260, 1970.
25. Varsani, F. and Dieke, G.H, *Phys. Rev. Lett.*, 7, 442, 1961.
26. Dorman, E., *J. Chem. Phys.*, 44, 2910, 1966.
27. Feofilov, P.P. and Trifimov, A.K., *Opt. Spect.*, 27, 538, 1969.
28. Nakazawa, E., *J. Luminesc.*, 12, 675, 1976.
29. Nakazawa, E. and Shionoya, S., *Phys. Rev. Lett.*, 25, 1710, 1982.
30. Rand, S.C., Lee, L. S., and Schawlow, A.L., *Opt. Commun.*, 42, 179, 1982.
31. Dexter, D.L., *Phys. Rev.*, 126, 1962, 1962.
32. Shinagawa, K., *J. Phys. Soc. Japan*, 23, 1057, 1967.
33. Kushida, T., *J. Phys. Soc. Japan*, 34, 1318, 1973; 34, 1327, 1973; 34, 1334, 1973.

Fundamentals of luminescence

Hajime Yamamoto

Contents

1.9   Excitation mechanism of luminescence by cathode-ray and ionizing radiation

1.9.1   Introduction

Luminescence excited by an electron beam is called *cathodoluminescence* and luminescence excited by energetic particles, i.e., α-ray, β-ray or a neutron beam, or by γ-ray, is called either *radioluminescence* or *scintillation.**

The excitation mechanism of cathodoluminescence and of radioluminescence can be discussed jointly because these two kinds of luminescence have a similar origin. In solids, both the electron beam and the high-energy radiation induce ionization processes, which in turn generate highly energetic electrons. These energetic electrons can be further multiplied in number through collisions, creating "secondary" electrons, which can then migrate in the solid with high kinetic energy, exciting the light-emitting centers. The excitation mechanism primarily relevant to cathodoluminescence is discussed here.

1.9.2   Collision of primary electrons with solid surfaces

Energetic electrons incident on a solid surface in vacuum are called *primary electrons* and are distinct from the secondary electrons mentioned above. A small fraction of the electrons is scattered and reflected back to the vacuum, while most of the electrons penetrate into

---

* The word originally means flash, as is observed under particle beam excitation.

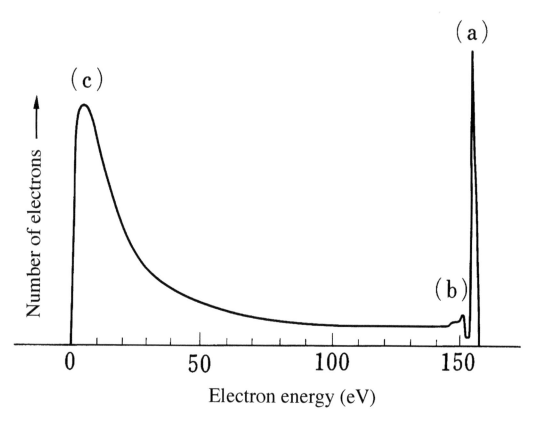

Figure 70   The energy distribution of electrons emitted from the Ag surface exposed by the primary electrons of 153 eV: (a)electrons emitted by elastic scattering, (b) electrons by inelastic scattering and (c) secondary electrons. (From Dekker, A.J., *Solid State Physics*, Prentice-Hall, Tokyo, 1960, 418-420. With permission.)

the solid. The reflected electrons can be classified into three types: (a) elastically scattered primary electrons, (b) inelastically scattered primary electrons, and (c) secondary electrons.[1] The secondary electrons mentioned here are those electrons generated by the primary electrons in the solid and are energetic enough to overcome the work function of the solid surface. This phenomenon, i.e., the escape of secondary electrons from the solid, is similar to the photoelectric effect. The relative numbers of the three types of scattered electrons observed for the Ag surface are shown in Figure 70.[1,2] As shown in this figure, the inelastically scattered primary electrons are much smaller in number than the other two types.

The ratio of the number of the emitted electrons to the number of the incident electrons is called the *secondary yield* and usually denoted as δ. With this terminology, δ should be defined only in terms of the secondary electrons (c), excluding (a) and (b). However, in most cases, δ is stated for all the scattered electrons—(a), (b), and (c)—for practical reasons.

For an insulator, δ depends on the surface potential relative to the cathode as is schematically shown in Figure 71. For $\delta < 1$, the insulator surface is negatively charged; as a consequence, the potential of a phosphor surface is not raised above V_{II} shown in Figure 71, even for an accelerating voltage higher than V_{II}. In other words, the surface potential stays at V_{II} and is called the sticking potential. To prevent electrical charging of surface, an aluminizing technique is employed in cathode-ray tubes (CRTs). Negative charging of a phosphor is also a problem for vacuum fluorescent tubes and some of field emission displays, which use low-energy electrons at a voltage below V_I. The aluminizing technique cannot be used in these cases, however, because the low-energy primary

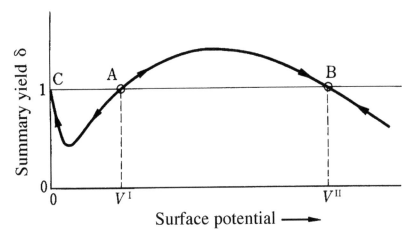

Figure 71 A schematic illustration of secondary yield as a function of the surface potential of an insulator. The secondary yield δ is unstable at point A, while it is stable at B and C. At these points, the state shifts toward the direction of the given arrows with a change in the potential. Near point C, where the potential is in a region of a few to several tens of volts, the yield approaches 1 because the incident primary electrons are reflected. (From Kazan, B. and Knoll, M., *Electron Image Storage*, Academic, New York, 1968, 22. With permission.)

electrons, for example a few ten or hundred eV, cannot go through an aluminum film, even if it is as thin as 100 nm, which is practically the minimum thickness required to provide sufficient electrical conductivity and optical reflectivity. It is, therefore, required to make the phosphor surface electrically conductive.

To evaluate a cathodoluminescence efficiency, one must exclude the scattered primary electrons (a) and (b) in Figure 70. The ratio of the electrons (a) and (b) to the number of the incident electrons is called *back-scattering factor*, denoted by η_0. Actually, the electrons (b) are negligible compared with the electrons (a) as shown in Figure 70. The value of η_0 depends weakly on the primary electron energy but increases with the atomic number of a solid. η_0 obeys an empirical formula, with the atomic number or the number of electrons per molecule being Z_m; that is:

$$\eta_0 = (1/6)\ln Z_m - (1/4) \tag{157}$$

The value calculated by this formula agrees well with experimental results obtained for single-crystal samples. For example, the calculated values for ZnS with Z_m of 23 is 0.25 and for YVO$_4$ with Z_m of 15.7 is 0.21, while the observed values for single crystals of these compounds are 0.25 and 0.20, respectively.[4] In contrast, a smaller value of η_0 is found for a powder layer because some of the reflected electrons are absorbed by the powder through multiple-scattering. The observed values of η_0 are 0.14,[4] both for ZnS and YVO$_4$ in powder form. It has also been reported that η_0 varies by several percent depending on the packing density of a powder layer.[6]

1.9.3 Penetration of primary electrons into a solid

The penetration path of an electron in a solid has been directly observed with an optical microscope by using a fine electron beam of 0.75 μm diameter (Figure 72). This experiment shows a narrow channel leading to a nearly spherical region for electron energy higher than 40 keV, while it shows a semispherical luminescent region for lower electron energies.

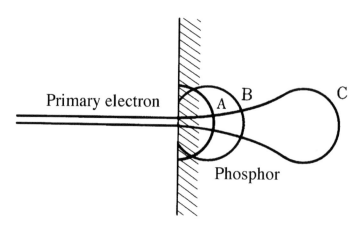

Figure 72 A schematic illustration of a region excited by an electron beam. This region can be visualized as the luminescent profile of the solid phosphor particle as seen through a microscope. The energy of the primary electrons increases in the order of A, B, and C. For A, the energy is several keV and for C, 40 keV or higher. (From Ehrenberg, W. and Franks, J., *Proc. Phys. Soc.*, B66, 1057, 1953; Garlick, G.F.J., *Br. J. Appl. Phys.*, 13, 541, 1962. With permission.)

The former feature is found also for high-energy particle excitation, i.e., the excitation volume is confined to a narrow channel until the energy is dissipated by ionization processes. This result indicates that the scattering cross-section of an electron or a particle in a solid is larger for lower electron energy. The energy lost by a charged particle passing through a solid is expressed by Bethe's formula[9]:

$$dE/dx = \left(2\pi N Z_m e^4 / E\right)\ln\left(E/E_i\right) \tag{158}$$

where E denotes the energy of a primary electron at distance x from the solid surface, N the electron density (cm^{-3}) of the solid, Z_m the mean atomic number of the solid, and E_i the mean ionization potential averaged over all the electrons of the constituent atoms.

Various formulas have been proposed to give the relation between E and x. Among them, the most frequently used is Thomson-Whiddington's formula,[10] which we can derive from Eq. 158 simply by putting $\ln(E/E_i)$ = constant.

$$E = E_0\left(1 - x/R\right)^{1/2} \tag{159}$$

Here, E_0 is the primary electron energy at the surface and R is a constant called as the range, i.e., the penetration depth at $E = 0^*$. It is to be noted that an incremental energy loss, $-dE/dx$, increases with x according to Eq. 159.

In a range of E_0 = 1–10 keV, the dependence of R on E_0 is given by[11]:

$$R = 250\left(A/\rho\right)\left(E_0 / Z_m^{1/2}\right)^n \tag{160}$$

where $n = 1.2/(1 - 0.29\log Z_m)$, ρ is the bulk density, A the atomic or molecular weight, Z_m the atomic number per molecule, and E_0 and R are expressed in units of keV and Å, respectively. When E_0 = 10 keV, Eq. 160 gives R = 1.5 μm for ZnS and R = 0.97 μm for $CaWO_4$. The experimental values agree well with the calculated values.

---

* Other formulas define the range as the penetration depth at $E = E_0/e$.

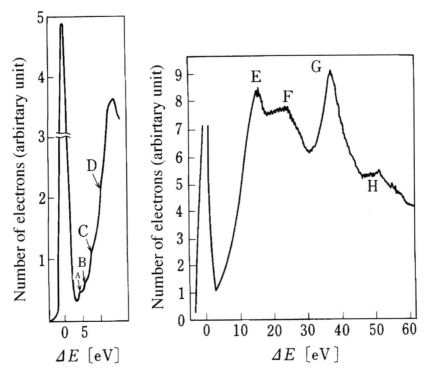

Figure 73   Electron energy loss spectra of YVO_4: (a) peaks A to D originate in the electronic transitions of the VO_4^{-3} complex; (b) peak E can be assigned to plasmon excitation. Peak G is due to a transition from Y $4p$ orbital to the conduction band and, peak H from V $3p$ to the conduction band. The origin of peak F is not identified. The strong peak at 0 eV indicates the incident electrons with no energy loss. (From Tonomura, A., Endoh, J., Yamamoto, H., and Usami, K., *J. Phys. Soc. Japan*, 45, 1654, 1978. With permission.)

When E_0 is decreased at a fixed electron beam current, luminescence vanishes at a certain positive voltage, called the dead voltage. One of the explanations of the dead voltage is that, at shallow R, the primary electron energy is dissipated within a dead layer near the surface, where nonradiative processes dominate as a result of a high concentration of lattice defects.[12] It is also known, however, that the dead voltage decreases with an increase in electrical conductivity, indicating that the dead voltage is affected by electrical charging as well.

1.9.4   Ionization processes

A charged particle, such as an electron, loses its kinetic energy through various modes of electrostatic interaction with constituent atoms when it passes through a solid. Elementary processes leading to energy dissipation can be observed experimentally by the electron energy loss spectroscopy, which measures the energy lost by a primary electron due to inelastic scattering (corresponding to the electrons (b) in Figure 70). Main loss processes observed by this method are core-electron excitations and creation of plasmons, which are a collective excitation mode of the valence electrons. Core-electron excitation is observed in the range of 10 to 50 eV for materials having elements of a large atomic number, i.e., rare-earth compounds or heavy metal oxides such as vanadates or tungstates.[13] The plasmon energy is found in the region of 15 to 30 eV. Compared with these excitation modes, the contribution of the band-to-band transition is small. As an example exhibiting various modes of excitation, the electron energy loss spectrum of YVO_4 is shown in Figure 73.[14]

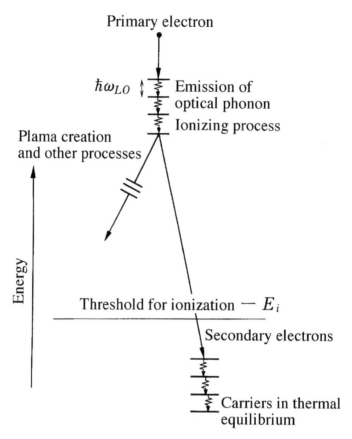

Figure 74 A schematic illustration of excitation processes by a high-energy electron, which penetrates into a solid. (From Robbins, D.J., *J. Electrochem. Soc.*, 127, 2694, 1980. With permission.)

Plasmons are converted to single-electron excitations in an extremely short period of time, ~10^{-15} s. As a consequence, electrons with energies of 10 to 50 eV are created every time an energetic primary electron is scattered in a solid as a result of core-electron excitation or plasmon creation. This results in a series of ionization processes in a solid. Most of the electrons generated by the scattering events, or the secondary electrons, are still energetic enough to create other hot carriers by Auger processes. Secondary electron multiplication can last until the energy of the electron falls below the threshold to create free carriers. All through this electron energy loss process, scattering is accompanied by phonon creation, as schematically shown in Figure 74.[15] Secondary electron multiplication is essentially the same as the photoexcitation process in the vacuum ultraviolet region.

The average energy required to create an electron-hole pair near the band edges, E_{av}, is given by the following empirical formula.[16]

$$E_{av} = 2.67E_g + 0.87 \ [\text{eV}] \tag{161}$$

where E_g is the bandgap energy either for the direct or the indirect gap. This formula was originally obtained for elements or binary compounds with tetrahedral bonding, but it is applied often to phosphors with more complex chemical compositions and crystal structures. It is not, however, straightforward to define the bandgap energy for a material having low-lying energy levels characteristic of a molecular group, e.g., vanadates or tungstates. Therefore, one must be careful in applying the above formula to some phosphors.

As described above, the average creation energy of an electron-hole pair is closely related to the cathodoluminescence efficiency (see also Section 1.9.6). There is, however, another way to consider the luminescence efficiency; it focuses on phonon emission,[16] which competes with the electron-hole pair creation in the ionization processes. The phonon emission probability, denoted as R_p here, is proportional to the interaction of an electron with an optical phonon, and is expressed as:

$$R_p \propto (\hbar\omega_{LO})^{1/2}(1/\varepsilon_\infty - 1/\varepsilon_0) \tag{162}$$

where ω_{LO} is the energy of a longitudinal optical phonon interacting with an electron, and ε_∞ and ε_0 are high-frequency and static dielectric constants, respectively. When multiplied with the phonon energy, the probability R_p contributes to the pair creation energy E_{av} as a term independent of E_g, e.g., the second term 0.87 eV in Eq. 161.

The luminescence excited by energetic particles is radioluminescence.[17] The excitation mechanism of radioluminescence has its own characteristic processes, though it involves ionization processes similar to the cathodoluminescence processes.

For example, the energy of γ-rays can be dissipated by three processes: (1) the Compton effect, (2) the photoelectric effect directly followed by X-ray emission and Auger effect, and (3) the creation of electron-positron pairs. Subsequent to these processes, highly energetic secondary electrons are created, followed by the excitation of luminescence centers, as is the case with cathodoluminescence.

A characteristic energy loss process of neutrons, which has no electric charges but much larger mass than an electron, is due to the recoil of hydrogen atoms. If the neutron energy is large enough, a recoiled hydrogen is ionized and creates secondary electrons. It must be added, however, that hydrogen atoms are not contained intentionally in inorganic phosphors.

1.9.5   Energy transfer to luminescence centers

The final products of the secondary-electron multiplication are free electrons and free holes near the band edge, i.e., so-called *thermalized* electrons and holes. They recombine with each other, and a part of the recombination energy may be converted to luminescence light emission.

The process in which either a thermalized electron-hole pair or the energy released by their recombination is transferred to a luminescence center is called *host sensitization* because the luminescence is sensitized by the optical absorption of the host lattice. This process is analogous to the optical excitation near the band edge. Detailed studies were made on the optical excitation of luminescence in IIb-VIb and IIIb-Vb compounds, as described in 2.7 and 2.8. Luminescence of rare-earth ions and Mn^{2+} ions arises because these ions capture electrons and holes by acting as isoelectronic traps.[18,19] In inorganic compounds having complex ions and organic compounds, the excitation energy is transferred to the luminescence centers through the molecular energy levels.

1.9.6   Luminescence efficiency

The cathodoluminescence energy efficiency η, for all the processes described above can be expressed by[20]:

Table 2   Examples of Cathodoluminescence Efficiency

Chemical composition	WTDS designation	Energy efficiency (%)	Peak wavelength (nm)	Luminescence color
$Zn_2SiO_4{:}Mn^{2+}$	GJ	8	525	Green
$CaWO_4{:}Pb$	BJ	3.4	425	Blue
$ZnS{:}Ag,Cl$	X	21	450	Blue
$ZnS{:}Cu,Al$	X	23, 17	530	Green
$Y_2O_2S{:}Eu^{3+}$	X	13	626	Red
$Y_2O_3{:}Eu^{3+}$	RF	8.7	611	Red
$Gd_2O_2S{:}Tb^{3+}$	GY	15	544	Yellowish green
$CsI{:}Tl^+$	—	11	—	Green
$CaS{:}Ce^{3+}$	—	22	—	Yellowish green
$LaOBr{:}Tb^{3+}$	—	20	544	Yellowish green

Note: The phosphor screen designation by WTDS (Worldwide Phosphor Type Designation System) is presented.

Many data are collected in Alig, R.C. and Bloom, S., *J. Electrochem. Soc.*, 124, 1136, 1977.

$$\eta = \left(1 - \eta_0\right)\eta_x\left(E_{em}/E_g\right)q \tag{163}$$

where η_0 is the back-scattering factor given by Eq. 157, η_x the mean energy efficiency to create thermalized electrons and holes by the primary electrons or E_g/E_{av}, q the quantum efficiency of the luminescence excited by thermalized electron-hole pairs, and E_{em} the mean energy of the emitted photons. Thus,

$$\eta < \eta_x\left(E_{em}/E_g\right) \tag{164}$$

and also $\eta_x < 1/3$ according to Eq. 161.

The energy efficiency, luminescence peak wavelength and color are shown in Table 2 for some efficient phosphors. For the commercial phosphors, $ZnS{:}Ag,Cl$; $ZnS{:}Cu,Al$; $Y_2O_2S{:}Eu^{3+}$; and $Y_2O_3{:}Eu^{3+}$, we find $\eta_x \simeq 1/3$ from Eq. 163 by assuming that $\eta_0 = 0.1$ and $q = 0.9$–1.0. This value of η_x suggests that the energy efficiency is close to the limit predicted by Eq. 163 for these phosphors. It is to be emphasized, however, that this estimate does not exclude a possibility for further improvement in the efficiency of these phosphors, for example by 10 or 20%, since the calculated values are based on a number of approximations and simplifying assumptions. It should also be noted that the bandgap energy is not known accurately for the phosphors given in Table 2, except for ZnS, CsI, and CaS. For the other phosphors, the optical absorption edge must be used instead of the bandgap energy, leaving the estimation of η approximate. For CaS, the indirect bandgap, 4.4 eV, gives $\eta_x = 0.21$, while the direct bandgap, 5.3 eV, gives the value exceeding the limit predicted by Eq. 163.

References

1. Dekker, A.J., *Solid State Physics*, Prentice-Hall, Maruzen, Tokyo, 1960, 418-420.
2. Rudberg, E., *Proc. Roy. Soc. (London)*, A127, 111, 1930.
3. Tomlin, S.G., *Proc. Roy. Soc. (London)*, 82, 465, 1963.
4. Meyer, V.G., *J. Appl. Phys.*, 41, 4059, 1970.
5. Kazan, B. and Knoll, M., *Electron Image Storage*, Academic Press, New York, 1968, 22.
6. Kano, T. and Uchida, Y., *Jpn. J. Appl. Phys.*, 22, 1842, 1983.
7. Ehrenberg, W. and Franks, J., *Proc. Phys. Soc.*, B66, 1057, 1953.

8. Garlick, G.F.J., *Br. J. Appl. Phys.*, 13, 541, 1962.
9. Bethe, H.A., *Ann. Physik*, 13, 541, 1930.
10. Whiddington, R., *Proc. Roy. Soc. (London)*, A89, 554, 1914.
11. Feldman, C., *Phys. Rev.*, 117, 455, 1960.
12. Gergley, Gy., *J. Phys. Chem. Solids*, 17, 112, 1960.
13. Yamamoto, H. and Tonomura, A., *J. Luminesc.*, 12/13, 947, 1976.
14. Tonomura, A., Endoh, J., Yamamoto, H., and Usami, K., *J. Phys. Soc. Japan*, 45, 1654, 1978.
15. Robbins, D.J., *J. Electrochem. Soc.*, 127, 2694, 1980.
16. Klein, C.A., *J. Appl. Phys.*, 39, 2029, 1968.
17. For example, Brixner, L.H., *Materials Chemistry and Physics*, 14, 253, 1987; Derenzo, S.E., Moses, W.W., Cahoon, J.L., Perera, R.L.C., and Litton, J.E., *IEEE Trans. Nucl. Sci.*, 37, 203, 1990.
18. Robbins, D.J. and Dean, P.J., *Adv. Phys.*, 27, 499, 1978.
19. Yamamoto, H. and Kano, T., *J. Electrochem. Soc.*, 126, 305, 1979.
20. Garlick, G.F.J., *Cathodo- and Radioluminescence in Luminescence of Inorganic Solids*, Goldberg, P., Ed., Academic Press, New York, 1966, 385-417.
21. Alig, R.C. and Bloom, S., *J. Electrochem. Soc.*, 124, 1136, 1977.

chapter one — section ten

Fundamentals of luminescence

**Shosaku Tanaka, Hiroshi Kobayashi, Hiroshi Sasakura,
and Noboru Miura**

Contents

1.10   Inorganic electroluminescence

1.10.1   Introduction

Electroluminescence (EL) is the generation of light by the application of an electric field to crystalline materials, or resulting from a current flow through semiconductors. The EL of inorganic materials is classified into the two groups: injection EL and high electric field EL. The high- field EL is further divided into two types: powder phosphor EL and thin-film EL. The classification of EL with regard to typical device applications is summarized as follows:

EL ──→ Injection EL ──────→ Light-emitting diodes (LED), Laser diodes (LD)
   └─→ High-field EL ──→ Powder phosphor EL ──────→ EL illumination panels
                   └─→ Thin-film EL ──────────→ EL display panels

Historically, the EL phenomenon was first observed by Destriau[1,2] in 1936, who observed luminescence produced from ZnS powder phosphors suspended in castor oil when a strong electric field was applied. This type of EL is, today, classified as powder phosphor EL. Later on, in the early 1960s, polycrystalline ZnS thin films were prepared and used as EL materials. This type of EL is typical of thin-film EL.

On the other hand, in 1952 Haynes and Briggs[3] reported infrared EL from forward-biased *p-n* junctions in Ge and Si diodes. This type of EL is classified as injection EL. Visible EL is observed in diodes made of wide bandgap semiconductors, such as GaP. These diodes are called light-emitting diodes (LEDs) and have been widely used since the late 1960s. Semiconductor lasers, first demonstrated in 1962 using GaAs diodes, operate by stimulating injection EL light in an appropriate optical cavity. As will be described below, the mechanisms of light generation in injection EL and high-field EL are quite different from each other. In addition, the applications of these EL phenomena to electronic devices are different.

Usually, the term EL is used, in a narrow sense, to mean high-field EL. In this section, therefore, the description will focus on the basic processes of the high-field EL, in particular on the excitation mechanisms in thin-film EL. The mechanisms of injection EL are described only briefly.

1.10.2   Injection EL

The term "injection EL" is used to explain the phenomenon of luminescence produced by the injection of minority carriers. Energy band diagrams for p-n junction at thermal equilibrium and under forward biased conditions (p-type side:positive) are shown in Figures 75(a) and (b), respectively. At thermal equilibrium, a depletion layer is formed and a diffusion potential V_d across the junction is produced. When the p-n junction is forward-biased, the diffusion potential V_d decreases to $(V_d - V)$, and electrons are injected from the n-region into the p-region while holes are injected from the p-region into the n-region; that is, minority carrier injection takes place. Subsequently, the minority carriers diffuse and recombine with majority carriers directly or through trapping at various kinds of recombination centers, producing injection EL. The total diffusion current on p-n junction is given by:

$$J = J_p + J_n = J_s \left(\exp\left(\frac{qV}{nkT} \right) - 1 \right)$$

$$J_s = q \left(\frac{D_p p_{n0}}{L_p} + \frac{D_n n_{p0}}{L_n} \right)$$

(165)

where D_p and D_n are diffusion coefficients for holes and electrons, p_{no} and n_{po} are the concentrations of holes and electrons as minority carriers at thermal equilibrium, and L_p and L_n are diffusion lengths given by $\sqrt{D\tau}$, where τ is the lifetime of the minority carriers.

The LEDs that became commercially available in the late 1960s were the green-emitting GaP:N and the red-emitting GaP:Zn,O diodes. GaP is a semiconductor having an indirect bandgap; the N and (Zn,O) centers in GaP are isoelectronic traps that provide efficient recombination routes for electrons and holes to produce luminescence in this material (See 2.8.2). Very bright LEDs used for outdoor displays were developed using III-V compound alloys in the late 1980s to early 1990s; these alloys all have a direct bandgap. Green-, yellow-, orange-, and red-emitting LEDs with high brightness are fabricated using InGaAlP, GaAsP, or GaAlAs (See 2.8.3). In 1993 to 1994, GaInN (another alloy with a direct bandgap) was developed, leading to very bright blue and green LEDs (See 2.8.5). Thus, LEDs covering the entire visible range with high brightness are now commercially available.

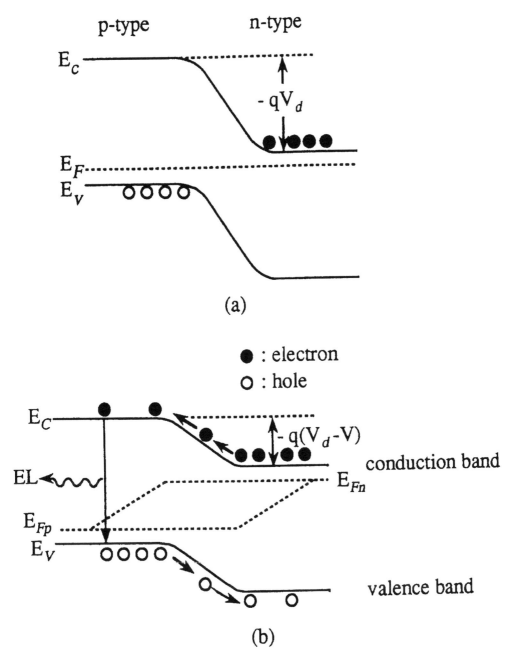

Figure 75 Energy band diagrams for the p-n junction under (a) thermal equilibrium and under (b) forward-biased conditions.

1.10.3 High-field EL

In the case of thin-film EL used for display panels, a high electric field of the order of 10^6 V cm^{-1} is applied to the EL materials. Electrons, which are the majority carriers in this case, are injected into the EL emitting layers. These electrons are accelerated by the electric field until some of them reach energies sufficient to cause impact excitation of luminescent centers generating EL light. The most common luminescent centers in ZnS and other EL hosts are Mn^{2+} and the rare-earth ions; these activators offer a wide variety of emission colors. As noted above, the mechanism of high-field thin-film EL is quite different from that of injection EL. Here, the basic processes of high-field EL—that is, (1) the injection of carriers, (2) the carrier energy distribution in the high electric field region, and (3) the excitation mechanism of the luminescence centers—are discussed.

In the case of high-field, powder phosphor EL, electrons and holes are injected by tunnel emission (field emission) induced by high electric field (10^6 V cm^{-1}) applied to a conductor-phosphor interface. The excitation mechanism is similar to that of thin-film EL, and is also discussed in this section.

1.10.3.1 Injection of carriers

Injection of majority carriers through a Schottky barrier.[4]   When a semiconductor is in contact with metal, a potential barrier, called the Schottky barrier, is formed in the contact region. Before interpreting the Schottky effect in the metal-semiconductor system, one can consider this effect in a metal-vacuum system, which will then be extended to the metal-semiconductor barrier.

The minimum energy necessary for an electron to escape into vacuum from its position within the Fermi distribution is defined as the work function $q\phi_m$, as shown in Figure 76. When an electron is located at a distance x from the metal, a positive charge will be induced on the metal surface. The force of attraction between the electron and the induced positive charge is equivalent to the force that would exist between the electron and the positive image charge located at a distance of $-x$. The attractive force is called the image force, and is given by: $F = -q^2/16\pi\varepsilon_0x^2$, where ε_0 is a permittivity in vacuum. The potential is given

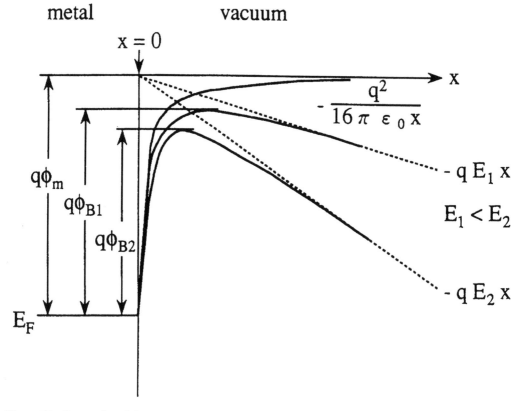

Figure 76   Energy band diagram representing the Schottky effect between metal surface and vacuum. The barrier lowering under reverse bias is $q(\phi_m - \phi_B)$.

by $U = q^2/16\pi\varepsilon_0 x$. When an external field E is applied, the total potential energy is given by: $U_T = q^2/16\pi\varepsilon_0 x + qEx$, as shown in the figure. Thus, at high fields, the potential barrier is lowered considerably, and the effective metal work function for thermoionic emission $q\phi_B$ is reduced. This lowering of the potential barrier induced by the image charge is known as the Schottky effect.

The energy band diagrams for an *n*-type semiconductor in contact with a metal are shown for the case of thermal equilibrium and under reverse-biased conditions (semiconductor side: positive) in Figures 77(a) and (b), respectively. When an electric field is applied to the metal-semiconductor contact region, the potential energy is lowered in the semiconductor by the image force or Schottky effect. The barrier height $q\Phi_B$ is lowered as discussed and electrons can be thermally injected into the semiconductor. The current density for this process is expressed as:

$$J = AT^2 \exp\left(\frac{-q\Phi_B - \left(qE/4\pi\varepsilon_s\right)^{1/2}}{kT}\right)$$

$$\sim T^2 \exp\left(\frac{aV^{1/2}}{T} - \frac{q\Phi_B}{kT}\right)$$

(166)

where the permittivity in the semiconductor ε_s is used instead of that in vacuum, ε_0. The injected electrons are then accelerated by the electric field and excite the luminescent centers by impact.

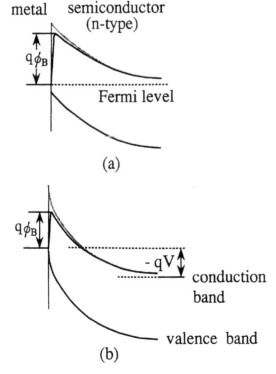

Figure 77 Energy band diagrams for the *n*-type semiconductor in contact with metal (a) under thermal equilibrium and (b) under reverse-biased conditions.

In the case of ZnS:Cu EL powder phosphors, a Schottky barrier is thought to be formed between the *n*-ZnS semiconductor and the Cu metal or the conducting Cu$_x$S microparticles found in the phosphors. In the latter, electron injection occurs from the conductive phase through the Schottky barrier and causes the electroluminescence.

Injection of carriers due to Poole-Frenkel emission.[4]   Semiconductors with fairly wide gaps of 3.5 to 4.5 eV (such as ZnS, CaS, and SrS) are used as EL materials. In these compounds, a large number of electrons are usually trapped in traps caused by lattice defects. When an electric field is applied, trapped electrons are released into the conduction band, as shown in Figure 78. This process is known as the Poole-Frenkel emission process and is due to field-enhanced thermal excitation of trapped electrons into the conduction band. For an electron trapped by a Coulomb-like potential U ∝ $1/r^n$, the expression for this process is identical to that for Schottky emission. With the barrier height qϕ_B reduced by the electric field as shown in the figure, the current density due to Poole-Frenkel emission is expressed as:

$$J \simeq E \exp\left(\frac{-q\Phi_B - \left(qE/\pi\varepsilon_s\right)^{1/2}}{kT} \right)$$

(167)

In the case of thin-film EL devices, a fraction of the initial electrons is injected by this process.

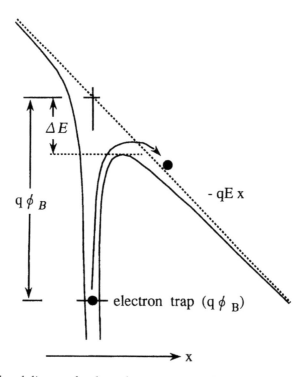

Figure 78   Energy band diagram for deep electron traps under high electric field. Electron injection due to Poole-Frenkel effect is illustrated.

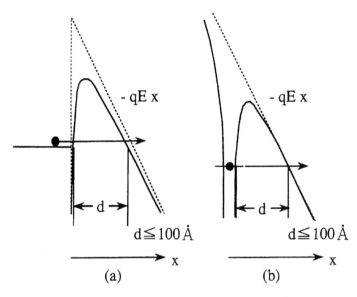

Figure 79 Energy band diagram for (a) Schottky barrier and for (b) deep electron traps under high electric field. Electron injection due to the tunneling effect is illustrated.

Injection of carriers due to tunnel emission (field emission).[4] When an extremely high electric field of over 10^6 V cm^{-1} is applied to a Schottky barrier or to electron traps, the barrier width d becomes very thin, with a thickness in the neighborhood of 100 Å. In this case, electrons tunnel directly into the conduction band, as illustrated in Figure 79. The current density due to this process depends only on the electric field and does not depend on temperature, and is described by:

$$J \simeq E^2 \exp\left(\frac{-4(2m^*)^{1/2}(q\Phi_B)^{3/2}}{3q\hbar E}\right)$$

$$\simeq V^2 \exp\left(-\frac{b}{V}\right)$$

(168)

where m^* is the effective electron mass.

Since the average electric field within thin phosphor films used in EL panels is nearly 10^6 V cm^{-1}, it is possible to conclude that electron injection due to tunneling takes place in addition to Schottky and Poole-Frenkel emission, with tunneling emission becoming predominant at high electric field conditions.

For powder-type EL devices, it is known that thin embedded Cu$_x$S conducting needles are formed in the ZnS microcrystals. Although the average applied electric field in the devices is about 10^4–10^5 V cm^{-1}, the electric field is concentrated at the tips of these microcrystals and the local electric field can be 10^6 V cm^{-1} or more. Electrons are injected by tunneling from one end of the needle and holes from the other end. This mechanism is known as the bipolar field-emission model. The injected electrons recombine with holes, which were injected by the same process and were trapped at centers previously, thus producing EL.

Injection of carriers from interfacial state.[5] When semiconductors are in contact with insulators (dielectric materials), states are formed at the interface having energy levels

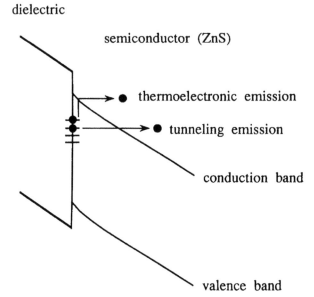

dielectric

semiconductor (ZnS)

● thermoelectronic emission

● tunneling emission

conduction band

valence band

Figure 80 Energy band diagram of a dielectric material and a semiconductor in contact. Electron injection from interfacial states under high electric field is illustrated.

distributed in the forbidden bandgap of the semiconductors, as illustrated in Figure 80. The density of the interfacial states is of the order of 10^{12}–10^{13} cm^{-2}. When an electric field is applied, electrons trapped in these states are injected into the conduction band due to tunneling and/or Poole-Frenkel emission. A typical ac-thin-film EL panel has a doubly insulating structure consisting of glass substrate/ITO (indium-tin oxide) transparent electrodes/insulating layer/EL phosphor layer/insulating layer/metal electrodes. For this type of EL device, the dominant electron injection mechanism into the EL phosphor layer is field emission from the insulator/EL phosphor interfacial states.

1.10.3.2   Electron energy distribution in high electric field

At thermal equilibrium, electrons in semiconductors emit and absorb phonons, but the net rate of energy exchange between the electrons and the lattice is zero. The energy distribution of electrons at thermal equilibrium is expressed by the Maxwell-Boltzmann distribution function as:

$$f(\varepsilon) = \exp\left(-\frac{\varepsilon}{kT}\right), \qquad \varepsilon = \frac{m^* v^2}{2} \tag{169}$$

This distribution function is spherical in momentum space, as illustrated in Figure 81(a).

In the presence of an electric field, the electrons acquire energy from the field and lose it to the lattice by emitting more phonons. Simultaneously, the electrons move with the drift velocity v_d, proportional to and in the direction of the electric field. In this case, the energy distribution of the electrons changes to a displaced Maxwell-Boltzmann distribution function (see Figure 81(b)) given by:

$$f(\varepsilon) = \exp\left(-\frac{\varepsilon}{kT}\right), \qquad \varepsilon = \frac{m^* (v - v_d)^2}{2} \tag{170}$$

At moderately high electric field (~10^3 V cm^{-1}), the most frequent scattering event is the emission of optical phonons. The electrons acquire on the average more energy than

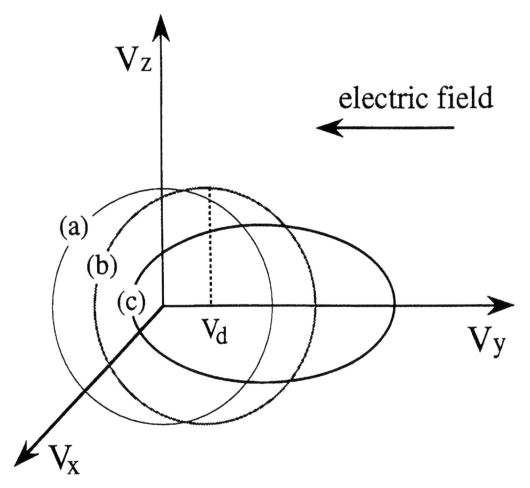

Figure 81 Electron energy distributions is in the momentum space as a function of electron velocity: (a) Maxwell-Boltzmann, (b) displaced Maxwell-Boltzmann, and (c) Baraff's distribution function.

they have at thermal equilibrium describable by an effective temperature T_e higher than the lattice temperature T_L. These electrons, therefore, are called *hot electrons*. However, the energy of hot electrons is still too low to excite luminescent centers or to ionize the lattice; the thermal energy of hot electrons is only 0.05 eV, even at $T_e = 600K$, while an energy of at least 2 to 3 eV is required for the impact excitation of luminescent centers.

When the electric field in semiconductors is increased above 10^5 V cm^{-1}, electrons gain enough energy to excite luminescent centers by impact excitation and also to create electron-hole pairs by impact ionization of the lattice. The energy distribution of these hot electrons can be expressed by Baraff's distribution function[6] (see Figure 81(c)), given by:

$$f(\varepsilon) = \varepsilon^{-a+0.5} \exp(-b\varepsilon)$$

$$a = \frac{E_O - qE\lambda}{2E_O + qE\lambda} \tag{171}$$

$$b^{-1} = \frac{3}{2}qE\lambda + \frac{1}{3}\frac{(qE\lambda)^2}{E_O}$$

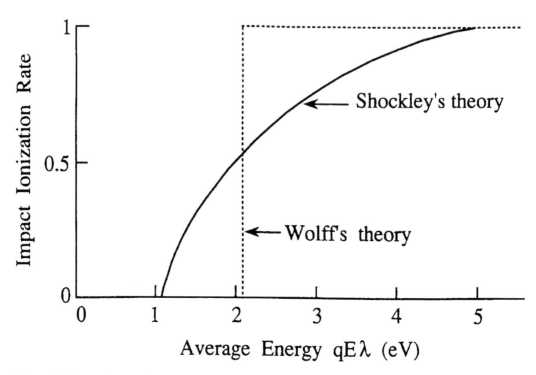

Figure 82 Dependence of impact ionization rate on the average electron energy calculated from Shockley's and Wolff's theories.

where E_O is the optical phonon energy and λ is the mean free path of electrons.

In the case of moderately high electric field ($qE\lambda < E_o$). When the electric field is moderately high, and the average electron energy, $qE\lambda$, is smaller than the optical phonon energy E_o, Eq. 171 can be reduced to the following form:

$$f(\varepsilon) \propto \varepsilon^{-1} \exp\left(-\frac{\varepsilon}{qE\lambda}\right) \tag{172}$$

This function agrees with Shockley's distribution function,[7] and implies that some electrons with very high energy can exist, even in the case of relatively low electric fields. This model is, therefore, called the lucky electron model. The impact ionization rate increases when the average electron energy is increased, as illustrated by the solid curve in Figure 82. The threshold energy—in other words, the threshold field—for impact ionization is relatively low in this model.

In the case of extremely high electric field ($qE\lambda \gg E_o$). When the electric field is extremely high, and the average electron energy, $qE\lambda$, is larger than the optical phonon energy E_O, Eq. 171 can be rearranged into the following form:

$$f(\varepsilon) \propto \exp\left(-\frac{3\varepsilon E_0}{(qE\lambda)^2}\right) \tag{173}$$

This function agrees with Wolff's distribution function derived using the diffusion approximation[8]; Eq. 173 gives a threshold energy for the impact ionization that is higher than that for the lucky electron model, as shown in Figure 82.

Recently, Bringuier[9,10] investigated electron transport in ZnS-type, thin-film EL. Two basic transport modes in the lucky-drift theory are considered. First, the ballistic regime, which is defined in terms of the optical-phonon mean free path λ and the electron-phonon collision rate $1/\tau_m$. This regime implies a collision-free (ballistic) mode. Second is the drift regime, which is characterized by the length λ_e and the rate $1/\tau_e$ of the energy relaxation. This mode predominates after the electron has suffered one collision since, once it has collided, it is deflected and the probability of other collisions is greatly increased. In the ballistic mode, an electron travels with a group velocity $v_g(\varepsilon)$, so that $\lambda = v_g\tau_m$; while in the drift mode, the motion is governed by a field-dependent drift velocity $v_d(\varepsilon)$ and $\lambda = v_d\tau_e$. The lucky-drift model may be applied to the case where $\tau_e \gg \tau_m$ and $\lambda_e \gg \lambda$, which should hold true for wide-gap semiconductors in the high-field regime. When these two inequalities are fulfilled, each collision results in an appreciable momentum loss for the electron, with little energy loss. Over the energy relaxation length, an electron drifting in the field loses its momentum and direction, but conserves much of its energy.

The energy exchange between electrons and phonons is described by the electron-phonon interaction Hamiltonian, where electrons can emit or absorb one phonon at a time. Because a phonon is a boson, the probability of the phonon occupation number changing from n to (n+1) is proportional to (n+1), while a change from n to (n–1) is proportional to n. Therefore, the ratio of the phonon emission $r_e(n \rightarrow n+1)$ to the phonon absorption $r_a(n \rightarrow n-1)$ rates is given by (n+1)/n. Because $r_e > r_a$, an electron experiences a net energy loss to the lattice, tending to stabilize the electron drift. Hot electrons in high electric field lose energy mostly to optical phonons and also to zone-edge acoustic phonons, though somewhat less effectively. At temperature T, the phonon occupation number $n(\omega)$ is given as $n(\omega) = 1/(\exp(\omega/kT)-1)$. For ZnS, the optical phone energy ω is 44 meV. Thus, one obtains an occupation number, $n(\omega) = 0.223$ at 300K. The analytical expression for the saturated drift velocity v_s in the lucky-drift theory is given by:

$$v_s = \left(\frac{\hbar\omega}{(2n+1)m^*} \right)^{1/2} \tag{174}$$

which yields 1.38×10^7 cm s^{-1} at 300K for electrons in the energy minimum Γ point at k = (000) of the conduction band.

In order to assess the electron-phonon coupling, the electron-phonon scattering rate $1/\tau$ (= $r_e + r_a$, r_e/r_a = (n + 1)/n) needs to be determined. From these rates, the average energy loss per unit time of an electron can be derived; in the steady state, this loss offsets the energy gained by drifting in the field, yielding:

$$\hbar\omega(r_e - r_a) = \frac{\hbar\omega}{(2n+1)\tau} = qEv_s \simeq 10^{13} \text{ eV s}^{-1} \tag{175}$$

By substituting n = 0.223 and ω = 44 meV into Eq. 175, one obtains $1/\tau \simeq 3.2 \times 10^{14}$ s^{-1}, or an electron mean free time of $\tau \simeq 3$ fs. The competition between heating by the field and cooling by a lattice scattering determines not only the average energy ε_{av} but also the nonequilibrium energy distribution function. The energy balance condition is obtained by setting the following equation to zero.

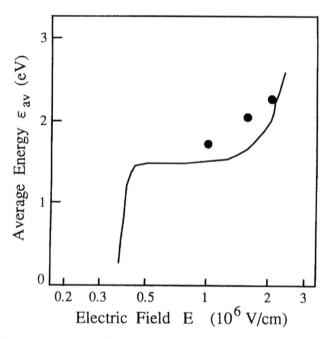

Figure 83 Solid line shows average electron energy ε_{av} as a function of electric field E in ZnS at 300K obtained from the energy balance condition. Solid circles are the Monte-Carlo calculated ε_{av}. (From Bringuier, E., *J. Appl. Phys.*, 75, 4291, 1994; Bhattacharyya, K., Goodnick, S.M., and Wager, J.F., *J. Appl. Phys.*, 73, 3390, 1993. With permission.)

$$\frac{d\varepsilon}{dt} = qEv_d - \frac{\hbar\omega}{(2n+1)\tau(\varepsilon)} \tag{176}$$

where $1/\tau(\varepsilon)$ is the energy-dependent scattering rate. The average electron energies ε_{av} obtained from this equation are plotted in Figure 83 as a function of the electric field E. It can be seen that the average electron energy ε_{av} increases sharply when the electric field exceeds 2×10^6 V cm^{-1}. An average electron energy ε_{av} exceeding 2 eV is sufficient for the impact excitation of luminescent centers, as described in the next section.

Recently, an ensemble Monte-Carlo simulation of electron transport in ZnS bulk at high electric fields was performed.[11] Scattering mechanisms associated with polar optical phonons, acoustic phonons, inter-valley scattering in the conduction band, and impurities were included into a nonparabolic multi-valley model. The average electron energy ε_{av} calculated in this way is also shown in Figure 83. Close agreement was obtained between the ε_{av} values calculated by the Monte-Carlo method and those obtained by the lucky-drift theory. Simulated results of the electron energy distribution are shown in Figure 84, together with the impact excitation cross-section for the Mn^{2+} center discussed in the next section. The results show that energetic electrons are available at field strengths exceeding 10^6 V cm^{-1} to cause impact excitation, and that transient effects such as ballistic transport can be disregarded in explaining the excitation mechanism of thin-film EL.

1.10.3.3 Excitation mechanism of luminescence centers

In EL phosphors presently used, there are two types of luminescent centers. One is the donor-acceptor pair type, and the other is the localized center type. For the latter, Mn^{2+} ions producing luminescence due to $3d^5$ intra-shell transitions are the most efficient centers used in ZnS thin-film EL devices. Some divalent and trivalent rare-earth ions emitting luminescence due to $4d^{n-1}5d \rightarrow 4f^n$ or $4f^n$ intra-shell transitions are also efficient luminescent

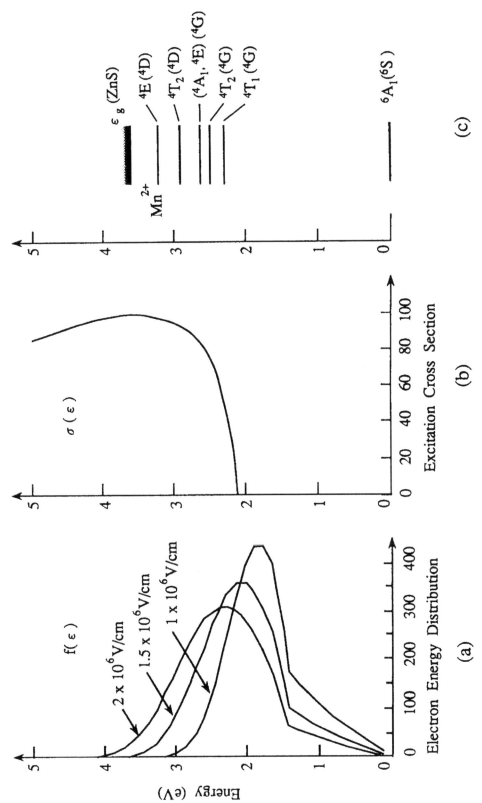

Figure 84 (a) Electron energy distribution f(ε), (b) Mn^{2+} impact excitation cross-section $\sigma(\varepsilon)$ as a function of energy, and (c) energy levels of Mn^{2+}. (From Bhattacharyya, K., Goodnick, S.M., and Wager, J.F., *J. Appl. Phys.*, 73, 3390, 1993. With permission.)

centers; these ions are potential candidates for color EL. The excitation processes of these luminescent centers are described in this section.[12]

Electron-hole pair generation by hot electron impact ionization. In the ZnS host lattice, a high electric field of 2×10^6 V cm^{-1} is enough to produce hot electrons. Consequently, these hot electrons ionize the ZnS lattice by collision, and by creating electron-hole pairs. This process is called impact ionization of the lattice. If impurities, donor and/or acceptor exist, they will also be ionized. The electron-hole pairs are recaptured by these ionized donors and acceptors, and luminescence is produced as a result of the recombination of electrons and holes. These processes are illustrated in Figure 85(a).

The ionization rate P_{ion} of the lattice is calculated using the following equation:

$$P_{ion} \propto \int_{\varepsilon_g}^{\infty} \sigma(\varepsilon) f(\varepsilon) d\varepsilon \tag{177}$$

where $\sigma(\varepsilon)$ is the ionization cross-section of the lattice, ε_g is the bandgap energy, and $f(\varepsilon)$ is the electron energy distribution function. $\sigma(\varepsilon)$ is proportional to the product of the density of states of the valence and conduction bands.

In cathode-ray tubes, luminescence due to donor-acceptor pair recombination is very efficient, and ZnS:Ag,Cl and ZnS:Cu,Al(Cl) phosphors are widely and commonly used as blue and green phosphors, respectively. ZnS:Cu,Al(Cl) phosphors are also used for powder-type EL. However, these phosphors are not efficient when used in thin-film EL devices. This is understood in terms of the reionization of the captured electrons and holes by the applied electric field prior to their recombination.

Direct impact excitation of luminescent centers by hot electrons. If hot electrons in the host lattice collide directly with localized luminescent centers, the ground-state electrons of the centers are excited to higher levels, so that luminescence is produced, as illustrated in Figure 85(b). EL of ZnS:Mn^{2+} is due to the impact excitation of the $3d^5$ intra-shell configuration of Mn^{2+} centers. Similarly, EL of trivalent rare-earth (RE)-doped ZnS is based on the impact excitation of the $4f^n$ intra-shell configurations. This excitation mechanism is thought to be dominant in thin-film EL device operation.

Assuming direct impact excitation, the excitation rate P of centers can be expressed by:

$$P \propto \int_{\varepsilon_0}^{\infty} \sigma(\varepsilon, \gamma) f(\varepsilon) d\varepsilon \tag{178}$$

where $\sigma(\varepsilon,\gamma)$ is the impact excitation cross-section to the excited state γ of the centers, $f(\varepsilon)$ is the energy distribution of hot electrons discussed above, and ε_0 is the threshold energy for the excitation.

Although calculations of impact excitation and ionization cross-sections in free atoms or ions are very sophisticated and accurate, they are still crude in solids. Allen[13] has pointed out that the problems lie in the form of the wavefunctions of the luminescent centers to be used, especially when covalent bonding with the host crystal is included. There is also a problem of dielectric screening. This screening should be properly taken as dependent on the energy and wave vector of carriers, or be taken approximately as a function of distance r using the screened Coulomb potential expressed by $\phi(r) = (-A/r)\exp(-r/\lambda_D)$, where λ_D is the potential decay coefficient. In addition, the carrier velocity is not a simple function of its energy. Allen[13] calculated a cross-section σ for impact excitation using a

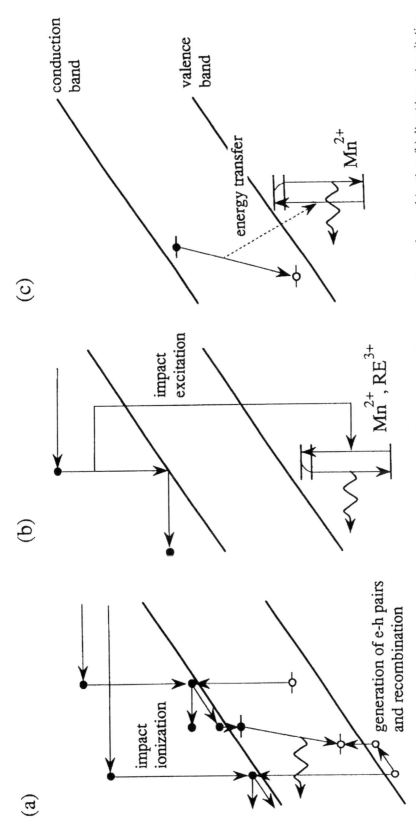

Figure 85 Three excitation processes of luminescent centers in thin-film EL devices: (a) impact ionization and recombination, (b) direct impact excitation, and (c) energy transfer.

simple Born-Bethe treatment of the direct Coulomb term and obtained the following expression.

$$\sigma = \frac{1}{\varepsilon^2 n_r \left(\dfrac{\varepsilon_{eff}}{\varepsilon_0}\right)} \times \frac{2\pi m_1^* e^2 \, \hbar c^3 S_{1u}^2}{v_u E_{ge}^2 K_u} \ln\left(\frac{k_u + k_1}{k_u - k_1}\right) \times \left(\frac{1}{\tau_{e-d}}\right) \tag{179}$$

Here, σ is explicitly written as the product of three terms. The first term describes the screening effect, where ε is the dielectric constant, n_r is the refractive index, and $(\varepsilon_{eff}/\varepsilon_0)$ is an effective field ratio. The second term is a function of the properties of the electron when the incident excited electron with the initial velocity v_u and with wave vector k_u in the conduction band is scattered to a lower state l with wave vector k_1, the electron loses kinetic energy E_{ge} corresponding to the energy difference between the ground and excited states of the center. In the second term, m_1^* is an electron effective mass, and S_{1u} represents the value of the overlap integral of the Bloch function for the electron with wave vector k_1 and that for the electron with k_u. The third term is the electric dipole radiative transition rate of the center. For centers with radiative lifetimes in the range of 10 μs to 1 ms, the cross-section is estimated to be 10^{-18} to 10^{-20} cm², which is too small to be useful.

There is an apparent difference between the nature of the electrons in vacuum and in a solid like ZnS. ZnS is a direct-bandgap semiconductor having both the bottom of the conduction band and the top of the valence band at the Γ point k = (000). In vacuum, the velocity of electrons increases monotonically with the increase in its energy; whereas, in ZnS, it is possible to have electrons with high energy but low velocity in the upper minima of the conduction band at the L [k = (111)] and X [k = (001)] valleys (see 2.7.2.3). As seen in Eq. 179, if the velocity of the incident electron v_u is low when it has sufficient energy E_{ge} for impact excitation, the excitation cross-section is considerably enhanced. Such a situation is realized in ZnS:Mn^{2+} when incident electrons in the X or L valley collide with Mn^{2+} centers and are scattered to the Γ valley; simultaneously, $3d^5$ electrons of Mn^{2+} are excited from the ground to the excited state.

Exchange effects are expected to dominate because of a resonance between the energy spacings of Mn^{2+} centers and those within the conduction band. The Born-Bethe treatment—i.e., the use of the Fermi Golden Rule—is not appropriate for exchange processes. Although there is no simple approximation to calculate rates of exchange processes, some qualitative conclusions can be drawn from our knowledge of what happens in the impact excitation of free atoms. The long-range part of the interaction is no longer dominant; so instead of the perturbing Hamiltonian in the dipole approximation, one must use the full Coulomb term. The exchange interaction is predominantly at the short range. Hence, in a crystal of high dielectric constant, the direct interaction is screened much more effectively than the exchange interaction. This is likely to be the cause for the large cross-section.

In ZnS:Mn^{2+}, the cross-section for the impact excitation is then doubly enhanced. One can, therefore, conclude that ZnS:Mn^{2+} is a suitable combination of host and center that produces efficient EL by the impact excitation. This is because ZnS satisfies the need for a host in which hot carriers have a suitable energy distribution, and Mn^{2+} has an unusually large cross-section near the threshold.

In Figures 84(a) and (b), the electron energy distribution $f(\varepsilon)$ and excitation cross-section of Mn^+ centers $\sigma(\varepsilon)$ are illustrated.[11] The energy levels of Mn^{2+} centers are also shown in Figure 84(c). The peak in $f(\varepsilon)$ near 2 eV implies that a large electron population exists in the X and L valleys, where electrons have sufficient energy to excite Mn^{2+} centers, but have relatively low velocities. The threshold energy for the excitation is a little smaller than the lowest excited state $^4T_1$ of Mn^{2+}. This results from the broadening due to the uncertainty principle. As seen from the Figures 84(a) and (b), at electric fields of the order of 1×10^6 V cm^{-1} and larger, a significant fraction of the total electron population exists at energies exceeding the threshold excitation energy of 2.1 eV for Mn^{2+}. A good match of the hot electron distribution with the Mn^{2+} excitation energy brings about the relatively high EL efficiencies in this system; the efficiencies are of the order of 4 to 6 lm W^{-1}.

It was demonstrated[14] that $ZnS:Tb^{3+}O^{2-}F^-$ thin-film EL devices show efficient green EL with an efficiency of the order of 1 to 2 lm W^{-1}. It has been shown that in TbOF complex centers, Tb^{3+} substitutes into the Zn^{2+} site, O^{2-} substitutes into the S^{2-} site, and F^- is located at an interstitial site to compensate for the charge difference. Therefore, TbOF centers seem to form centers isoelectronic with ZnS. $ZnS:Tb^{3+}F^-$ EL films with a Tb:F ratio of unity, with Tb^{3+} at Zn^{2+} sites and interstitial F^- ions, also form isoelectric centers that show efficient EL. It is believed that these isoelectronic centers have larger cross-sections for impact excitation than those for isolated rare-earth ions, so that high EL efficiencies result.

Energy transfer to luminescent centers.   In AC powder EL phosphors such as ZnS:Cu,Cl, donor (Cl)–acceptor (Cu) (D-A) pairs are efficient luminescent centers (see 2.7.4), and the EL emission is caused by the radiative recombination of electron-hole pairs through D-A pairs. Electrons and holes are injected into the ZnS lattice by bipolar field emission. By further incorporating Mn^{2+} centers in ZnS:Cu,Cl, yellow emission due to Mn^{2+} is observed. In this case, the excitation of Mn^{2+} centers is due to the nonradiative resonant energy transfer from D-A pairs to Mn^{2+}, as illustrated in Figure 85(c).

Ionization of centers and recapture of electrons to produce luminescence.   In thin-film EL of rare-earth-doped IIa-VIb compounds such as blue/green-emitting $SrS:Ce^{3+}$ and red-emitting $CaS:Eu^{2+}$, the transient behavior of the EL emission peaks under pulse excitation exhibits emission peaks when the pulsed voltage is turned on and turned off; in other words, the second peak appears when the electric field is reversed in the direction due to polarization charge trapped in the phosphor-insulator interfaces.[15] The luminescence of these phosphors is due to the $4f^{n-1}5d \rightarrow 4f^n$ transition. It is probable that the $4f^n$ ground-state level is located in the forbidden gap, while the $4f^{n-1}5d$ excited state is close to the bottom of the conduction band. The EL excitation mechanism is as follows: the luminescence centers are excited by the impact of the electron accelerated by the pulsed voltage and then ionized by the applied pulsed field. Electrons released to the conduction band are captured by traps. This process has been experimentally confirmed by measurements of the excitation spectra of the photoinduced conductivities.[15] When the voltage is turned back to zero, the trapped electrons are raised by the reversed field to the conduction band again and are recaptured by the ionized centers to produce luminescence.

References

1. Destriau, G., *J. Chim. Phys.*, 33, 620, 1936.
2. Destriau, G., *Phil. Mag.*, 38, 700, 1947; 38, 774, 1947; 38, 880, 1947.
3. Haynes, J.R. and Briggs, H.B., *Phys. Rev.*, 99, 1892, 1952.
4. Sze, S.M., *Physics of Semiconductor Devices*, 2nd edition, John Wiley & Sons, New York, 1981, chap. 5 and 6.
5. Kobayashi, H., *Optoelectronic Materials and Devices, Proc. 3rd Int. School*, Cetniewo, 1981, PWN-Polish Scientific Publishers, Warszawa, 1983, chap. 13.
6. Baraff, G.A., *Phys. Rev.*, 133, A26, 1964.
7. Shockley, W., *Solid State Electron.*, 2, 35, 1961.

8. Wolff, P.A., *Phys. Rev.*, 95, 1415, 1954.
9. Bringuier, E., *J. Appl. Phys.*, 66, 1314, 1989.
10. Bringuier, E., *J. Appl. Phys.*, 75, 4291, 1994.
11. Bhattacharyya, K., Goodnick, S.M., and Wager, J.F., *J. Appl. Phys.*, 73, 3390, 1993.
12. Kobayashi, H., *Proc. SPIE*, 1910, 15, 1993.
13. Allen, J.W., *Springer Proc. in Physics 38, Proc. 4th Int. Workshop on Electroluminescence*, Springer-Verlag, Heidelberg, 1989, p. 10.
14. Okamoto, K., Yoshimi, T., and Miura, S., *Springer Proc. in Physics 38, Proc. 4th Int. Workshop on Electroluminescence*, Springer-Verlag, Heidelberg, 1989, p. 139.
15. Tanaka, S., *J. Crystal Growth*, 101, 958, 1990.

chapter one — section eleven

Fundamentals of luminescence

Pieter Dorenbos

Contents

1.11   Lanthanide level locations and its impact on phosphor performance

1.11.1   Introduction

The lanthanide ions either in their divalent or trivalent charge state form a very important class of luminescence activators in phosphors and single crystals.[1] The fast 15–60 ns 5d–4f emission of Ce^{3+} in compounds like $LaCl_3$, $LaBr_3$, Lu_2SiO_5, and Gd_2SiO_5 is utilized in scintillators for γ-ray detection.[2] The same emission is utilized in cathode ray tubes and electroluminescence phosphors. The photon cascade emission involving the $4f^2$ levels of Pr^{3+} has been investigated for developing high quantum efficiency phosphors excited by means of a Xe discharge in the vacuum-UV.[3] The narrow-line $4f^3$ transitions in Nd^{3+} are used in laser crystals like $Y_3Al_5O_{12}$:Nd^{3+}. Sm^{3+} is utilized as an efficient electron trap and much research has been devoted to its information storage properties. For example, MgS:Ce^{3+};Sm^{3+} and MgS:Eu^{2+};Sm^{3+} were studied for optical memory phosphor applications,[4] Y_2SiO_5:Ce^{3+};Sm^{3+} was studied for X-ray imaging phosphor applications,[5] and $LiYSiO_4$:Ce^{3+};Sm^{3+} for thermal neutron imaging phosphor applications.[6] The famous $^5D_0 \rightarrow {}^7F_j$, $4f^6$ redline emissions of Eu^{3+} and the blue to red 5d–4f emission of Eu^{2+} are both used in display and lighting phosphors.[1] The $4f^8$ line emission of Tb^{3+} is often responsible for the green component in tricolor tube lighting.[1] Dy^{3+} plays an important role in the persistent luminescence phosphor $SrAl_2O_4$:Eu^{2+};Dy^{3+}.[7,8] Er^{3+} and Tm^{3+} are, like Pr^{3+}, investigated for possible photon cascade emission phosphor applications.

   This brief and still incomplete summary illustrates the diversity of applications involving the luminescence of lanthanide ions. It also illustrates that we can distinguish two types of lanthanide luminescent transitions. (1) Transitions between levels of the $4f^n$ configuration. In this chapter, the energy of each $4f^n$ excited state relative to the lowest $4f^n$ state will be regarded as invariant with the type of compound. One may then use the Dieke diagram with the extension provided by Wegh et al.[9] to identify the many possible luminescence emission and optical absorption lines. (2) Transitions between the $4f^{n-1}$ 5d and the $4f^n$ configurations. The energy of 5d levels, contrary to the 4f levels, depends very strongly on the type of compound. For example, the wavelength of the 5d–4f emission of Ce^{3+} may range from the ultraviolet region in fluorides like that of $KMgF_3$ to the red region in sulfides like that of Lu_2S_3.[10]

   In all phosphor applications the color of emission and the quantum efficiency of the luminescence process are of crucial importance as is the thermal stability of the emission in some applications. These three aspects are related to the relative and absolute location of the lanthanide energy levels. For example, the position of the host-sensitive lowest 5d state relative to the host-invariant 4f states is important for the quenching behavior of both 5d–4f and 4f–4f emissions by multiphonon relaxation. The absolute position of the 4f and 5d states relative to valence band and conduction band states also affects luminescence quenching and charge-trapping phenomena. Although it was realized long ago that absolute location is crucial for phosphor performance, the experimental and theoretical understanding of the placement of energy levels relative to the intrinsic bands of the host has been lacking.

   In this section, first, a survey is provided on how relative and absolute locations of lanthanide energy levels affect phosphor performance. Next, methods and models to determine relative and absolute locations are treated. After discussing the energy levels of the free (or gaseous) lanthanide ions, the influence of the host compound on the location of the 5d levels relative to the 4f levels is presented. Next, the influence of the host compound on the absolute location of the lowest $4f^n$ state above the top of the valence band is explained. This forms the basis for drawing schemes for the absolute placement of both the 4f and 5d states of all the divalent and trivalent lanthanide ions.

1.11.2   *Level position and phosphor performance*

The importance of the relative and absolute positions of the energy levels of lanthanide ions is illustrated in Figure 86. We distinguish occupied states that can donate electrons and empty states that can accept electrons. Let us start with the "occupied states." Figure 86(a) illustrates the downward shift of the lowest-energy 5d level when a lanthanide is brought from the gaseous state (free ion) into the crystalline environment of a compound (A). Due to the interaction with the neighboring anion ligands (the crystal field interaction), the degenerate 5d levels of the free ion split (crystal field splitting), depending on the site symmetry. In addition, the whole 5d configuration shifts (centroid shift) toward lower energy. The crystal field splitting combined with the centroid shift lowers the lowest 5d level with an amount known as the redshift or depression D. Clearly the value of D determines the color of emission and wavelength of absorption of the 4f–5d transitions.

   Figure 86(b) illustrates the importance of lowest-energy 5d level location relative to $4f^2$ levels in Pr^{3+}. With the 5d level above the $^1S_0$ level of Pr^{3+}, multiphonon relaxation from the lowest 5d state to the lower lying $^1S_0$ level takes place. A cascade emission of two photons may result, which leads to quantum efficiency larger than 100%. However, with the lowest 5d state below $^1S_0$, broad-band 5d–4f emission is observed. Much research is devoted toward the search for Pr^{3+} quantum-splitting phosphors and for finding efficient 5d–4f-emitting Pr^{3+}-doped materials for scintillator applications. Depending on the precise

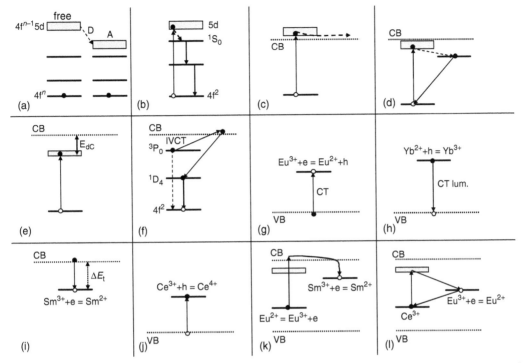

Figure 86 Illustration of influence of level location on phosphor properties: (a) the redshift *D* of the 5d state, (b) photon cascade emission in Pr^{3+}, (c) 5d–4f emission quenching by autoionization, (d) anomalous 5d emission, (e) thermal quenching by ionization, (f) quenching by intervalence charge transfer, (g) valence band charge transfer, (h) charge transfer luminescence, (i) electron trapping by Sm^{3+}, (j) hole trapping by Ce^{3+}, (k) electron transfer from Eu^{2+} to Sm^{3+}, (l) luminescence quenching by lanthanide to lanthanide charge transfer.

location of the lowest 5d state in Nd^{3+}, Eu^{2+}, and Sm^{2+}, either broad-band 5d–4f or narrow-line 4f–4f emissions can be observed.[11]

Figure 86(c), (d), and (e) show the interplay between the localized 5d electron and the delocalized conduction band states. If the lowest 5d state is above the bottom of the conduction band as in Figure 86(c), autoionization occurs spontaneously and no 5d–4f emission is observed. This is the case for $LaAlO_3$:Ce^{3+}, rare-earth sesquioxides Ln_2O_3:Ce^{3+},[1] and also for Eu^{2+} on trivalent rare-earth sites in oxide compounds.[12] Figure 86(d) illustrates the situation with 5d just below the conduction band. The 5d electron delocalizes but remains in the vicinity of the hole left behind. The true nature of the state, which is sometimes called an impurity trapped exciton state, is not precisely known. The recombination of the electron with the hole leads to the so-called anomalous emission characterized by a very large Stokes shift.[13,14] Finally, Figure 86(e) shows the situation with the 5d state well below the conduction band, leading to 5d–4f emission. The thermal quenching of this emission by means of ionization to conduction band states is controlled by the energy E_{dC} between the 5d state (d) and the bottom of the conduction band (C).[15,16] A review on the relationship between E_{dC} for Eu^{2+} and thermal quenching of its 5d–4f emission recently appeared.[16] Knowledge on such relationships is important for developing temperature-stable Eu^{2+}-doped light-emitting diode (LED) phosphors or temperature-stable Ce^{3+}-doped scintillators. For electroluminescence applications, E_{dC} is an important parameter to discriminate the mechanism of impact ionization against the mechanism of field ionization.[17]

Figure 86(f) shows a typical situation for Pr^{3+} in a transition metal complex compound like $CaTiO_3$. The undesired blue emission from the Pr^{3+} $^3P_0$ level is quenched by intervalence charge transfer (IVCT).[18] The electron transfers from the $^3P_0$ level to the transition metal (Ti^{4+}). The electron is transferred back to the red emitting Pr^{3+} $^1D_4$ level. The position of the $^3P_0$ level relative to the transition metal-derived conduction band controls the quenching process, and thereby the color of emission.

So far we have discussed examples of absolute location of "occupied states." However, a trivalent lanthanide ion may accept an electron to form a divalent lanthanide ion. The location of the occupied ground-state level of a divalent lanthanide ion is therefore the same as the unoccupied electron-accepting state of the corresponding trivalent lanthanide ion. The accepted electron may originate from the valence band, the conduction band, or another lanthanide ion. Figure 86(g) pertains to a Eu^{3+}-doped compound. Eu^{3+} introduces an unoccupied Eu^{2+} state in the forbidden gap. The excitation of an electron from the valence band to the unoccupied state creates the ground state of Eu^{2+}. This is a dipole-allowed transition that is used, for example, to sensitize Y_2O_3:Eu^{3+} phosphors to the 254 nm Hg emission in tube lighting.[1] Recombination of the electron with the valence band hole leaves the Eu^{3+} ion in the $^5D_0$ excited state resulting in red $4f^6$–$4f^6$ emission. Figure 86(h) shows a similar situation for Yb^{3+}. In the case of Yb^{3+} the recombination with the hole in the valence band produces a strong Stokes-shifted charge transfer (CT) luminescence. This type of luminescence gained considerable interest for developing scintillators for neutrino detection.[19] Clearly, the absolute location of the divalent lanthanide ground state is important for CT excitation and CT luminescence energies.

Figure 86(i) shows the trapping of an electron from the conduction band by Sm^{3+} to form the ground state of Sm^{2+}. The absolute location of an "unoccupied" divalent lanthanide ground state determines the electron trapping depth provided by the corresponding trivalent lanthanide ion. On the other hand, the absolute location of an "occupied" lanthanide ground state determines the valence band hole trapping depth provided by that lanthanide ion. Figure 86(j) illustrates trapping of a hole from the valence band by Ce^{3+}. This hole trapping is an important aspect of the scintillation mechanism in Ce^{3+}-doped scintillators. Similarly, Eu^{2+} is an efficient hole trap of importance for the X-ray storage phosphor BaFBr:Eu^{2+}.

Phosphor properties become more complicated when we deal with "double lanthanide-doped systems." Figure 86(k) shows the situation in Eu^{2+} and Sm^{3+} double-doped compounds like SrS and MgS that were studied for optical data storage applications.[4,11] The ultraviolet write pulse excites an electron from Eu^{2+} to the conduction band, which is then trapped by Sm^{3+}. Eu^{3+} and Sm^{2+} are created in the process. An infrared read pulse liberates the electron again from Sm^{2+}, resulting, eventually, in Eu^{2+} 5d–4f emission. Similar mechanisms apply for Y_2SiO_5:Ce^{3+};Sm^{3+} and $LiYSiO_4$:Ce^{3+};Sm^{3+} compounds that were developed for X-ray and thermal neutron storage phosphor applications, respectively.[5,6] The true mechanism in the persistent luminescence phosphor $SrAl_2O_4$:Eu^{2+};Dy^{3+} is still disputed. One needs to know the absolute level energy locations to arrive at plausible mechanisms or to discard implausible ones.[8]

As a last example, Figure 86(l) shows quenching of emission in Ce^{3+} and Eu^{3+} co-doped systems. The Ce^{3+} electron excited to the lowest 5d state can jump to Eu^{3+} when the unoccupied Eu^{2+} ground state is located at a lower energy than the occupied lowest Ce^{3+} 5d excited state. After the jump, Eu^{2+} and Ce^{4+} are formed. The Eu^{2+} electron can jump back to Ce^{4+} if the unoccupied Ce^{3+} ground state is located below the occupied Eu^{2+} ground state. The original situation is restored without emission of a photon. Similar quenching routes pertain to Ce^{3+} in Yb-based compounds, and with appropriate level schemes, other "killing" combinations can be found as well.

The above set of examples shows the importance of energy level locations for the performance of phosphors. This importance was realized long ago, but not until recently methods and models became available that allow the determination of these absolute positions. In the following sections, the historic developments and current status of absolute level positioning are briefly reviewed. For detailed information, original literature should be consulted.

1.11.3 The free (gaseous) lanthanide ions

The previous section illustrated the importance of lanthanide level locations for phosphor performance. To understand and predict these locations we first need to understand the properties of the free (gaseous) lanthanide ions. Figure 87 shows the data available on the energy (E_{fd}) needed to excite an electron from the lowest level of the $4f^n5d^06s^m$ configuration to the lowest level of the $4f^{n-1}5d^16s^m$ configuration in the gaseous free lanthanide ions or atoms. The data are from Brewer[20] and Martin[21] together with later updates.[11] Data are most complete for the neutral atoms ($m = 2$, curve c), the monovalent lanthanides ($m = 1$, curve b), and the divalent lanthanides ($m = 0$, curve a). A universal curve, curve a in Figure 87, can be constructed. By shifting the energy of this universal curve, the 4f–5d energies as a function of n can be reproduced irrespective of the charge of the lanthanide ion (0, +1, +2, or +3) or the number, m, of electrons in 6s ($m = 0, 1,$ or 2). This remarkable phenomenon is due to the inner-shell nature of the 4f orbital. Apparently, the occupation number of electrons in the 6s shell has no influence on the universal behavior. The main features of this universal variation have been known for a long time and understood in terms of Jörgensens spin pairing theory for the binding of 4f electrons.[22] The energy is large when the 4f configuration is half- ($n = 7$) or completely ($n = 14$) filled, and the energy is small when it is occupied by one or eight electrons.

Figure 88 shows the binding energy (or ionization energy) of the 4f and 5d electrons in the free divalent and free trivalent lanthanide ions with $m = 0$. When we add the corresponding energies, E_{fd}, from Figure 87 to curves b and d in Figure 88, we obtain the binding energies for the 5d electron (see curves a and c). The stronger binding of the 4f and 5d electrons in the trivalent lanthanides than in the divalent ones is due to a stronger Coulomb attraction. Clearly, the binding of the 4f electron is responsible for the universal behavior in the 4f–5d transitions. The binding energy of the 5d electron is rather constant with n which indicates that the nature of the 5d state is relatively invariant with the type of lanthanide ion.

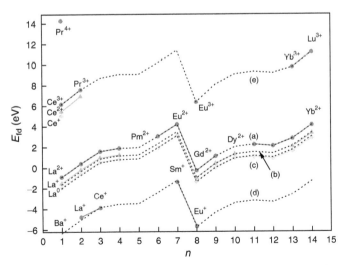

Figure 87 Experimentally observed energies E_{fd} for the transition between the lowest $4f^n5d^06s^m$ and the lowest $4f^{n-1}5d^16s^m$ states of free (gaseous) lanthanide ions and atoms. A shift of the dashed curve (a) by –0.71 eV, –1.09 eV, –5.42 eV, and +7.00 eV gives curves (b), (c), (d), and (e), respectively.

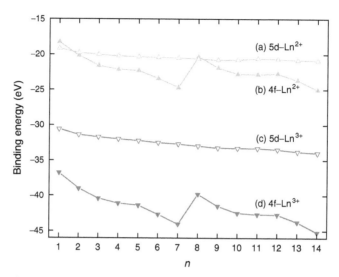

Figure 88 The binding energy in eV of the 5d (curves a and c) and 4f electron (curves b and d) in the free divalent (curves a and b) and free trivalent lanthanide ions (curves c and d).

1.11.4 4f–5d energy differences of lanthanide ions in compounds

Figure 87 indicates that the variation of E_{fd} with n does not depend on the charge of the lanthanide ion or on the number of electrons in the 6s orbital. It is also well established that the Dieke diagram of 4f energy levels is almost invariant with the type of compound.

The situation is completely different for the 5d states. Their energies are influenced 50 times stronger by the host compound than those of 4f states. Due to crystal field splitting of the 5d states and a shift (centroid shift) of the average energy of the 5d configuration, the lowest level of the 5d configuration decreases in energy as illustrated in Figure 89 for Ce^{3+} in $LiLuF_4$ (see also Figure 86(a)). The decrease is known as the redshift or depression $D(n,Q,A) \approx D(Q,A)$ where n, Q, and A stand for the number of electrons in the $4f^n$ ground state, the charge of the lanthanide ion, and the name of the compound, respectively. The redshift depends very strongly on A but appears, to good first approximation, independent of n, i.e., the type of lanthanide ion. This implies that both the crystal field splitting and the centroid shift of the 5d levels depend on the type of compound but to a good first approximation are the same for each lanthanide ion.

Figure 90 shows this principle. It is an inverted Dieke diagram where the zero of energy is at the lowest 5d state of the free trivalent lanthanide ion. When the lanthanide ions are present in a compound, one simply needs to shift the 5d levels down by the redshift $D(3+,A)$ to find the appropriate diagram for that compound. Figure 90 illustrates this for $LiLuF_4$. The 4f–5d transition energy of each lanthanide ion can be read from the diagram. In equation form this is written as:

$$E_{fd}(n,3+,A) = E_{fd}(n,3+,\text{free}) - D(3+,A) \tag{180}$$

where $E_{fd}(n,3+,\text{free})$ is the energy for the first $4f^n$–$4f^{n-1}$ 5d transition in the trivalent (3+) free lanthanide ion.[23] In addition to 4f–5d energies in $LiLuF_4$, the diagram also predicts that the lowest 5d state of Pr^{3+} is below the $^1S_0$ state, and broad-band 5d–4f emission and

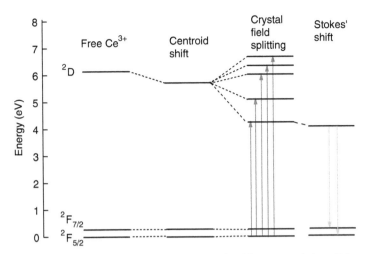

Figure 89 The effect of the crystal field interaction on the (degenerate) free Ce^{3+} energy states in $LiLuF_4$. The combination of centroid shift and crystal field splitting decreases the lowest 5d state with a total energy D. On the far right the Stokes shifted 5d–4f emission transitions are shown.

not narrow-band $^1S_0$ line emission will be observed (see Figure 86(b)). The lowest-energy Nd^{3+} 5d state in $LiLuF_4$ is predicted to be stable enough against multiphonon relaxation to the $^2G_{7/2}$ level. Indeed Nd^{3+} 5d–4f emission has been observed.

Redshift values are known for many hundreds of different compounds.[10] Figure 91 summarizes the redshift values $D(3+,A)$ for the trivalent lanthanide ions.[10] It is by definition zero for the free ions, and for the halides it increases from F to I in the sequence F, Cl, Br, I. For the chalcogenides, an increase in the sequence O, S, Se, and presumably Te is observed. This is directly connected with the properties of the anions that affect the centroid shift. The origin of the centroid shift is very complicated and related with covalency and polarizability of the anions in the compound.[24-26] The crystal field splitting is

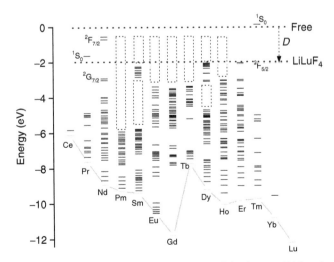

Figure 90 The inverted Dieke diagram where the energy of the lowest 5d level of the free trivalent lanthanide ions are defined as the zero of energy. A downward shift of the 5d levels with the redshift value $D = 1.9$ eV provides the relative position of the lowest 5d level for the trivalent lanthanides in $LiLuF_4$.

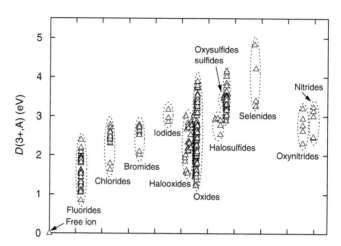

Figure 91 The redshift $D(3+,A)$ for trivalent lanthanide ions in compounds. The parameter along the horizontal axis groups the data depending on the type of compound.

related with the shape and size of the first anion coordination polyhedron.[26,27] The small fluorine and oxygen anions provide the largest values for the crystal field splitting and this is the main reason for the large spread in redshift values for these two types of compounds.[26] With the compiled redshift values, one may predict the 4f–5d transition energies for each of the 13 trivalent lanthanide ions in several hundreds of different compounds using the very simple relationship of Eq. 180. Eq. 180 equally well applies for the 5d–4f emission because the Stokes shift between absorption and emission is to first approximation also independent of the lanthanide ion.

For the divalent lanthanide ions in compounds the story is analogous.[11,12] Again one can introduce a redshift $D(2+,A)$ with a similar relationship

$$E_{fd}(n,2+,A) = E_{fd}(n,2+,\text{free}) - D(2+,A) \qquad (181)$$

and construct figures like Figure 90 and Figure 91.

With all data available on $D(3+,A)$ and $D(2+,A)$, the redshift in divalent lanthanides can be compared with that in the trivalent ones; a roughly linear relationship is found.[28]

$$D(2+,A) = 0.64D(3+,A) - 0.233\,\text{eV} \qquad (182)$$

Investigations also show a linear relationship between crystal field splitting, centroid shift, and Stokes' shift.[28]

Combining Eqs. 180, 181, and 182 with the available data on $D(2+,A)$ and $D(3+,A)$, it is now possible to predict 4f–5d energy differences for all 13 divalent and all 13 trivalent lanthanides in about 500 different compounds, i.e., about 13000 different combinations! Usually, the accuracy is a few 0.1 eV but deviations occur. The work by van Pieterson et al.[29,30] on the trivalent lanthanides in LiYF$_4$, YPO$_4$, and CaF$_2$ shows that the crystal field splitting decreases slightly with the smaller size of the lanthanide ion. In these cases the redshift may not be the same for all lanthanide ions. A study on the crystal field splitting in Ce^{3+} and Tb^{3+} also revealed deviations of the order of a few tenths of eV from the idealized situation expressed by Eqs. 180, 181, and 182.[31]

1.11.5   Methods to determine absolute level locations

The experimental basis for the results in the previous sections are from the 4f–5d energy differences, which are easily measured by means of luminescence, luminescence excitation, or optical absorption techniques. We deal with a dipole-allowed transition from a localized ground state to a localized excited state involving one and the same lanthanide ion. Both states have a well-defined energy.

To determine the location of energy levels relative to the valence band or to the conduction band is not straightforward. Again one may use information from optical spectroscopy. Figure 86(g) shows the transition of an electron from the top of the valence band (an anion) to Eu^{3+}. The final state is the $4f^7$ ground state of Eu^{2+}. The energy needed for this CT provides then a measure for the energy difference between the valence band and the Eu^{2+} ground state. Wong et al.[32] and Happek et al.[33] assume that the CT energy provides the location of the ground state of the electron-accepting lanthanide relative to the top of the valence band directly. However, this is not so trivial. The transferred electron and the hole left behind are still Coulomb attracted to each other, and this reduces the transition energy by perhaps as much as 0.5 eV. On the other hand Eu^{2+} is about 18 pm larger than Eu^{3+}, and the optical transition ends in a configuration of neighboring anions that is not yet in its lowest-energy state. Both these effects tend to compensate each other, and fortuitously the original assumption by Wong et al., and later by Happek et al., appears quite plausible.[34] The location of occupied 4f states relative to the occupied valence band states can also be probed by X-ray or UV photoelectron spectroscopy (XPS or UPS).[35]

With the techniques mentioned in the preceding paragraph, the localized level positions of lanthanide ions relative to the valence band states can be probed. The level locations relative to the conduction band can be determined with other techniques. Various methods rely on the ionization of 5d electrons to conduction band states. The thermal quenching of 5d–4f emission in Ce^{3+} or Eu^{2+} is often due to such ionization processes.[15,16] By studying the quenching of intensity or the shortening of decay time with temperature, the energy difference, E_{dC}, between the (lattice relaxed) lowest 5d state and the bottom of the conduction band can be deduced from their Arrhenius behavior.[15] Such studies were done by Lizzo et al.[36] for Yb^{2+} in $CaSO_4$ and SrB_4O_7, by Bessière et al.[17] for Ce^{3+} in $CaGa_2S_4$, and by Lyu and Hamilton.[15] Also, the absence of Ce^{3+} emission due to a situation sketched in Figure 86(c) or the presence of anomalous emission as in Figure 86(d) provides qualitative information on 5d level locations.[14] One may also interpret the absence or presence of vibronic structures in 5d excitation bands as indicative of 5d states contained within the conduction band.[29] One- or two-step photoconductivity provides information on the location of 4f ground states relative to the bottom of the conduction band.[37–40] Another related technique is the microwave conductivity method developed by Joubert and coworkers that was applied to $Lu_2SiO_5:Ce^{3+}$.[41]

1.11.6   Systematic variation in absolute level locations

The previous section provides an explanation on the techniques that have been used to obtain information on level positions. But often these techniques were applied to a specific lanthanide ion in a specific compound with the aim of understanding properties of that combination. Furthermore, each of these techniques provides its own source of unknown systematic errors. These individual studies do not provide us with a broad overview on how level energies change with the type of lanthanide ion and the type of compound. Such an overview is needed to predict phosphor properties and to guide the researcher in the quest for new and better materials.

One of the first systematic approaches was by Pedrini et al. who undertook photo-conductivity measurements to determine the location of the 4f ground state of divalent lanthanides in the fluorite compounds CaF_2, SrF_2, and BaF_2 relative to the bottom of the conduction band.[39] They also provide a model to explain the observed variation in 4f ground-state energy with n.

The first systematic approach to determine the levels of trivalent lanthanides was undertaken by Thiel and coworkers using XPS.[42,43] They studied the trivalent lanthanides in $Y_3Al_5O_{12}$ and determined the 4f ground-state energies relative to the valence band of the host crystal. They also combined their findings with the systematic in 4f–5d energy difference found in Ref. 23 to locate the 5d states in the band gap. The absolute energy of the lowest 5d state appears relatively constant with the type of lanthanide ion.

Both XPS and photoconductivity experiments have drawbacks. The oscillator strength for the transition of the localized 4f ground state to the delocalized conduction band states is very small and photoconductivity is rarely observed due to such direct transitions. Two-step photoconductivity is observed more frequently. After a dipole-allowed excitation to the 5d state, it is either followed by autoionization (see Figure 86(c)) or thermally assisted ionization (see Figure 86(e)). For the XPS experiments, high Ln^{3+}-concentrated samples are needed,[42,44] and one has to deal with uncertain final state effects to obtain reliable data.[45] At this moment the amount of information obtained with these two methods is scarce. Although they provide us with very valuable ideas and insight on how level energies change with the type of lanthanide ion, there is not enough information to obtain detailed insights into the effect of type of compound.

Another method to obtain the systematic variation in level position with the type of lanthanide is CT spectroscopy. It appears that the energy of CT to Sm^{3+} is always (at least in oxide compounds) a fixed amount higher than that for the CT to Eu^{3+}. The same applies for Tm^{3+} and Yb^{3+}. This was noticed long ago[22,46,47] and reconfirmed by more recent studies.[48–50] An elaborate analysis of data on CT retrieved from the literature revealed that the systematic behavior in CT energies holds for all lanthanides in all types of different compounds.[34]

Figure 92 illustrates the method to construct diagrams with absolute level location of the divalent lanthanide in $CaGa_2S_4$. The top of the valence band is defined as zero of energy. The arrows numbered 1 through 6 show the observed energies for CT to trivalent lanthanide ions, and they provide us with the location of the ground state of the corresponding divalent lanthanides (see Figure 86(g)). Using these data we can construct precisely the same universal curve, but in an inverted form, as found for the energy E_{fd} of 4f–5d transitions in the free lanthanide ions and atoms of Figure 87. Arrow 7 shows the energy of the first 4f–5d transition in Eu^{2+}. Using Eq. 181, the absolute location of the lowest 5d state for each divalent lanthanide ion can be drawn in the scheme. It appears constant with n.

The universal behavior in the energy of the lowest 4f state with n is determined by the binding of 4f electrons, similar to that depicted in Figure 88, but modified by the Madelung potential at the lanthanide site in the compound. This Madelung potential increases with smaller size of the lanthanide ion due to the inward relaxation of the neighboring negatively charged anions.[14,34,39,43] The increase in 5d electron binding energy by 1–2 eV, as observed for the free divalent lanthanides in Figure 88, is absent in $CaGa_2S_4$ where the binding of the 5d electron is found independent of n. This fortuitous situation for $CaGa_2S_4$, which is also expected for other sulfide compounds, does not apply to oxides and fluorides. For these compounds it was found that from Eu^{2+} to Yb^{2+} the binding of the levels gradually decrease by about 0.5 eV.[14,34] In other words, the 5d state of Yb^{2+} is found 0.5 eV closer to the bottom of the conduction band than that of Eu^{2+}, which is

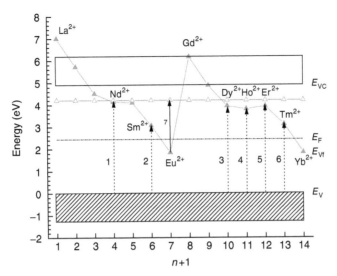

Figure 92 The location of the lowest 4f and lowest 5d states of the divalent lanthanide ions in $CaGa_2S_4$. Arrows 1 through 6 show observed energies of charge transfer to Ln^{3+}. Arrow 7 shows the observed energy for the first 4f–5d transition in Eu^{2+}.

consistent with the observation that Yb^{2+} in oxides and fluorides is more susceptible to anomalous emission than Eu^{2+} in these compounds.[14]

The universal behavior in both 4f–5d energy differences and CT energies forms the basis for a construction method of the diagrams as seen in Figure 92. Only three host-dependent parameters, i.e., E^{CT} (6,3+,A), D(2+,A), and the energy E_{VC} (A) between the top of the valence band (*V*) and the bottom of the conduction band, are needed. These parameters are available for many different compounds.[51] Figure 93 shows the energy E^{CT} (6,3+,A) of CT to Eu^{3+} (with $n = 6$) in compound (A), and Figure 94 shows the energy of the first excitonic absorption maximum. The mobility band gap, i.e., the energy of the bottom of the conduction band at E_{VC}, is assumed to be 8% higher in energy.[51]

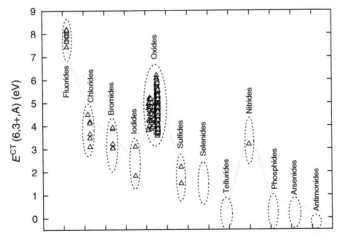

Figure 93 The energy E^{CT} (6,3+,A) of charge transfer to Eu^{3+} in inorganic crystalline compounds. The parameter along the horizontal axis groups the data depending on the type of compound. The solid curve is given by $E^{CT} = 3.72\eta(X) - 2.00\,eV$ where $\eta(X)$ is the Pauling electronegativity of the anion.

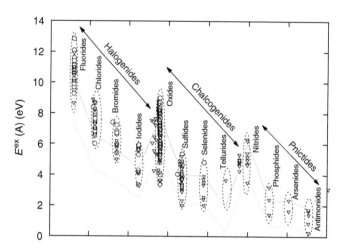

Figure 94   The optical band gap E^{ex} of inorganic compounds. The parameter along the horizontal axis groups the data depending on the type of compound. The solid curve is given by $E^{ex} = 4.34\eta-$ 7.15 eV where η is the Pauling electronegativity of the anion. The dashed curve is the same as in Figure 93.

With the richness of data on E^{CT} (6,3+,A), $D(2+,A)$, $D(3+,A)$, and E_{VC} (A) pertaining to hundreds of different compounds, one may now study the relationship between level location and type of compound in detail. Figure 93 shows already the effect of the type of anion on the position of the Eu^{2+} ground state above the top of the valence band. The location is at around 8 eV in fluorides and a clear pattern emerges when the type of anion varies. The energy decreases for the halides from F to I in the sequence F, Cl, Br, I, and for the chalcogenides from O to Se in the sequence O, S, Se, and presumably Te. This pattern is not new and has been interpreted with the Jörgensen model of optical electronegativity.[52]

$$E^{CT} = 3.72\big(\eta(X) - \eta_{opt}(Eu)\big) \qquad\qquad (183)$$

where $\eta(X)$ is the Pauling electronegativity of the anion X and $\eta_{opt}(Eu)$ is the optical electronegativity of Eu, a value that must be determined empirically from observed CT energies. With $\eta_{opt}(Eu) = 2$ the curve through the data in Figure 93 was constructed.[51] The curve reproduces the main trend with the type of anion. It also predicts where we can expect the Eu^{2+} ground state in the pnictides; a decrease in the sequence N, O, As, Sb is expected. However, the wide variation of CT energies within, for example, the oxide compounds is not accounted for by the Jörgensen model. Parameters like lanthanide site size and anion coordination number are also important and need to be considered for a refined interpretation of CT data.[51]

An equation similar to Eq. 183 can be introduced for E^{ex} to illustrate the main trend in the band gap with the type of anion.[51] The band gap follows the same pattern as the energy of CT with changing the type of anion. Interestingly, one may also notice a similar behavior in the values for the redshift in Figure 91 with changing the type of anion. This shows that the parameter values of our model (E^{CT} (6,3+,A), $D(2+,A)$, $D(3+,A)$, and E_{VC} (A)) are not entirely independent from one another.

Analogous to the systematic behavior of the lowest 4f energies for divalent lanthanides with changing n, a systematic behavior of the lowest 4f energies for trivalent lanthanides has been proposed.[34] One may then use, in principle, the same method as used for the divalent lanthanides to construct absolute level diagrams for trivalent lanthanide ions.

For the divalent lanthanides, the "anchor point" of construction is the CT energy to Eu^{3+} (see Figure 93). The CT to Ce^{4+} might play the role of such an anchor point for the trivalent lanthanide level positions. However, information on CT to the Ce^{4+} ion is only sparsely available, insufficient to routinely construct level diagrams. We, therefore, need another anchor point. The energy difference E_{dC} (1,3+,A) between the lowest 5d state of Ce^{3+} and the bottom of the conduction band may serve as the required anchor point. Its value (see Figure 86(d)) can be obtained from two-step photoconductivity experiments or from luminescence-quenching data.

Figure 95 demonstrates the level positions of both divalent and trivalent lanthanides in the same compound YPO_4. The scheme can be compared with that of the free ions in Figure 88. Note that the binding energy difference of more than 10 eV between the free trivalent and free divalent 5d levels is drastically reduced to about 0.8 eV in YPO_4. Energy differences of 0.5–1.0 eV are commonly observed when constructing diagrams for other compounds. The binding energy difference of almost 20 eV between the 4f states of the free ions is reduced to about 7.5 eV in YPO_4. This value also appears fairly constant for different host materials.

The full potential of schemes, similar to those for YPO_4, is demonstrated by comparing Figure 95 with the situations sketched in Figure 86. Actually, each of the 12 situations in Figure 86 can be found in the scheme of YPO_4. The arrows marked 86g, 86h, 86i, 86j, and 86l show the same type of transitions as in Figure 86(g), (h), (i), (j), and (l), respectively. Various other types of transitions, quenching routes, and charge-trapping depths can be read directly from the diagram. To name a few: (1) The lowest 5d states of all the divalent lanthanide ions are between E^{ex} and the bottom of the conduction band. In this situation, the 5d–4f emission is always quenched due to autoionization processes (see Figure 86(c) and (d)). (2) The 5d states of the trivalent lanthanides are well below E^{ex}, and for Ce^{3+}, Pr^{3+}, Nd^{3+}, Er^{3+}, and Tm^{3+} 5d–4f emissions are observed.[29,30] (3) Apart from Eu^{3+}, Gd^{3+}, Yb^{3+}, and Lu^{3+}, all the trivalent lanthanides form valence band hole traps. The trap is

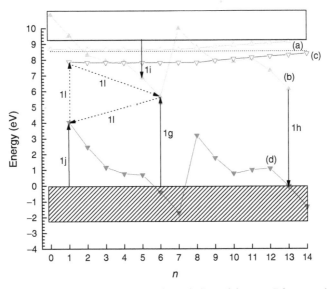

Figure 95 The location of the lowest 4f (curves b and d) and lowest 5d states (curves a and c) of the divalent (curves a and b) and trivalent (curves c and d) lanthanide ions in YPO_4. n and $n + 1$ are the number of electrons in the 4f shell of the trivalent and divalent lanthanide ion, respectively. Arrows indicate specific transitions that were also discussed in Figure 86. The horizontal dashed line at 8.55 eV is E^{ex}.

deepest for Ce^{3+} followed by Tb^{3+}. (4) The ground-state energies for the divalent lanthanides are high above the top of the valence band. In practice this means that even for Eu and Yb it is not possible to stabilize the divalent state during synthesis. (5) The trivalent lanthanides create stable electron traps because the ground states of the corresponding divalent lanthanides are well below the conduction band. (6) The ground states of Sm^{2+}, Eu^{2+}, Tm^{2+}, and Yb^{2+} are below the 5d state of the trivalent lanthanides. This means that Sm^{3+}, Eu^{3+}, Tm^{3+}, and Yb^{3+} can quench the 5d emission of trivalent lanthanide ions.

1.11.7 Future prospects and pretailoring phosphor properties

With the methods described in this section one can construct level schemes for all the lanthanide ions with few parameters. These parameters are available for hundreds of different compounds. At this stage, the schemes still contain systematic errors. Often the bottom of the conduction band is not well defined or known or levels may change due to charge-compensating defects and lattice relaxation which may result in (systematic) errors that are estimated at around 0.5 eV. Such errors are still very important for phosphor performance because a few tenths of eV shift of absolute energy level position may change the performance of a phosphor from very good to useless. The level schemes are, however, already very powerful in predicting 4f–5d and CT transition energies.

We may deduce trends in the energy difference between the lowest 5d state and the bottom of the conduction band, and then use these trends to guide the search for finding better temperature stable phosphors.[16] We may deduce trends in the absolute location of the lanthanide ground state that determines its susceptibility to oxidization or reduction.[53] For example, oxidation of Eu^{2+} is believed to play an important role in the degradation of $BaMgAl_{10}O_{17}$:Eu^{2+} phosphors,[54] and knowledge on level energies may provide us ideas to further stabilize Eu^{2+}. The level schemes are particularly useful when more than one lanthanide ion is present in the same compound. CT reactions and pathways from one lanthanide to the other can be read from the level schemes. For permanent information storage deep charge traps are required and for persistent luminescence shallow traps are needed. The level schemes provide very clear ideas on what combination of lanthanide ions are needed to obtain the desired properties. Perhaps even more importantly at this stage is that the level schemes provide very clear ideas on what combination not to choose for a specific application.

This chapter has surveyed where we are today with our knowledge and experimental techniques on the prediction and determination of absolute location of lanthanide ion energy levels in phosphors. Currently we have a basic model, but it needs to be more accurate. Aspects like lattice relaxation, charge-compensating defects, intrinsic defects, the nature of the bottom of the conduction band, dynamic properties involved in charge localization and delocalization processes, and theoretical modeling all need to be considered to improve our knowledge further. It will be the next step on the route for the tailoring of phosphor properties beforehand.

References

1. Blasse, G., and Grabmaier, B.C., *Luminescent Materials*, Springer-Verlag, Berlin, 1994.
2. Weber, M.J., Inorganic scintillators: Today and tomorrow, *J. Lumin.*, 100, 35, 2002.
3. van der Kolk, E., et al., Vacuum ultraviolet excitation and emission properties of Pr^{3+} and Ce^{3+} in MSO_4 (M = Ba, Sr, and Ca) and predicting quantum splitting by Pr^{3+} in oxides and fluorides, *Phys. Rev.*, B64, 195129, 2001.
4. Chakrabarti, K., Mathur, V.K., Rhodes, J.F., and Abbundi, R.J., Stimulated luminescence in rare-earth-doped MgS, *J. Appl. Phys.*, 64, 1362, 1988.

5. Meijerink, A., Schipper, W.J., and Blasse, G., Photostimulated luminescence and thermally stimulated luminescence of Y_2SiO_5-Ce,Sm, *J. Phys. D: Appl. Phys.*, 24, 997, 1991.
6. Sidorenko, A.V., et al., Storage effect in $LiLnSiO_4$:Ce^{3+},Sm^{3+},Ln = Y,Lu phosphor, *Nucl. Instrum. Methods*, 537, 81, 2005.
7. Matsuzawa, T., Aoki, Y., Takeuchi, N., and Murayama, Y., A new long phosphorescent phosphor with high brightness, $SrAl_2O_4$:Eu^{2+},Dy^{3+}, *J. Electrochem. Soc.*, 143, 2670, 1996.
8. Dorenbos, P., Mechanism of persistent luminescence in Eu^{2+} and Dy^{3+} co-doped aluminate and silicate compounds, *J. Electrochem. Soc.*, 152, H107, 2005.
9. Wegh, R.T., Meijerink, A., Lamminmäki, R.-J., and Hölsä, J., Extending Dieke's diagram, *J. Lumin.*, 87–89, 1002, 2000.
10. Dorenbos, P., The 5d level positions of the trivalent lanthanides in inorganic compounds, *J. Lumin.*, 91, 155, 2000.
11. Dorenbos, P., f → d transition energies of divalent lanthanides in inorganic compounds, *J. Phys.: Condens. Matter*, 15, 575, 2003.
12. Dorenbos, P., Energy of the first $4f^7$→$4f^65d$ transition in Eu^{2+}-doped compounds, *J. Lumin.*, 104, 239, 2003.
13. McClure, D.S. and Pedrini, C., Excitons trapped at impurity centers in highly ionic crystals, *Phys. Rev.*, B32, 8465, 1985.
14. Dorenbos, P., Anomalous luminescence of Eu^{2+} and Yb^{2+} in inorganic compounds, *J. Phys.: Condens. Matter*, 15 2645, 2003.
15. Lyu, L.-J. and Hamilton, D.S., Radiative and nonradiative relaxation measurements in Ce^{3+}-doped crystals, *J. Lumin.*, 48&49, 251, 1991.
16. Dorenbos, P., Thermal quenching of Eu^{2+} 5d–4f luminescence in inorganic compounds, *J. Phys.: Condens. Matter*, 17, 8103, 2005.
17. Bessière, A., et al., Spectroscopy and lanthanide impurity level locations in $CaGa_2S_4$:Ln (Ln = Ce, Pr, Tb, Er, Sm), *J. Electrochem. Soc.*, 151, H254, 2004.
18. Boutinaud, P., et al., Making red emitting phosphors with Pr^{3+}, *Opt. Mater.*, 28, 9, 2006.
19. Guerassimova, N., et al., X-ray excited charge transfer luminescence of ytterbium-containing aluminium garnets. *Chem. Phys. Lett.*, 339, 197, 2001.
20. Brewer, L., *Systematics and the Properties of the Lanthanides*, edited by S.P. Sinha, D. Reidel Publishing Company, Dordrecht, The Netherlands, 1983, 17.
21. Martin, W.C., Energy differences between two spectroscopic systems in neutral, singly ionized, and doubly ionized lanthanide atoms, *J. Opt. Soc. Am.*, 61, 1682, 1971.
22. Jörgensen, C.K., Energy transfer spectra of lanthanide complexes, *Mol. Phys.*, 5, 271, 1962.
23. Dorenbos, P., The $4f^n$↔$4f^{n-1}5d$ transitions of the trivalent lanthanides in halogenides and chalcogenides, *J. Lumin.*, 91, 91, 2000.
24. Andriessen, J., Dorenbos, P., and van Eijk, C.W.E., Ab initio calculation of the contribution from anion dipole polarization and dynamic correlation to 4f–5d excitations of Ce^{3+} in ionic compounds, *Phys. Rev.*, B72, 045129, 2005.
25. Dorenbos, P., 5d-level energies of Ce^{3+} and the crystalline environment. I. Fluoride compounds, *Phys. Rev.*, B62, 15640, 2000.
26. Dorenbos, P., 5d-level energies of Ce^{3+} and the crystalline environment. IV. Aluminates and simple oxides, *J. Lumin.*, 99, 283, 2002.
27. Dorenbos, P., 5d-level energies of Ce^{3+} and the crystalline environment. II. Chloride, bromide, and iodide compounds, *Phys. Rev.*, B62, 15650, 2000.
28. Dorenbos, P., Relation between Eu^{2+} and Ce^{3+} f→d transition energies in inorganic compounds, *J. Phys.: Condens. Matter*, 15, 4797, 2003.
29. van Pieterson, L., et al., $4f^n$→$4f^{n-1}5d$ transitions of the light lanthanides: Experiment and theory, *Phys. Rev.*, B6, 045113, 2002.
30. van Pieterson, L., Reid, M.F., Burdick, G.W., and Meijerink, A., $4f^n$→$4f^{n-1}5d$ transitions of the heavy lanthanides: Experiment and theory, *Phys. Rev.*, B65, 045114, 2002.
31. Dorenbos, P., Exchange and crystal field effects on the $4f^{n-1}5d$ levels of Tb^{3+}, *J. Phys.: Condens. Matter*, 15, 6249, 2003.

32. Wong, W.C., McClure, D.S., Basun, S.A., and Kokta, M.R., Charge-exchange processes in titanium-doped sapphire crystals. I. Charge-exchange energies and titanium-bound excitons, *Phys. Rev.*, B51, 5682, 1995.

33. Happek, U., Choi, J., and Srivastava, A.M., Observation of cross-ionization in $Gd_3Sc_2Al_3O_{12}$:Ce^{3+}, *J. Lumin.*, 94–95, 7, 2001.

34. Dorenbos, P., Systematic behaviour in trivalent lanthandie charge transfer energies, *J. Phys.: Condens. Matter*, 15, 8417, 2003.

35. Sato, S., Optical absorption and X-ray photoemission spectra of lanthanum and cerium halides, *J. Phys. Soc. Jpn.*, 41, 913, 1976.

36. Lizzo, S., Meijerink, A., and Blasse, G., Luminescence of divalent ytterbium in alkaline earth sulphates, *J. Lumin.*, 59, 185, 1994.

37. Jia, D., Meltzer, R.S., and Yen, W.M., Location of the ground state of Er^{3+} in doped Y_2O_3 from two-step photoconductivity, *Phys. Rev.*, B65, 235116, 2002.

38. van der Kolk, E., et al., 5d electron delocalization of Ce^{3+} and Pr^{3+} in Y_2SiO_5 and Lu_2SiO_5, *Phys. Rev.*, B71, 165120, 2005.

39. Pedrini, C., Rogemond, F., and McClure, D.S., Photoionization thresholds of rare-earth impurity ions. Eu^{2+}:CaF_2, Ce^{3+};YAG, and Sm^{3+}:CaF_2, *J. Appl. Phys.*, 59, 1196, 1986.

40. Fuller, R.L. and McClure, D.S., Photoionization yields in the doubly doped SrF_2:Eu,Sm system, *Phys. Rev.*, B43, 27, 1991.

41. Joubert, M.F., et al., A new microwave resonant technique for studying rare earth photoionization thresholds in dielectric crystals under laser irradiation, *Opt. Mater.*, 24, 137, 2003.

42. Thiel, C.W., Systematics of 4f electron energies relative to host bands by resonant photoemission of rare-earth ions in aluminum garnets, *Phys. Rev.*, B64, 085107, 2001.

43. Thiel, C.W., Sun, Y., and Cone, R.L., Progress in relating rare-earth ion 4f and 5d energy levels to host bands in optical materials for hole burning, quantum information and phosphors, *J. Mod. Opt.*, 49, 2399, 2002.

44. Pidol, L., Viana, B., Galtayries, A., and Dorenbos, P., Energy levels of lanthanide ions in a $Lu_2Si_2O_7$:Ln^{3+} host, *Phys. Rev.*, B72, 125110, 2005.

45. Poole, R.T., Leckey, R.C.G., Jenkin, J.G., and Liesegang, J., Electronic structure of the alkaline-earth fluorides studied by photoelectron spectroscopy, *Phys. Rev.*, B12, 5872, 1975.

46. Barnes, J.C. and Pincott, H., Electron transfer spectra of some lanthanide (III) complexes, *J. Chem. Soc.* (a), 842, 1966.

47. Blasse, G. and Bril, A., Broad-band UV excitation of Sm^{3+}-activated phosphors, *Phys. Lett.*, 23, 440, 1966.

48. Krupa, J.C., Optical excitations in lanthanide and actinide compounds, *J. of Alloys and Compounds*, 225, 1, 1995.

49. Nakazawa, E., The lowest 4f-to-5d and charge-transfer transitions of rare earth ions in YPO_4 hosts, *J. Lumin.*, 100, 89, 2002.

50. Krupa, J.C., High-energy optical absorption in f-compounds, *J. Solid State Chem.*, 178, 483, 2005.

51. Dorenbos, P., The Eu^{3+} charge transfer energy and the relation with the band gap of compounds, *J. Lumin.*, 111, 89, 2004.

52. Jörgensen, C.K., *Modern Aspects of ligand Field Theory*, North-Holland Publishing Company, Amsterdam, 1971.

53. Dorenbos, P., Valence stability of lanthanide ions in inorganic compounds, *Chem. Mater.*, 17, 2005, 6452.

54. Howe, B., and Diaz, A.L., Characterization of host-lattice emission and energy transfer in $BaMgAl_{10}O_{17}$:Eu^{2+}, *J. Lumin.*, 109, 51, 2004.

chapter two — section one

Principal phosphor materials and their optical properties

Shinkichi Tanimizu

Contents

2.1 Luminescence centers of ns²-type ions

Ions with the electronic configuration ns^2 for the ground state and $nsnp$ for the first excited state (n = 4, 5, 6) are called ns^2-type ions. Table 1 shows 15 ions with the outer electronic configuration s^2. Luminescence from most of these ions incorporated in *alkali halides* and other crystals has been observed. Among these ions, luminescence and related optical properties of Tl^+ in KCl and other similar crystals have been most precisely studied,[1-5] so s^2 ions are also called Tl^+-like ions. As for powder phosphors, excitation and emission spectra of Sn^{2+}, Sb^{3+}, Tl^+, Pb^{2+}, and Bi^{3+} ions introduced into various oxygen-dominated host lattices have been reported,[6,7] though the analyses of these spectra have not yet been completed due to structureless broad-band spectra and unknown site symmetries. In this section, therefore, experimental and theoretical works on s^2 ions mainly in alkali halides will be summarized.

2.1.1 Optical spectra of s² ions in alkali halides

2.1.1.1 Absorption spectra

The intrinsic absorption edge of a pure KCl crystal is located at about 7.51 eV (165 nm) at room temperature. When Tl^+ is incorporated as a substitutional impurity in the crystal with concentrations below 0.01 mol%, four absorption bands appear below 7.51 eV, as shown in Figure 1(a). They have been labeled A, B, C, and D bands in order of increasing energy. Similar bands are observed by the incorporation of Pb^{2+} or Ag^- ions, as shown in Figures 1(b), (c).[8-10] One or two D bands lying near the absorption edge are due to charge-transfer transitions from Cl^- to s^2 ions or to perturbed excitons, and are not due to $s^2 \rightarrow sp$ transitions. The following discussion will, therefore, be restricted to the A, B, and C bands.

First, a model based on free Tl^+ ions following the original work of Seitz[1] will be discussed. The $6s^2$ ground state is expressed by $^1S_0$. The $6s6p$ first excited state consists of a triplet ^3P_J and a singlet $^1P_1$. The order of these states is $^3P_0$, $^3P_1$, $^3P_2$, and $^1P_1$ from the low-energy side. When a Tl^+ ion is introduced into an alkali halide host and occupies a cation site, it is placed in an octahedral (O_h) crystal field. The energy levels of the Tl^+ ion are labeled by the irreducible representation of the O_h point group. The labeling is made as follows: for the ground state $^1S_0 \rightarrow {}^1A_{1g}$, and for the excited state $^3P_0 \rightarrow {}^3A_{1u}$, $^3P_1 \rightarrow {}^3T_{1u}$, $^3P_2 \rightarrow {}^3E_u + {}^3T_{2u}$, and $^1P_1 \rightarrow {}^1T_{1u}$.

The $^1A_{1g} \rightarrow {}^1T_{1u}$ transition is dipole- and spin-allowed, while the $^1A_{1g} \rightarrow {}^3A_{1u}$ transition is strictly forbidden. The $^1A_{1g} \rightarrow {}^1T_{1u}$ transition is partially allowed by singlet-triplet spin-orbit mixing, and $^1A_{1g} \rightarrow ({}^3E_u + {}^3T_{2u})$ is also allowed due to vibronic mixing of ^3E_u and $^3T_{2u}$ with $^3T_{1u}$.

Then, the observed absorption bands shown in Figure 1 can be assigned as follows:

$$A \text{ bands}: {}^1A_{1g} \rightarrow {}^3T_{1u} \qquad \left({}^1S_0 \rightarrow {}^3P_1\right)$$

$$B \text{ bands}: {}^1A_{1g} \rightarrow {}^3E_u + {}^3T_{2u} \qquad \left({}^1S_0 \rightarrow {}^3P_2\right)$$

$$C \text{ bands}: {}^1A_{1g} \rightarrow {}^1T_{1u} \qquad \left({}^1S_0 \rightarrow {}^1P_1\right)$$

Focusing on the characteristics of the A, B, and C absorption bands, the centers of the gravity of the energies of these bands are given by[11]:

$$\bar{E}_A = F - \zeta/4 - \sqrt{\left(G + \zeta/4\right)^2 + \left(\lambda\zeta\right)^2 / 2}$$

$$\bar{E}_B = F - G + \zeta/2$$

$$\bar{E}_C = F - \zeta/4 + \sqrt{\left(G + \zeta/4\right)^2 + \left(\lambda\zeta\right)^2 / 2}$$

Here, F and G are the parameters of Coulomb and exchange energies as defined by Condon and Shortley.[11] ζ is the spin-orbit coupling constant. λ for the A and C bands is called the King-Van Vleck factor,[12] and is a parameter expressing the spatial difference between the $^1T_{1u}$ and $^3T_{1u}$ wavefunctions. The values of ζ and λ can be obtained from the values of $\bar{E}_A$ and $\bar{E}_C$ extrapolated to $T = 0K$, as shown in Figure 2.[13] The oscillator strength ratio of the C to A bands is given by[14]:

$$f_C / f_A = \left(\bar{E}_C / \bar{E}_A\right) \cdot R(\lambda, x)$$

where

$$R(\lambda, x) = \frac{1 + \lambda^2(1-x) + \sqrt{1 + 2\lambda^2 x(1-x)}}{1 + \lambda^2 x - \sqrt{1 + 2\lambda^2 x(1-x)}} \tag{1a}$$

Table 1   Ions with the ns^2 Configuration in the Ground State

Atomic No.	Element	$(ns)(np)$	Ion species
29	Cu	$(4s)^1$	Cu^-
30	Zn	$(4s)^2$	Zn^0
31	Ga	$(4s)^2(4p)^1$	Ga^+
32	Ge	$(4s)^2(4p)^2$	Ge^{2+}
33	As	$(4s)^2(4p)^3$	As^{3+}
47	Ag	$(5s)^1$	Ag^-
48	Cd	$(5s)^2$	Cd^0
49	In	$(5s)^2(5p)^1$	In^+
50	Sn	$(5s)^2(5p)^2$	$^*Sn^{2+}$
51	Sb	$(5s)^2(5p)^3$	$^*Sb^{3+}$
79	Au	$(6s)^1$	Au^-
80	Hg	$(6s)^2$	Hg^0
81	Tl	$(6s)^2(6p)^1$	$^*Tl^+$
82	Pb	$(6s)^2(6p)^2$	$^*Pb^{2+}$
83	Bi	$(6s)^2(6p)^3$	$^*Bi^{3+}$

* Luminescence is observed also in powder phosphors. (See 2.1.2)

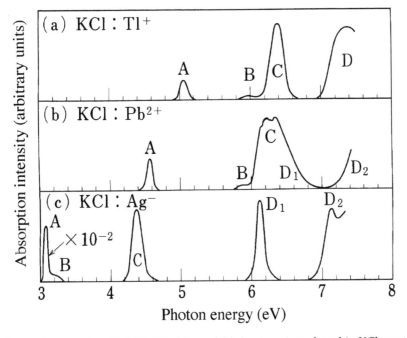

Figure 1   Absorption spectra of (a) Tl^+, (b) Pb^{2+}, and (c) Ag^- ions introduced in KCl crystals at 77K. (From Fukuda, A., *Science of Light (Japan)*, 13, 64, 1964; Kleeman, W., *Z. Physik*, 234, 362, 1970; Kojima, K., Shimanuki, S., and Kojima, T., *J. Phys. Soc. Japan*, 30, 1380, 1971. With permission.)

and

$$x = \left(\overline{E}_B - \overline{E}_A\right)/\left(\overline{E}_C - \overline{E}_A\right) \tag{1b}$$

Values of important parameters mentioned above are listed in Table 2[9] for various ns^2-type ions.

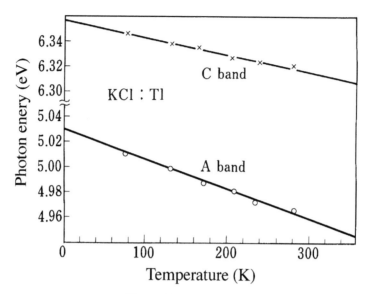

Figure 2 Temperature dependence of $\overline{E}_A$ and $\overline{E}_C$ for the A and C absorption bands in KCl:Tl[+]. (From Homma, A., *Science of Light (Japan)*, 17, 34, 1968. With permission.)

Table 2 Various Parameters Related to the A, B, and C Absorption Bands of ns²-Type Ions in Alkali Halide Crystals

ns²	Phosphors	R	$\overline{E}_A$ (eV)	$\overline{E}_B$ (eV)	$\overline{E}_C$ (eV)	λ	ζ (eV)	G (eV)	F (eV)
	KCl:Sn²⁺	18	4.36	4.94	5.36	0.599	0.527	0.316	4.992
	KCl:In⁺	54	4.343	4.630	5.409	0.754	0.268	0.447	4.943
	CsI:Ag⁻	360	2.780	2.865	3.770	0.897	0.082	0.472	3.296
5 s²	KCl:Ag⁻	435	3.100	3.250	4.349	0.575	0.147	0.585	3.762
	Cd⁰	478	3.80	3.87	5.41	0.762	0.142	0.769	4.643
	KI:Ag⁻	525	2.878	2.981	3.985	0.663	0.101	0.516	3.457
	KBr:Ag⁻	570	3.005	3.132	4.180	0.556	0.125	0.554	3.624
	KCl:Pb²⁺	4.2	4.57	5.86	6.33	1.03	0.951	0.304	5.688
6 s²	KCl:Tl⁺	5.4	5.031	5.930	6.357	0.984	0.692	0.283	5.867
	KCl:Au⁻	14.0	4.08	4.37	5.44	2.412	0.199	0.540	5.258
	Hg⁰	34.2	4.89	5.11	6.70	0.758	0.529	0.731	5.92

Note: R: see text, $\overline{E}_A$, $\overline{E}_B$, $\overline{E}_C$: The centers of gravity of the energies of A, B, and C absorption bands, λ: King-Van Vleck factor, ζ: Spin-orbit coupling constant, G: Exchange energy, F: Coulomb energy.

From Kleeman, W., *Z. Physik*, 234, 362, 1970. With permission.

If the $^1T_{1u}$ and $^3T_{1u}$ wavefunctions are identical, λ becomes 1. Assuming that λ = 1, Eq. 1a becomes:

$$R(x) = \frac{4 - 2x + \sqrt{6 - 2(2x-1)^2}}{2 + 2x - \sqrt{6 - 2(2x-1)^2}} \qquad (2)$$

This equation is known as Sugano's formula.[14] Figure 3[9] shows a plot of Eq. 2 and the experimental data obtained for various ns²-type ions in alkali halide crystals. Deviations

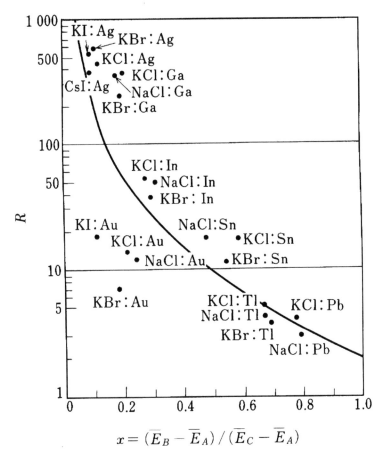

$$x = (\overline{E}_B - \overline{E}_A) / (\overline{E}_C - \overline{E}_A)$$

Figure 3 Experimentally obtained R values plotted against x for various ns^2-type ions in alkali halide crystals. The drawn curve is Sugano's formula, Eq. 2. (From Kleeman, W., *Z. Physik*, 234, 362, 1970. With permission.)

from the curve reflect the deviation of λ from 1. Figure 3 and Table 2 show that the observed R values for the same s^2-type ions are nearly the same magnitude for different alkali halide hosts, whereas the values for cationic s^2-type ions and for anionic s^2-type ions differ markedly for the same hosts; for example, R is 5.4 for KCl:Tl$^+$, and 435 for KCl:Ag$^-$. In the case of anionic Ag$^-$, the energy separation between the A and B absorption bands is as small as 0.15 eV, and their intensities are about one-hundredth of that of the C band because of the weak spin-orbit interaction of Ag$^-$.

It may be worth mentioning at this point that Sugano's formula was derived from molecular orbital approximation, but it uses the experimentally determined values for both G and ζ. The formula should, therefore, be considered as a special case of the atomic orbital approximation.

2.1.1.2 Structure of the A and C absorption bands

The C absorption band of KCl:Pb^{2+} has a triplet structure as shown in Figure 1(b). This structure is explained as a result of the splitting of the excited states due to the interaction with lattice vibrations, i.e., due to the dynamical Jahn-Teller effect.[15] The lattice vibrational modes interacting with the excited states of s^2 ions in O_h symmetry consist of A_{1g}, E_g, and T_{2g}. The symmetric triplet structure of the C band appears when the potential curves of the ground and excited states in the configurational coordinate model have the same

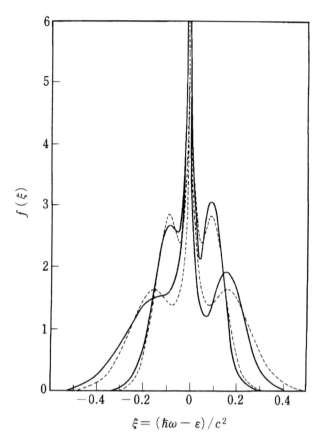

Figure 4   Calculated spectra of the C absorption band ($^1A_{1g} \rightarrow {}^1T_{1u}$) for two different (high and low) temperatures. Dotted curves represent symmetric cases. Solid curves represent the case that the $^1T_{1u}$ excited state has the curvature that is half as small as that of the $^1A_{1g}$ ground state. (From Fukuda, A., *J. Phys. Soc. Japan*, 27, 96, 1969. With permission.)

curvature within the framework of the Franck-Condon approximation, while the asymmetric triplet structure of the C band appears when they have different curvatures.

Figure 4 shows examples of calculated spectra of the C band for two different temperatures by taking account of the T_{2g} interaction mode.[15,16] The parameter c^2 appearing in the horizontal axis is that representing the coupling constant between s^2-type ions and lattice vibrational modes. The value of c^2 becomes smaller as the host lattice constant becomes larger, and becomes larger if the charge number of the ion becomes larger in the same host lattices. For example, the values of c^2 are 1.2 eV for NaCl:Tl$^+$, 0.82 eV for KCl:Tl$^+$, and 1.82 eV for KCl:Pb^{2+}.[15]

The A band, on the other hand, theoretically has a doublet structure, because two components consisting of the above-mentioned triplet structure have coalesced together due to the interaction between the A and B bands. Figure 5 shows an example of the calculated A absorption bands for two different temperatures θ. However, it is noted that the observed A bands shown in Figure 1 have no clear-cut doublet structure, in disagreement with the calculated bands in Figure 5, and appear as structureless bands. It is also noted that the doublet structure can be observed for KCl:Sn^{2+} and KCl:In$^+$ (see p. 836-837 in Reference 4).

Define the calculated splitting energy of the A doublet band as δ_A and that of the C triplet band as δ_C. The ratio of the two is given by[15]:

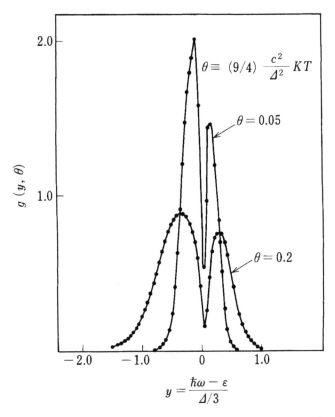

Figure 5 Calculated spectra of the A absorption band ($^1A_{1g} \rightarrow {}^3T_{1u}$) for two different (high and low) temperatures. Δ is a normalized energy parameter of the adiabatic potentials for $^3T_{1u}$ interacting with the T_{2g} mode. (From Toyozawa, Y. and Inoue, M., *J. Phys. Soc. Japan*, 21, 1663, 1966. With permission.)

$$\delta_A/\delta_C = 0.85 \cdot (R-2)/(R-1/2) \tag{3}$$

where R is the parameter of Eq. 2. It is understood that the values of δ_A are smaller than those of δ_C for heavy ions such as Tl$^+$ and Pb^{2+} because of their smaller R values, as shown in Table 2. This is considered as one reason that the doublet structure of the A band is not observed experimentally.

2.1.1.3 Temperature dependence of the A, B, and C absorption bands

The intensity of the C band is rather constant up to about 150K, and then slightly increases between 150 and 300K. The triplet structure of this band has a tendency to be prominent at higher temperatures. As for the B band, the intensity increases as temperature increases, because the band originates from vibration-allowed transitions. In some cases, temperature-dependent structure is observed in this band, but it is not precisely studied because of the small intensity of this band. The intensity of the A band varies with temperature similar to the C band. In KCl:Tl$^+$, however, the increase of the B band intensity is counterbalanced by the decrease of the A band intensity, which suggests a mixing of the excited states of the A and B bands.

The above-mentioned characteristics of the A, B, and C absorption bands are prominent features of s^2 ions in alkali halide host lattices. In hosts other than alkali halides, these features are also observed. The appearance of these features is useful for the identification of observed absorption and excitation bands.

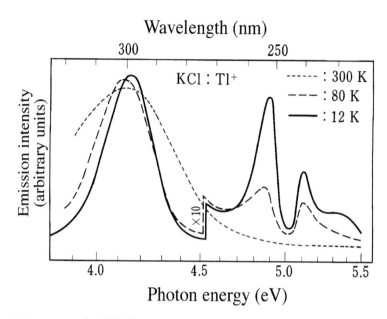

Figure 6 Emission spectra for KCl:Tl$^+$ at 300, 80, and 12K. (From Edgerton, R. and Teegarden, K., *Phys. Rev.*, 129, 169, 1963. With permission.)

2.1.1.4 Emission spectra

Figure 6 shows emission spectra of KCl:Tl$^+$ (0.01 mol%) as an example.[17] At 300K (dotted curve), excitation in any of the A, B, or C bands produces the same emission spectrum, i.e., the A emission band peaking at 4.12 eV (300 nm) and having a width at half-maximum of 0.56 eV (40 nm). At low temperatures, excitation in the A absorption band produces the emission at 4.13–4.17 eV, similar to the case at 300K; whereas excitation in the B or C bands produces another emission band located at about 5 eV in addition to the A band. This emission band has a large dip at 5 eV because of the overlap with the A absorption band. The 5-eV emission observed below 80K is assigned to the C emission ($^1P_1 \rightarrow {}^1S_0$).

Although the A emission band in KCl:Tl$^+$ has a simple structure, the A band in most other cases of s^2-type ion luminescence is composed of two bands: the high-energy band labeled A$_T$ and the low-energy band labeled A$_X$. Table 3[18] shows energy positions of the A$_T$ and A$_X$ bands for various monovalent s^2-type ions at temperatures in the range of 4.2 to 20K. In Group I, A$_T$ is much stronger than A$_X$ at 4.2K. With increasing temperature, the A$_T$ intensity decreases while the A$_X$ intensity increases, maintaining the sum of both intensities as constant. Above 60K, only A$_X$ is observed. In Group II, there is no temperature region in which A$_X$ is mainly observed. In Group III, the only band observed is assigned to A$_T$.

The mechanism that the A emission band is composed of two bands is ascribed to the spin-orbit interaction between the A band emitting state (i.e., the triplet $^3T_{1u}$ state) and the upper singlet $^1T_{1u}$ state.[18] This is explained by the configurational coordinate model as shown in Figure 7.[3-5] If the spin-orbit interaction is strong enough, the $^3T_{1u}$ state and $^1T_{1u}$ states repel each other, so that the lower triplet state is deformed to a relaxed excited state with two minima as shown in Figure 7(b). Thus, the two emission bands are produced from the two minima T and X.

As for decay kinetics of the A emission in KCl:Tl$^+$, readers are referred to Reference 19.

2.1.2 s^2-Type ion centers in practical phosphors

Some of the ns^2-type ions listed in Table 1 have long been known as luminescence centers of fluorescent lamp phosphors. In oxygen-dominated host lattices, the emissions

Table 3   Classification of the Observed A Emission Peaks
at 4.2–20K and Their Assignments

Group	Phosphor	A_T (eV)	A_X (eV)
I	KI:Ga$^+$	2.47	2.04
	KBr:Ga$^+$	2.74	2.24
	KCl:Ga$^+$	2.85	2.35
	NaCl:Ga$^+$	3.10	2.45
	KI:In$^+$	2.81	2.20
	KI:Tl$^+$	3.70	2.89
II	KBr:In$^+$	2.94	2.46
	KBr:Tl$^+$	4.02	3.50
III	KCl:In$^+$	2.95	—
	NaCl:In$^+$	3.05	—
	KCl:Tl$^+$	4.17	—

From Fakuda, A., *Phys. Rev.*, B1, 4161, 1970. With permission.

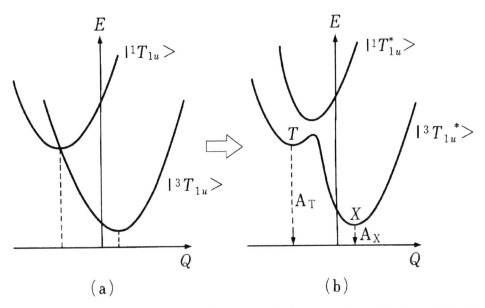

Figure 7   Configurational coordinate model to account for the A_T and A_X emission bands: (a) without spin-orbit interaction, (b) with spin-orbit interaction. (From Farge, Y. and Fontana, Y.P., *Electronic and Vibrational Properties of Point Defects in Ionic Crystals*, North-Holland Publishing, Amsterdam, 1974, 193; Ranfagni, A., Magnai, D., and Bacci, M., *Adv. Phys.*, 32, 823, 1983; Jacobs, P.W.M., *J. Phys. Chem. Solids*, 52, 35, 1991. With permission.)

from Sn^{2+}, Sb^{3+}, Tl$^+$, Pb^{2+}, and Bi^{3+} are reported. These ions are marked with asteriks in the table.

Luminescence features of the above five ions are as follows.

1. The luminescence is due to the A band transition ($^3P_1 \rightarrow {}^1S_0$).
2. The luminescence is usually associated with a large Stokes' shift, and the spectra are considerably broad, especially in case of Sn^{2+} and Sb^{3+}.
3. The luminescence decay is not very fast and of the order of microseconds. This is because the luminescence transition is spin-forbidden.

Spectral data[20] and $1/e$ decay times of practical phosphors activated with s^2-type ions at room temperature under 230–260 nm excitation are given below.

$Sr_2P_2O_7:Sn^{2+}$ (Ref. 21,22)
Excitation bands: 210, 233, and 250 nm.
Emission band: 464 nm with halfwidth 105 nm.

$SrB_6O_{10}:Sn^{2+}$ (Ref. 23)
Excitation bands: 260 and 325 nm.
Emission band: 420 nm with halfwidth 68 nm.
Decay time: 5 μs.

$Ca_5(PO_4)_3F:Sb^{3+}$ (Ref. 24, 25)
Excitation bands: • 175,[26] 202, 226, 235, 250, and 281 nm for O_2-compensated samples.
                     • 190, 200, 225, 246, and 267 nm for Na-compensated samples.
Emission bands: • 480 nm with halfwidth 140 nm.
                     • 400 nm with halfwidth 96 nm.
Decay times: • 7.7 μs for 480 nm emission.
                     • 1.95 μs for 400 nm emission.

The behavior of Sb^{3+} in fluorapatite $[Ca_5(PO_4)_3F]$ host lattice is not so simple, because of the existence of two different Ca sites and charge compensation. The low-lying excited states of Sb^{3+} with and without O_2 compensation were calculated by a molecular orbital model.[25] However, the reason why the decay times for 480 and 400 nm emission bands differ noticeably has not yet been elucidated.

$YPO_4:Sb^{3+}$ (Ref. 27, 28)
Excitation bands: 155 nm, 177–202 nm, 230 nm, and 244 nm.
Emission bands: 295 nm with halfwidth 46 nm, and 395 nm with halfwidth 143 nm.
Decay time: Below 1 μs.

$(Ca,Zn)_3(PO_4)_2:Tl^+$ (Ref. 29)
Excitation bands: 200 and 240 nm.
Emission band: 310 nm with halfwidth 41 nm.

The emission peaks vary with Zn contents.

$BaMg_2Al_{16}O_{27}:Tl^+$ (Ref. 30)
Excitation bands: • 200 nm and 245 nm for 1% Tl.
                     • Unknown for 3 and 10% Tl.
Emission bands: • 1% Tl: 295 nm with halfwidth 30 nm.
                     • 3% Tl: 420 nm with halfwidth 115 nm.
                     • 10% Tl: 460 nm with halfwidth 115 nm.
Decay times: • 0.2 μs for 295 nm emission.
                     • 0.6 μs for 460 nm emission.

$BaSi_2O_5:Pb^{2+}$ (Ref. 31, 32)
Excitation bands: 187 and 238 nm.
Emission band: 350 nm with halfwidth 39 nm.

In $BaO\text{-}SiO_2$ systems, Ba_2SiO_4, $BaSiO_3$, and $BaSi_3O_8$, are also known. $Ba_2SiO_4:Pb^{2+}$ reveals two emissions peaked at 317 and 370 nm. The excitation bands lie at 180, 202, and 260 nm.

Pb^{2+} in another host; $SrAl_{12}O_{19}:Pb^{2+}$ (Ref. 30)

Excitation bands:	Below 200 nm, and 250 nm for 1% Pb.
	• Unknown for 25 and 75% Pb.
Emission bands:	• 1% Pb: 307 nm with halfwidth 40 nm.
	• 25% Pb: 307 nm with halfwidth 46 nm, and 385 nm with half-width 75 nm.
	• 75% Pb: 405 nm with halfwidth 80 nm.
Decay time:	• 0.4 µs for 307 nm emission.

As for the spectral data and decay times of Bi^{3+} activated phosphors, readers are referred to References 33, 34, 35, and 36.

$YPO_4:Bi^{3+}$ (Ref. 33, 36)

Excitation bands:	156, 169, 180, 220, 230, and 325 nm (for a Bi-Bi pair)
Emission bands:	241 nm
Decay time:	0.7 s

References

1. Seitz, F., *J. Chem. Phys.*, 6, 150, 1938.
2. Fowler, W.B., Electronic States and Optical Transitions of Color Centers, in *Physics of Color Centers*, Fowler, W.B., Ed., Academic Press, New York, 1968, 133.
3. Farge, Y. and Fontana, M.P., *Electronic and Vibrational Properties of Point Defects in Ionic Crystals*, North-Holland Publishing Co., Amsterdam, 1974, 193.
4. Ranfagni, A., Magnai, D., and Bacci, M., *Adv. Phys.*, 32, 823, 1983.
5. Jacobs, P.W.M., *J. Phys. Chem. Solids*, 52, 35, 1991.
6. Butler, K.H., *Fluorescent Lamp Phosphors*, Pennsylvania State University Press, 1980, 161.
7. Blasse, G. and Grabmaier, B.C., *Luminescent Materials*, Springer Verlag, Berlin, 1994, 28.
8. Fukuda, A., *Science of Light (Japan)*, 13, 64, 1964.
9. Kleemann, W., *Z. Physik*, 234, 362, 1970.
10. Kojima, K., Shimanuki, S., and Kojima, T., *J. Phys. Soc. Japan*, 30, 1380, 1971.
11. Condon, E.U. and Shortley, G.H., *The Theory of Atomic Spectra*, Cambridge University Press, London, 1935.
12. King, G.W. and Van Vleck, J.H., *Phys. Rev.*, 56, 464, 1939.
13. Homma, A., *Science of Light (Japan)*, 17, 34, 1968.
14. Sugano, S., *J. Chem. Phys.*, 36, 122, 1962.
15. Toyozawa, Y. and Inoue, M., *J. Phys. Soc. Japan*, 21, 1663, 1966; Toyozawa, Y., *Optical Processes in Solids*, Cambridge University Press, London, 53, 2003.
16. Fukuda, A., *J. Phys. Soc. Japan*, 27, 96, 1969.
17. Edgerton, R. and Teegarden, K., *Phys. Rev.*, 129, 169, 1963.
18. Fukuda, A., *Phys. Rev.*, B1, 4161, 1970.
19. Hlinka, J., Mihokova, E., and Nikl, M., *Phys. Stat. Sol.*, 166(b), 503, 1991.
20. See Table 10 and 10a in 5.6.2.
21. Ropp, R.C. and Mooney, R.W., *J. Electrochem. Soc.*, 107, 15 1960.
22. Ranby, P.W., Mash, D.H., and Henderson, S.T, *Br. J. Appl. Phys.*, Suppl. 4, S18, 1955.
23. Leskela, M., Koskentalo, T., and Blasse, G., *J. Solid State Chem.*, 59, 272, 1985.
24. Davis, T.S., Kreidler, E.R., Parodi, J.A., and Soules, T.F., *J. Luminesc.*, 4, 48, 1971.
25. Soules, T.F., Davis, T.S., and Kreidler, E.R., *J. Chem. Phys.*, 55, 1056, 1971; Soules, T.F., Bateman, R.L., Hewes, R.A., and Kreidler, E.R., *Phys. Rev.*, B7, 1657, 1973.

26. Tanimizu, S. and Suzuki, T., *Electrochem. Soc.*, Extended Abstr., 74-1, No. 96, 236, 1974.
27. Grafmeyer, J., Bourcet, J.C., and Janin, J., *J. Luminesc.*, 11, 369, 1976.
28. Omen, E.W.J.L., Smit, W.M.A., and Blasse, G., *Phys. Rev.*, B37, 18, 1988.
29. Nagy, R., Wollentin, R.W., and Lui, C.K., *J. Electrochem. Soc.*, 97, 29, 1950.
30. Sommerdijk, J.L., Verstegen, J.M.P.J., and Bril, A., *Philips Res. Repts.*, 29, 517, 1974.
31. Clapp, R.H. and Ginther, R.J., *J. Opt. Soc. Am.*, 37, 355, 1947.
32. Butler, K.H., *Trans. Electrochem. Soc.*, 91, 265, 1947.
33. Blasse, G. and Bril, A., *J. Chem. Phys.*, 48, 217, 1968.
34. Boulon, G., *J. Physique*, 32, 333, 1971.
35. Blasse, G., *Prog. Solid State Chem.*, 18, 79, 1988.
36. J-Stel, T., Huppertz, P., Mayr, W., Wiechert, D.U. *J. Lumin.*, 106, 225, 2004.

chapter two — section two

Principal phosphor materials and their optical properties

Masaaki Tamatani

Contents

2.2 Luminescence centers of transition metal ions

2.2.1 Crystal field theory[1-7]

The $3d$ transition metal ions utilized in commercial powder phosphors have three electrons (in the case of Cr^{3+} and Mn^{4+}) or five electrons (Mn^{2+} and Fe^{3+}) occupying the outermost $3d$ electron orbitals of the ions. When the $3d$ ions are incorporated into liquids or solids, spectroscopic properties (such as spectral positions, widths, and intensities of luminescence and absorption bands) are considerably changed from those of gaseous free ions.

These changes are explained in terms of *crystal field theory*, which assumes anions (ligands) surrounding the metal ion as point electric charges. When the theory is extended to take into consideration the overlap of electron orbitals of the metal ion and ligands, it is called *ligand field theory.* In the following, these theories will be described briefly. For more details, the reader is referred to Reference 1.

2.2.1.1 The simplest case: 3d¹ electron configuration

First, take the case of an ion that has the $3d^1$ electron configuration, such as Ti^{3+}. Table 4 shows the wavefunctions for the five $3d$ electron orbitals, and Figure 8 the electron distributions for these orbitals. For a free ion, the energies of the five $3d$ orbitals are identical, and are determined by an electron kinetic energy and a central field potential caused by the inner electron shell.* In cases where different orbitals have the same energy, the orbitals are said to be degenerate.

When this ion is incorporated in a crystal, surrounding anions affect it. Consider the case where there are six anions (negative point charges) at a distance R from a central cation nucleus located at $\pm x$, $\pm y$, and $\pm z$ as shown by open circles in Figure 8. This ligand arrangement is called the octahedral coordination. These anions induce an electrostatic potential V on a $3d$ electron of the central cation, which is expressed by

$$V = \sum_{i}^{i=6} \frac{Ze^2}{|R_i - r|} \tag{4}$$

Here, R_i represents a position of the ith anion, r a position of the $3d$ electron (coordinates x, y, z), Z a valency of an anion, and e an electron charge.

When $|R_i| \gg |r|$, the following equation is obtained from Eq. 4 by the expansion on r up to 4th order.

$$V = \frac{6Ze^2}{R} + \frac{35Ze^2}{4R^5}\left(x^4 + y^4 + z^4 - \frac{3}{5}r^4\right) \tag{5}$$

The effect of the potential V on the $3d$ electron orbital energy is expressed by the following integration.

$$\int \psi(3d)V\psi(3d)d\tau = \langle 3d|V|3d \rangle \tag{6}$$

The first term of Eq. 5 increases the energy of all five orbitals by the same amount. It may be neglected in the field of optical spectroscopy, where only energy differences among electron states are meaningful. From the second term in Eq. 5, the following orbital energies are obtained.

$$\langle \xi|V|\xi \rangle = \langle \eta|V|\eta \rangle = \langle \varsigma|V|\varsigma \rangle = -4Dq \tag{7}$$

$$\langle u|V|u \rangle = \langle v|V|v \rangle = 6Dq \tag{8}$$

---

* Here, the spin-orbit interaction of an electron is neglected.

Table 4   Wavefunctions for a 3d Electron

$$\varphi_u = \sqrt{5/16\pi}\, R_{3d}(r)(1/r^2)(3z^2 - r^2)$$

$$\varphi_v = \sqrt{5/16\pi}\, R_{3d}(r)(1/r^2)(x^2 - y^2)$$

$$\varphi_\xi = \sqrt{15/4\pi}\, R_{3d}(r)(1/r^2)yz$$

$$\varphi_\eta = \sqrt{15/4\pi}\, R_{3d}(r)(1/r^2)zx$$

$$\varphi_\varsigma = \sqrt{15/4\pi}\, R_{3d}(r)(1/r^2)xy$$

Note: $R_{3d}(r)$ means the radial wavefunction of a 3d electron. There are many ways to construct five wavefunctions for a 3d electron. Here, they are constructed so as to diagonalize the matrix for the cubic crystal field V; that is, nondiagonal elements of the seqular equation (e.g., $\langle u|V|v \rangle$) are equal to zero.

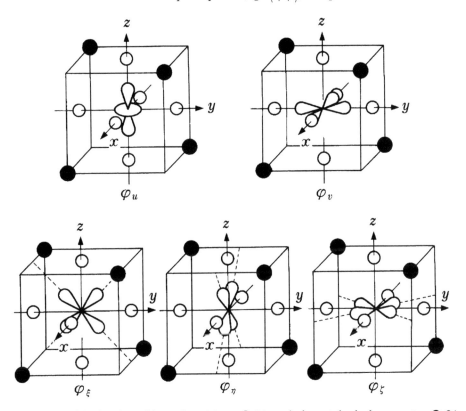

Figure 8   Shapes of d orbitals and ligand positions. ○: Ligands for octahedral symmetry. ●: Ligands for tetrahedral symmetry.

Here,

$$D = \frac{35Ze}{4R^5} \tag{9}$$

$$q = \frac{2e}{105} \int |R_{3d}(r)|^2 r^4 dr \tag{10}$$

Therefore, the fivefold degenerate 3d orbitals split into triply degenerate orbitals (ξ,η,ζ) and doubly degenerate orbitals (u,v). The former are called t_2 orbitals, and the latter e orbitals.* The energy difference between the t_2 and e orbitals is $10Dq$.** The splitting originates from the fact that u and v orbitals pointing toward anions in the x, y, and z directions suffer a larger electrostatic repulsion than ξ, η, and ζ orbitals, which point in directions in which the anions are absent.

Next, consider the case where four anions at a distance R from the central cation form a regular tetrahedron (tetrahedral coordination). The electrostatic potential caused by these anions at a 3d electron of the cation, V_t, is expressed as follows.

$$V_t = \frac{4Ze^2}{R} + eTxyz + eD_t\left(x^4 + y^4 + z^4 - \frac{3}{5}r^4\right) \tag{11}$$

Here,

$$T = \frac{10\sqrt{3}\,Ze}{3R^4} \tag{12}$$

$$D_t = -\frac{4}{9}D \tag{13}$$

The sign of the second term in Eq. 11 changes when the electron coordinates are inverted as $x \rightarrow -x$, $y \rightarrow -y$, and $z \rightarrow -z$, (that is, the term has "odd parity"), and the integrated value of Eq. 6 is zero. Since the third term of Eq. 11 has the same form as the second term in Eq. 5, values similar to Eqs. 7 and 8 are obtained for the 3d electrons, lifting the degeneration. However, as shown by Eq. 13, a t_2 orbital has a higher energy than an e orbital, and the splitting is smaller than that in octahedral coordination. These results reflect the facts that the t_2 orbitals point toward the anion positions and that the number of the ligands is smaller than that in the octahedral case.

In most crystals, each metal ion is surrounded by four or six ligands. So, the electrostatic effect from the ligands on the central cation (the crystal field) may be approximated by Eqs. 5 or 11, where all ligands are assumed to be located at an equal distance from the central cation, and to have a geometric symmetry of O_h or T_d in notation of the crystal point group. The crystal field with a slightly lower symmetry than the O_h or T_d may be treated by a perturbation method applied to Eq. 5 or 11. The energy levels split further in this case.

For the above procedures, group theory may be utilized based on the symmetry of the geometric arrangement of the central ion and ligands. This is based on the fact that a crystal field having a certain symmetry is invariant when the coordinates are transformed by elemental symmetry operations that belong to a point group associated with the symmetry; all terms other than the crystal field in the Hamiltonian for electrons are also not changed in form by the elemental symmetry operations. In addition, electron wavefunctions can be used as the basis of a representative matrix for the symmetry operations, and the eigenvalues (energies) of the Hamiltonian can be characterized by the reduced representations. Particularly when the Hamiltonian includes the inter-electron electrostatic and spin-orbit interactions in a multi-electron system, group theory is useful for obtaining

---

* They are sometimes called $d\varepsilon$ and $d\gamma$ orbitals in crystal field theory. Notation of t_2 and e is generally used more in ligand field theory.
** The energy difference of $10Dq$, a measure of the crystal field, is sometimes represented as Δ.

Table 5 Correlation of Reduced Representations

Point group	O_h	T_d	D_{4h}	C_{2v}	C_{3v}
	A_{1g}	A_1	A_{1g}	A_1	A_1
	A_{2g}	A_2	B_{1g}	A_2	A_2
Representation	E_g	E	A_{1g}, B_{1g}	A_1, A_2	E
	T_{1g}	T_1	A_{2g}, E_g	A_2, B_1, B_2	A_2, E
	T_{2g}	T_2	B_{2g}, E_g	A_1, B_1, B_2	A_1, E

Note: Subscript g means even parity. Odd parity representations,
$A_{1u}, A_{2u}, \ldots, T_{2u}$, are not shown.

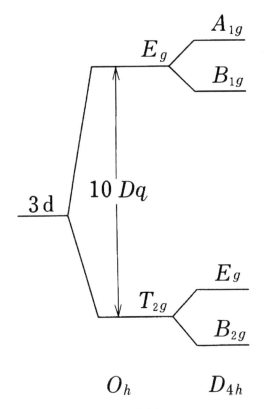

Figure 9 $3d$ level splitting caused by the crystal field.

energy level splitting and wavefunctions, calculating level energies, and predicting the
selection rule for transitions between energy levels. Wavefunctions for the t_2 and e orbitals
are the basis for the reduced representations T_{2g} and E_g, respectively, in the O_h group.

When the symmetry of the crystal site is lowered from O_h to D_{4h}, one obtains representations in the lower symmetry group contained in the original (higher) symmetry
representations from a correlation table of group representations.[8] Table 5 shows an example. From this table, the number (splitting) and representations of energy levels in the
lower symmetry can be seen. Figure 9 shows the energy level splitting due to symmetry
lowering.

2.2.1.2   The cases of more than one d electron

Strong crystal field.   When there are more than one electron, the electrons affect each
other electrostatically through a potential of $\sum_{i,j} e^2/r_{ij}$, where r_{ij} represents the distance

between the two electrons. When the contribution of the crystal field is so large that the electrostatic interaction can be neglected, energies of the states for the d^N electron configuration are determined by the number of electrons occupying the t_2 and e orbitals only. That is, $(N + 1)$ energy levels of e^N, $t_2 e^{N-1}$, ..., t_2^N configurations are produced with energies for $t_2^n e^{N-n}$ given by:

$$E(n,\ N-n) = \left(-4n + 6(N-n)\right)Dq \tag{14}$$

The energy difference between the neighboring two levels is $10Dq$.

When the electrostatic interaction is taken into consideration as a small perturbation, the lower symmetry levels split from the levels of these electron configurations. They are derived from the group theoretical concept of products of representations, applied together with the *Pauli principle*. The latter states that only one electron can occupy each electron orbital, inclusive of spin state. For example, in the case of d^2 (V^{3+} ion), the following levels can be derived:

$$t_2^2 \rightarrow\ ^3T_1,\ ^1A_1,\ ^1E,\ ^1T_2$$

$$t_2 e \rightarrow\ ^3T_1,\ ^3T_2,\ ^1T_1,\ ^1T_2$$

$$e^2 \rightarrow\ ^3A_2,\ ^1A_1,\ ^1E$$

Here, each level of $^{2S+1}\Gamma$, which is $(2S+1)(\Gamma)$ degenerate, is called a multiplet. S stands for the total spin angular momentum of the electrons. (Γ) represents the degeneracy of the reduced representation Γ; it is 1 for A_1, A_2, B_1, and B_2, 2 for E, and 3 for T_1 and T_2. The energy for a multiplet is obtained as the sum given by Eq. 14 and the expectation value of e^2/r_{12} (e.g., $\left\langle t_2^2\ ^3T_1 \middle| e^2/r_{12} \middle| t_2^2\ ^3T_1 \right\rangle$). To distinguish the parent electron configuration, each multiplet is usually expressed in the form of $^{2S+1}\Gamma(t_2^n e^{N-n})$.

Medium crystal field. When the crystal field strength decreases, one cannot neglect the interaction between levels having the same reduced representation but different electron configurations; for example, $\left\langle t_2^2\ ^3T_1 \middle| e^2/r_{12} \middle| t_2 e\ ^3T_1 \right\rangle$. This interaction is called the configuration interaction. The level energies of the reduced representation are derived from the eigenvalues of a determinant or a secular equation that contains the configuration interaction.

Weak crystal field. When the crystal field energy is very small compared with that of the configuration interaction, total angular quantum numbers of L and S for orbitals and spins, respectively, determine the energy. In the case of $Dq = 0$, a level is expressed by ^{2S+1}L, with degeneracy of $(2S+1)(2L+1)$. Symbols S, P, D, F, G, H, ... have been used historically, corresponding to $L = 0, 1, 2, 3, 4, 5, ...$ For the d^2 configuration, there exist ^1S, ^1G, ^3P, ^1D, and ^3F levels. Levels split from these levels by a small crystal field perturbation are represented by $^{2S+1}\Gamma(^{2S+1}L)$.

In all three cases described above, integral values for e^2/r_{12} can be shown as linear combinations of a set of parameters: A, B, and C introduced by Racah (Racah parameters). Parameter A makes a common contribution to energies of all levels. Therefore, level energies are functions of Dq, B, and C for spectroscopic purposes, where the energy difference between the levels is the meaningful quantity.

2.2.1.3   Tanabe-Sugano diagrams[2]

Each crystal field and electron configuration interaction affects the level energies of the $3d$ transition metal ions by about 10^4 cm^{-1}. Tanabe and Sugano[2a] calculated the determinants of the electron configuration interaction described in Section 2.2.1.2 for the d^2 to d^8 configurations in an octahedral crystal field. They presented the solutions of the determinants in so-called Tanabe-Sugano diagrams.[2b] Figures 10 to 16 show the diagrams for the d^2 to d^8 configurations. These diagrams were prepared for the analysis of optical spectra.* The level energies (E) from the ground level are plotted against the crystal field energy (Dq), both in units of B. For $C/B = \gamma$, values of 4.2 to 4.9 obtained from the experimental spectra in free ions are used. Note that one can treat the configuration interaction for n electrons occupying 10 d orbitals in the same manner as that for (10–n) holes; the diagram for d^n is the same as that for d^{10-n} for $Dq = 0$. In addition, the sign of the Dq value for electrons becomes opposite for holes, so that the diagram for d^{10-n} in the octahedral field is also used for d^n in the tetrahedral field.

Optical absorption spectra for $[M(H_2O)_6]^{n+}$ complex ions of $3d$ metals can be well explained by the Tanabe-Sugano diagrams containing the two empirical parameters of Dq and B (about 1000 cm^{-1}).[3] The Dq values for metal ions are in the order:

$$Mn^{2+} < Ni^{2+} < Co^{2+} < Fe^{2+} < V^{2+} < Fe^{3+} < Cr^{3+} < V^{3+} < Co^{3+} < Mn^{4+} \tag{15}$$

They are about 1000 cm^{-1} for divalent metals, and about 2000 cm^{-1} for trivalent metals. For a metal ion, Dq is known to depend on ligand species in the order:

$$I^- < Br^- < Cl^- \sim SCN^- < F^- < H_2O < NH_3 < NO_2^- < CN^- \tag{16}$$

This ordering is called the *spectrochemical series*.[7] Dq values in liquid are not so different from those in crystal, but are governed by the ligand ion species directly bound to the central metal ion. Thus, the spectrochemical series may be rewritten as[7]:

$$I < Br < Cl < S < F < O < N < C \tag{17}$$

Tanabe-Sugano diagrams demonstrate that those configurations in which the lowest excited levels (light-emitting levels) are located in the visible spectral region are d^3 and d^5. For d^3 (Figure 11), the light-emitting levels are $^2E(^2G)$ and $^4T_2(^4F)$ above and below the crossover value of $Dq/B \sim 2.2$, respectively. As will be described later, luminescence bands from these two levels are observed for Cr^{3+} depending on the crystal field strength of host materials. For d^5 (Figure 13), $^4T_1(^4G)$ is the lowest excited level, which is located in the visible region at weak crystal field of $Dq/B < 1.5$. Mn^{2+} of this configuration, having the smallest Dq value among transition metal ions in Eq. 15, is a suitable activator for green- to red-emitting phosphors. The dependence of the $^2E(^2G)$ states for d^3 on Dq is almost parallel to that of the ground level. This suggests that the wavelength of the emitted light does not depend significantly on the crystal field strength of different host materials or on the temperature. Lattice vibrations also lead to instantaneous Dq variation, but the emitting level energy is insensitive to these variations and, consequently, the spectral band may be a sharp line. On the other hand, the curves of the $^4T_2(^4F)$ for d^3 and $^4T_1(^4G)$ for d^5 have steep slopes when plotted against Dq, suggesting that the position of

---

* Orgel[3] also presented diagrams of energy levels as a function of Dq for some transition metal ions such as $V^{3+}(d^2)$, $Ni^{2+}(d^8)$, $Cr^{3+}(d^3)$, $Co^{2+}(d^7)$, and $Mn^{2+}(d^5)$.

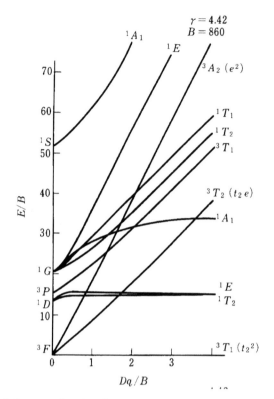

Figure 10 Energy level diagram for the d^2 configuration. (From Kamimura, H., Sugano, S., and Tanabe, Y., *Ligand Field Theory and its Applications*, Syokabo, Tokyo, 1969 (in Japanese). With permission.)

the emitting bands will depend strongly on host materials and that their bandwidths may be broad.

There have been extensive studies on solid-state laser materials doped with transition metal ions. In particular, for tunable lasers working in the far-red to infrared regions, the optical properties of various ions—d^1 (Ti^{3+}, V^{4+}); d^2 (V^{3+}, Cr^{4+}, Mn^{5+}, Fe^{6+}); d^3 (V^{2+}, Cr^{3+}, Mn^{4+}); d^4 (Mn^{3+}); d^5 (Mn^{2+}, Fe^{3+}); d^7 (Co^{2+}); and d^8 (Ni^{2+})—have been investigated in terms of Tanabe-Sugano diagrams with considerable success.

In the diagrams for d^4, d^5, d^6, and d^7 configurations, the ground levels are replaced by those of the lower spin quantum numbers when Dq/B exceeds 2 to 3. This gives an apparent violation of Hund's rule, which states that the ground state is the multiplet having the maximum orbital angular quantum number among those having the highest spin quantum number. It is known that the ion valency is unstable around the Dq/B values at which Hund's rule starts to break down.[4,5]

2.2.1.4   Spin-orbit interaction

For $3d$ transition metal ions, the contribution from the spin-orbit interaction in electrons ($\sum_i \xi l_i \cdot s_i$) is as small as 100 cm^{-1}, compared with that due to the crystal field (~10^4 cm^{-1}). Hitherto, this interaction has been neglected. Spin-orbit plays a role, however, in determining the splitting of sharp spectral lines and the transition probability between the levels.

2.2.1.5   Intensities of emission and absorption bands

The interaction between an oscillating electromagnetic field of light and an electron brings about a transition between different electronic states. Since the electric dipole (P)

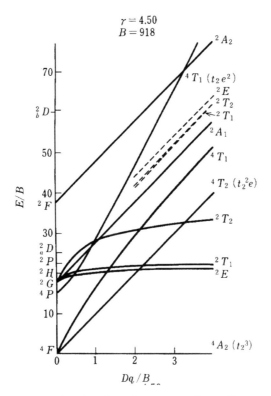

$\gamma = 4.50$
$B = 918$

Figure 11 Energy level diagram for the d^3 configuration. (From Kamimura, H., Sugano, S., and Tanabe, Y., *Ligand Field Theory and its Applications*, Syokabo, Tokyo, 1969 (in Japanese). With permission.)

component of the electric field of light has odd parity, and since all wavefunctions of pure d^n states have even parity, one obtains

$$\langle 3d^n\ f\,|P|3d^n\ i\rangle = 0 \tag{18}$$

This means that a transition between the states i and f having the same parity is forbidden (Laporte's rule). When a crystal field V_{odd} has no inversion symmetry, however, this expression may have small finite value, since wavefunctions having odd parity may be admixed with the $3d^n$ wavefunctions according to the following expression.

$$\psi = \psi_{3d^n} + \sum_u \frac{\psi_u \langle u|V_{odd}|3d^n\rangle}{\Delta E_u} \tag{19}$$

Here, ψ_u is a wavefunction for an odd parity state lying at higher energies; these could be $(3d)^{n-1}4p$ states and/or charge-transfer states which will be described later. ΔE_u is the energy difference between the ψ_{3d^n} and ψ_u states.

Even in the case of O_h having the inversion symmetry, V_{odd} may be produced instantaneously by lattice vibrations having odd parity, resulting in a slight violation of Laporte's rule. On the other hand, a magnetic dipole produced by the oscillating magnetic field of light has even parity, and transitions between d^n levels are allowed via this mechanism. In the above, it is assumed that multiplets involved in the transition have a same spin quantum number. Transitions between different spin states are forbidden by orthogonality

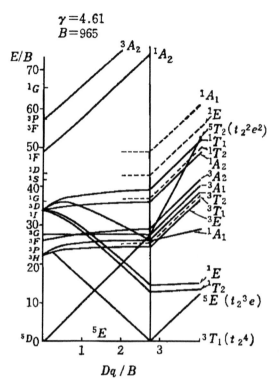

$\gamma = 4.61$
$B = 965$

Figure 12 Energy level diagram for the d^4 configuration. (From Kamimura, H., Sugano, S., and Tanabe, Y., *Ligand Field Theory and its Applications*, Syokabo, Tokyo, 1969 (in Japanese). With permission.)

of the spin wavefunctions (spin selection rule). However, in this case also, they can be partly allowed, since different spin wavefunctions may be slightly mixed by means of the spin-orbit interaction. Based on the above considerations, intensities of absorption bands in the visible region for metal complexes have been evaluated in terms of their oscillator strength f. Table 6 shows the results.*

The luminescence decay time, i.e., the time required for an emission from a level to reach $1/e$ of its initial intensity value after excitation cessation, τ (seconds), and the oscillator strength, f, have the following relations.[9] For electric dipole transitions,

$$f\tau = \frac{1.51\left(E_c/E_{eff}\right)^2 \lambda_0^2}{n} \tag{20}$$

and for magnetic dipole transitions,

$$f\tau = \frac{1.51\lambda_0}{n^3} \tag{21}$$

Here, E_c is the average electric field strength in a crystal, E_{eff} is the electric field strength at the ion position, λ_0 is the wavelength in vacuum (cm), and n is the refraction index. In Table 6, τ values estimated from these equations are also shown.

---

* Note that oscillator strength f for transitions allowed by odd lattice vibrations depends on temperature as $\coth\left(\hbar\omega/2kT\right)$. Here, $\hbar\omega$ means phonon energy.

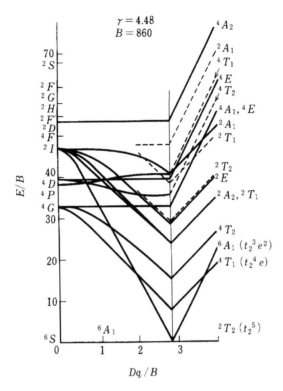

$\gamma = 4.48$
$B = 860$

Figure 13 Energy level diagram for the d^5 configuration. (From Kamimura, H., Sugano, S., and Tanabe, Y., *Ligand Field Theory and its Applications*, Syokabo, Tokyo, 1969 (in Japanese). With permission.)

2.2.2 Effects of electron cloud expansion

2.2.2.1 Nephelauxetic effect[7]

The Racah parameters, B and C, in a crystal are considerably smaller than those for a free ion, as shown in Tables 7, 8, 10, and 11. The reason is as follows. Some electrons of the ligands move into the orbitals of the central ion and reduce the cationic valency. Due to this reduction, the d-electron wavefunctions expand toward the ligands to increase the distances between electrons, reducing the interaction between them. This effect is called the *nephelauxetic effect*. In fact, some $3d$ electrons are known to exist even at the positions of the nuclei of the ligands as determined by ESR and NMR experiments. Therefore, the assumption in crystal field theory that expansion of the $3d$ orbitals may be negligibly small does not strictly hold. The reduction of B and C for various ligands is in the order:

$$F < O \sim N < Cl \sim C < Br < I \sim S \tag{22}$$

For central cations, it is:

$$Mn^{2+} < Ni^{2+} < Cr^{3+} < Fe^{3+} < Co^{3+} \tag{23}$$

This effect may be considered to increase with covalency between the cation and ligands. Note that the relation in Eq. 22 corresponds to a decreasing order in the electronegativity of elements.

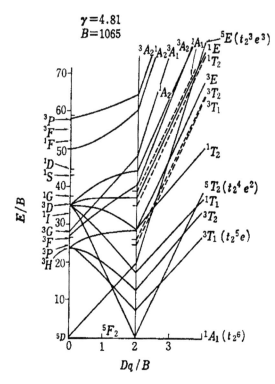

Figure 14 Energy level diagram for the d^6 configuration. (From Kamimura, H., Sugano, S., and Tanabe, Y., *Ligand Field Theory and its Applications*, Syokabo, Tokyo, 1969 (in Japanese). With permission.)

2.2.2.2 Charge-transfer band

In crystal field theory, transitions with higher energies than those within the d^n configuration entail $d^n \to d^{n-1}s$ or $d^n \to d^{n-1}p$ processes. However, in energy regions (e.g., 200 to 300 nm for oxides) lower than these interconfigurational transitions, strong absorption bands ($f \sim 10^{-1}$), called charge-transfer (CT) (or electron-transfer) bands, are sometimes observed.[4,7,9] These absorption bands are ascribed classically to electron transfers from the ligands to a central cation. It is argued that (1) the band energy is lower as the electronegativity of the ligands decreases, and (2) it is reduced as the valency increases for cations having the same number of electrons.[4,7] Charge-transfer states for $3d$ ions, however, are not fully understood, unlike those for $4d$ and $5d$ ion complexes.*

2.2.3   Cr³⁺ phosphors (3d³)

Luminescence due to Cr^{3+} is observed in the far-red to infrared region, and only limited applications have been proposed for Cr^{3+} phosphors.[12] This ion has attracted, however, the attention of spectroscopists since the 1930s, because Cr^{3+} brings about luminescence with an interesting line structure in the 680- to 720-nm spectral region in various host materials. In particular, the optical spectra of ruby (Al_2O_3:Cr^{3+}) were fully explained for the first time by applying crystal field theory (1958)[13]; ruby was utilized for the first solid-state laser (1960).[14]

Figures 17 and 18 show the luminescence[15] and absorption[1] spectra of ruby crystals, respectively. The two strong luminescence lines at 694.3 nm (= 14399 cm⁻¹) and 692.9

---

* See 2.4. For rare-earth phosphors, the effect of the charge-transfer bands is investigated in considerable detail with respect to the fluorescence properties of *f-f* transitions.[11]

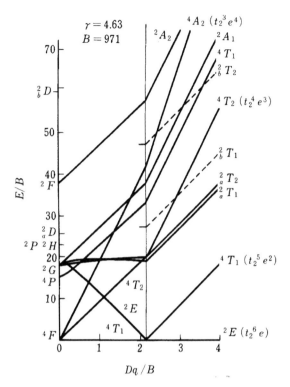

Figure 15   Energy level diagram for the d^7 configuration. (From Kamimura, H., Sugano, S., and Tanabe, Y., *Ligand Field Theory and its Applications*, Syokabo, Tokyo, 1969 (in Japanese). With permission.)

nm (= 14428 cm^{-1}) with width of ~10 cm^{-1} and decay time of 3.4 ms at room temperature are called R_1 and R_2 lines. They lie at the same wavelengths as lines observed in the absorption spectrum (zero-phonon lines). These lines correspond to the transition from $^2E(t_2^3) \rightarrow {}^4A_2(t_2^3)$ in Figure 11. The ^2E level splits into two levels due to a combination of the spin-orbit interaction and symmetry reduction in the crystal field from cubic to trigonal.[1] Two strong absorption bands at ~18000 cm^{-1} and ~25000 cm^{-1} correspond to the spin-allowed transitions from the ground level ($^4A_2(t_2^3)$) to the $^4T_2(t_2^2e)$ and $^4T_1(t_2^2e)$ levels, respectively. The spectral band shape differs, depending on the electric field direction of the incident light due to the axial symmetry in the crystal field (dichroism).

Many spin doublets originate from the t_2^2e configuration of Cr^{3+} in addition to the above two spin quartets.* Transitions from the ground level ($^4A_2$) to those spin doublets are spin-forbidden, the corresponding absorption bands being very weak to observe.** Strong spin-allowed absorption bands to those spin doublets, however, are observable from $^2E(t_2^3)$, when a number of Cr^{3+} ions are produced by an intense light excitation into this excited state (excited-state absorption).[16] For 11 multiplet levels, including those obtained through excited-state absorption studies, all the properties of the absorption bands—such as spectral position, absorption intensity, and dependence on the polarized light—have been found to agree very well with those predicted from crystal (ligand) field theory.[1,16]

As shown in Figure 17, with the increase in Cr^{3+} concentration, additional luminescence lines begin to appear at the longer wavelength side of the R lines, and grow up to be broad bands that become stronger than R lines; this is accompanied by the reduction

* In Figure 11, positions for these doublets are not shown clearly.
** In a strong crystal field, two-electron transitions such as $t_2^3 \rightarrow t_2e^2$ are forbidden.

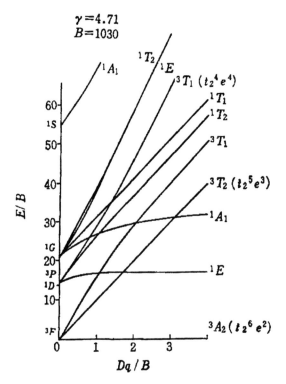

$\gamma = 4.71$
$B = 1030$

Figure 16 Energy level diagram for the d^8 configuration. (From Kamimura, H., Sugano, S., and Tanabe, Y., *Ligand Field Theory and its Applications*, Syokabo, Tokyo, 1969 (in Japanese). With permission.)

Table 6 Oscillator Strength and Luminescence Decay Time

		Laporte's rule allowed		Laporte's rule forbidden Electric dipole	
		Electric dipole	Magnetic dipole	V_{odd} allowed	Lattice vibration allowed
Spin-allowed	f	~1	~10^{-6}	~10^{-4}	~10^{-4}
	τ	~5 ns	~1 ms	~50 μs	~50 μs
Spin-forbidden	f	10^{-2}–10^{-3}	10^{-8}–10^{-9}	10^{-6}–10^{-7}	10^{-6}–10^{-7}
	τ	0.5–5 μs	10^2–10^3 ms	5–50 ms	5–50 ms

Note: 1. f values for the case of spin-allowed are estimated in Reference 1. f values for the case of spin-forbidden are assumed to be 10^{-2}–10^{-3} of those for spin-allowed.

2. Decay times are calculated from Eqs. 22 and 23, assuming $E_c/E_{eff} = (n^2 + 2)/3$ (Lorenz field), $n = 1.6$, and $\lambda_0 = 500$ nm.

in the luminescence decay time of R lines, in the case of Figure 17, from 3.5 ms to 0.8 ms at room temperature.[15] Additional lines are attributed to magnetically coupled Cr^{3+}-Cr^{3+} pairs and clusters. Luminescence lines are assigned to such pairs up to the fourth nearest neighbor; for example, the N_1 line is assigned to pairing to the third nearest neighbor, and N_2 to the fourth nearest.[17]

In compounds such as various gallium garnets in which Cr^{3+} ions are located in weak crystal fields, $^4T_2(^4F)$, instead of $^2E(^2G)$, is the emitting level.[18] As expected from Figure 11, the luminescence spectrum consists of a broad band in the near-infrared region, i.e., at a longer wavelength region than that in the ^2E case. The decay time is as short as ~0.1 ms because the transition is spin-allowed. These properties make them promising candidates

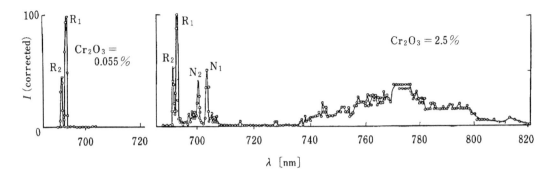

Figure 17 Luminescence spectra in rubies (at 77K). (Figure 1 in the source shows luminescence spectra and decay times for rubies containing 0.4, 0.86, 1.5, and 8% concentrations of Cr_2O_3, in addition to the above two examples.) (From Tolstoi, N.A., Liu, S., and Lapidus, M.E., *Opt. Spectrosc.*, 13, 133, 1962. With permission.)

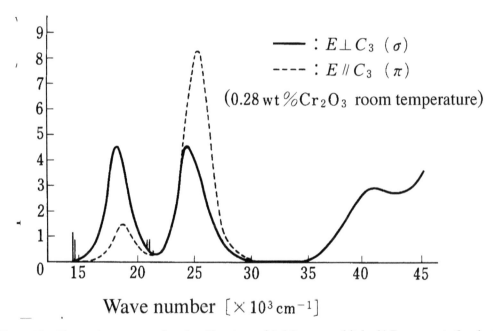

Figure 18 Absorption spectra of a ruby. (Courtesy of A. Misu, unpublished.) E represents the electric field direction of an incident light, and C_3 does a three-fold axis direction of the crystal. Spectrum at higher energies than 35000 cm^{-1} is for natural light. Absorption lines around 15000 and 20000 cm^{-1} are shown only in the case of the σ spectrum, qualitatively with respect to intensity and linewidth. (From Kamimura, H., Sugano, S., and Tanabe, Y., *Ligand Field Theory and its Applications*, Syokabo, Tokyo, 1969 (in Japanese). With permission.)

for tunable solid-state laser materials.[19,20] The change of the emitting state depending on the host materials is a good example of the importance of the crystal field in determining the optical properties of the transition-metal-doped compounds.

Table 7 shows the crystal field parameters obtained from absorption spectra and luminescence decay times for Cr^{3+} in several hosts. Most luminescence bands in $3d$ ions are caused by electric dipole transitions. In such materials as $MgAl_2O_4$ and MgO, in which a metal ion lies in the crystal field with the inversion symmetry, however, the R lines occur via a magnetic dipole process[21,22]; consequently, the decay times are long.

Table 7 Crystal Field Parameters for Cr^{3+}

Host	λ (nm)		Dq (cm^{-1})	B (cm^{-1})	C (cm^{-1})	τ (ms)		Ref.
α-Al_2O_3 (ruby)	694.3	692.9[14]	1630	640	3300	3	(R)	23
$Be_3Al_2Si_6O_{18}$ (Emerald)	682.1	679.2[26]	1630	780	2960			23
$MgAl_2O_4$	682.2	681.9	1825	700	3200	36.5	(N)	21
MgO		698[27]	1660	650	3200	12	(N)	22
$LiAl_5O_8$ [a]	715.8	701.6	1750	800	2900	3.7		24
$Y_3Al_5O_{12}$	688.7	687.7	1725	640	3200	1.5		28
$Gd_3Ga_5O_{12}$	745 (broad)[b]		1471	645		0.16		18
$Y_3Ga_5O_{12}$	730 (broad)[b]		1508	656		0.24		18
$Cr(H_2O)_6^{3+}$	Abs.		1720	765				3
Free ion	684.2	(^2G)[25]		918				3

Note: λ: peak wavelength of luminescence; τ: $1/e$ decay time; (R), room temperature; (N), 77K.
[a] Ordered type
[b] $^4T_2 \rightarrow {}^4A_2$ transition, otherwise $^2E \rightarrow {}^4A_2$ transition.

2.2.4 Mn^{4+} phosphors (3d^3)

Only $3.5MgO \cdot 0.5MgF_2 \cdot GeO_2$:$Mn^{4+}$ is now in practical use among the Mn^{4+} phosphors, though $6MgO \cdot As_2O_5$:Mn^{4+}, which has a performance almost equal to that of $3.5MgO \cdot 0.5MgF_2 \cdot GeO_2$:$Mn^{4+}$, was used previously,[29] and a number of titanate phosphors were developed between 1940 and 1950.[30]

Luminescence bands due to Mn^{4+} exist at 620 to 700 nm in most host materials. The spectrum has a structure consisting of several broad lines originating from transitions aided by lattice vibration. In Al_2O_3 and Mg_2TiO_4, it resembles the R lines of Cr^{3+}, and is assigned to the $^2E(t_2^3) \rightarrow {}^4A_2(t_2^3)$ transition.

Figure 19 shows the luminescence spectra for $3.5MgO \cdot 0.5MgF_2 \cdot GeO_2$:$Mn^{4+}$. It consists of more than six lines at room temperature; the intensity of the lines at the shorter wavelength side decreases at low temperatures. This behavior is explained by assuming that thermal equilibrium exists between two levels in the emitting state, and that there are more than two levels in the ground state.[31] As for the origin of the emitting and ground states, different assignments have been proposed.

Kemeny and Haake assigned the bands to the $^4T_2(t_2^2e) \rightarrow {}^4A_2(t_2^3)$ transition in Figure 11, assuming the Mn^{4+} site has octahedral coordination.[31] They propose that the $^4T_2$ level splits into two levels due to the low symmetry field, and that more than two vibronic levels accompany the ground state. Butler insisted that a $(MnO_4)^{4-}$ complex replaced $(GeO_4)^{4-}$, which is tetrahedrally coordinated.[32] In this case, the appropriate energy diagram is Figure 15 instead of Figure 11, and the luminescence originates from the $^2E(e^3) \rightarrow {}^4T_1(e^2t_2)$ transition.* The ^2E and $^4T_1$ levels split into two and three due to the low symmetry field, respectively. These proposals, however, could not account for such facts as the luminescence has a decay time of the order of milliseconds; in addition, no visible luminescence has been observed due to Mn^{4+} in solid-state materials in which the metal ions are tetrahedrally coordinated.

Ibuki's group assigned the lines to transitions from two excited levels of $^2E(t_2^3)$ and $^2T_1(t_2^3)$ to the ground state $^4A_2(t_2^3)$ in Figure 11, assuming Mn^{4+} has an octahedral coordination.[33] The main peak structure in the range 640 to 680 nm at room temperature originates from the lattice vibration associated with the $^2E \rightarrow {}^4A_2$ zero-phonon transition at 640 nm.

Blasse explained the spectral characteristics by assuming only one electronic transition of $^2E \rightarrow {}^4A_2$ in octahedrally coordinated Mn^{4+}.[34] Both the ground and excited states are

* See 2.2.1.3. The transition corresponds to $^2E(t_2^6e) \rightarrow {}^4T_1(t_2^5e^2)$ in Figure 15.

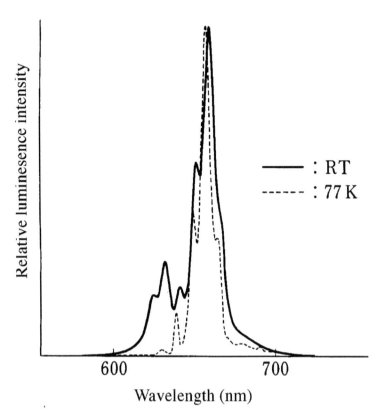

Figure 19 Luminescence spectra of $3.5MgO \cdot 0.5MgF_2 \cdot GeO_2:Mn^{4+}$. (Observed by the author.)

coupled with special vibration modes. The shorter wavelength peaks, which disappear at low temperatures, are ascribed to transitions from an excited vibronic level (anti-Stokes vibronic transitions).

Strong absorption bands due to Mn^{4+} exist, corresponding to the spin-allowed transitions of $^4A_2(t_2^3) \rightarrow {}^4T_1, {}^4T_2(t_2^2e)$ in the visible to near-UV region, and the body color of the phosphor is usually yellow. Table 8 shows the crystal field parameters and luminescence decay times for Mn^{4+} in several hosts. The larger valency leads to Dq/B values as large as 3, compared with those for Cr^{3+} (~2.5), and this, in turn, to the absorption bands at shorter wavelengths as expected from Figure 11. The charge-transfer band, on the other hand, lies at longer wavelength (~285 nm in Al_2O_3), resulting from the larger valency of Mn^{4+}.[10,35] (See 2.2.2.2.)

2.2.5 Mn^{2+} phosphors ($3d^5$)

2.2.5.1 Crystal field
Luminescence due to Mn^{2+} is known to occur in more than 500 inorganic compounds.[40] Of these, several are being used widely for fluorescent lamps and CRTs. The luminescence spectrum consists of a structureless band with a halfwidth of 1000 to 2500 cm^{-1} at peak wavelengths of 490 to 750 nm. Figure 20 shows the luminescence and excitation spectra due to Mn^{2+} in $La_2O_3 \cdot 11Al_2O_3$ as an example.[41] The energy level diagram for Mn^{2+} in both octahedral and tetrahedral coordinations is represented by Figure 13. In phosphors, Mn^{2+} ions are located in the weak crystal field of $Dq/B \simeq 1$, and the luminescence corresponds to the $^4T_1(^4G) \rightarrow {}^6A_1(^6S)$ transition.

When a metal ion occupies a certain position in a crystal, the crystal field strength that affects the ion increases as the space containing the ion becomes smaller, as expected

Table 8 Crystal Field Parameters for Mn^{4+}

Host	λ (nm)		Dq (cm^{-1})	B (cm^{-1})	C (cm^{-1})	τ (ms)		Ref.
α-Al$_2$O$_3$	676.3	672.6[39]	2170	700	2800	0.8	(N)	35
Mg$_2$TiO$_4$	655.6	653.2 vib	2096	848	3300	0.5[38]	(R)	36
LiAl$_5$O$_8$ [a]	716	702 vib	2014	725	2900	0.2	(N)	37
3.5MgO·0.5 MgF$_2$·GeO$_2$		623–664 str.	2375	709	3080	3.3[31]	(R)	33
Mg$_6$As$_2$O$_{11}$		620–665 str.	(2375)	(709)	(3080)	2.8[27]	(R)	33
Free ion	576.4 (^2G)[25]			1065	4919			3

Note: λ: peak wavelength of luminescence; τ: 1/e decay time; (R): room temperature; (N): 77K; vib: vibration structure, str: structured band.

[a] Ordered type.

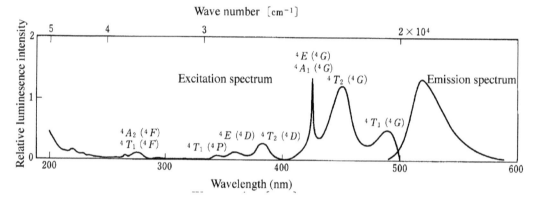

Figure 20 Luminescence and excitation spectra of La$_2$O$_3$·11Al$_2$O$_3$:Mn^{2+}. (From Tamatani, M., *Jpn. J. Appl. Phys.*, 13, 950, 1974. With permission.)

from Eq. 9. For increases in the field, the transition energy between the $^4T_1$ and $^6A_1$ levels is predicted to decrease (shift to longer wavelengths). (See Figure 13.) In fact, the peak wavelength of the Mn^{2+} luminescence band is known to vary linearly to longer wavelength (547 to 602 nm) with a decrease in Mn-F distance (2.26 to 1.99Å) in a group of fluorides already studied, including 10 perovskite lattices of the type A^{I}B^{II}F$_3$, ZnF$_2$, and MgF$_2$.[42] A similar relationship also holds for each group of oxo-acid salt phosphors having an analogous crystal structure; the wavelength is longer when Mn^{2+} replaces a smaller cation in each group, as seen in Table 9. On the other hand, a larger anion complex makes the cation space shrink, leading to longer-wavelength luminescence. For Ca$_{10}$(PO$_4$)$_6$F$_2$:Mn^{2+}, the crystal field at a Mn^{2+} ion produced by ions in eight unit cells around it was calculated theoretically. The result was consistent with the observed luminescence peak shift (100 cm^{-1}) to longer wavelength due to a lattice constant decrease (0.14%) when one Ca in each Ca$_{10}$(PO$_4$)$_6$F$_2$ is replaced by Cd.[43]

In spite of the fact that the ionic radius for Zn^{2+} (0.72 Å) is smaller than that for Ca^{2+} (0.99 Å), the luminescence wavelength in Zn$_2$SiO$_4$:Mn^{2+} is shorter than in CaSiO$_3$:Mn^{2+}. This is attributed to a smaller coordination number (4) in the former as compared with that (6) in the latter. (See Eq. 13.) In materials containing a spinel structure, Mn^{2+} can occupy either octahedral or tetrahedral sites. From the fact that the luminescence occurs in the shorter-wavelength (green) region, the tetrahedral site is expected to be occupied preferentially by Mn^{2+}. This is confirmed by ESR[44] and ion-exchange[45] studies for β-aluminas and supported by thermodynamic data.[45]

Table 9 Mn^{2+} Sites and Luminescence Properties

Host	Crystal symmetry	Site	Coordination number	Inversion symmetry	λ (nm)	τ (ms)
CaF_2	O_h	Ca	8	g	495	83[46]
ZnF_2	D_{4h}	Zn	6	g	587	100
$KMgF_3$	(O_h)	Mg	6	g	602[42]	104[62]
$ZnGa_2O_4$	O_h	(A site)	(4)	u	506	4
$ZnAl_2O_4$	O_h	(A site)	(4)	u	513	5
Zn_2SiO_4	C_{3i}	2Zn	4	u	525	12
Zn_2GeO_4	C_{3i}	2Zn	4	u	537	10
$Ca_5(PO_4)_3F$	C_{6h}	2Ca	6[63]	u	570[a]	14[66]
$Sr_5(PO_4)_3F$	C_{6h}	2Sr	6	u	558	
monocl-$CaSiO_3$	C_2	3Ca[64]	6	u	550 620	30
monocl-$MgSiO_3$	C_{2h}	2Mg[65]	6	u	660 740	
CaS	O_h	Ca	6	g	588	2.2–4.8[67]
hex–ZnS	T_d	Zn	4	u	591	0.25

Note: 1. 2Ca in the site column means existence of two different Ca sites. (A site) means larger probability for existence in A sites than for octahedral B sites.

2. Except for those referred, crystal symmetries follow those in Reference 61, and luminescence wavelengths and decay times in Reference 51.

3. In the inversion symmetry column, g and u correspond to existence and nonexistence of a center of symmetry, respectively.

[a] A value obtained in an Sb-Mn co-doped sample.

In CaF_2:Mn^{2+}, though Mn^{2+} occupies a cubic site with high coordination number, Dq is not so large because the anion valency of F^- is smaller than that of O^{2-}. In addition, B is large because of the smaller nephelauxetic effect.[46] Consequently, this compound yields the shortest luminescence wavelength (~495 nm) observed among Mn^{2+}-doped phosphors.[*]

Since every excited level of d^5 is either a spin quartet or a doublet, all transitions from the ground sextet to them are spin-forbidden. Optical absorption intensity is weak, and the phosphors are not colored (i.e., the powder body color is white). The $^4A_1$ and $^4E(^4G)$ levels have the same energy and are parallel to the ground level $^6A_1$ in Figure 13. The absorption band corresponding to $^6A_1 \rightarrow {}^4A_1,{}^4E(^4G)$ therefore has a narrow bandwidth, lying at ~425 nm, irrespective of the kind of host material.[48,49] One notices that this band splits into more than one line when carefully investigated. The splitting is considered to reflect the reduction of the crystal field symmetry.[48,49]

Table 10 shows the crystal field parameters for Mn^{2+} in representative phosphors. Note that Dq/B for the tetrahedral coordination is smaller (<1) than that (>1) for the octahedral one.

2.2.5.2   Different Mn^{2+} sites in crystals

Since the luminescence wavelength due to Mn^{2+} is sensitive to the magnitude of the crystal field, several emission bands are observed when different types of Mn^{2+} sites exist in a host crystal. In $SrAl_{12}O_{19}$, the bands at 515, 560, and 590 nm are considered to originate from Mn^{2+} ions replacing tetrahedrally coordinated Al^{3+}, fivefold coordinated Al^{3+}, and 12-fold coordinated Sr^{2+}, respectively.[45] In lanthanum aluminate, which has a layer structure of spinel blocks, a 680-nm band is observed due to Mn^{2+} in octahedral coordination, in addition to a green-emitting band due to tetrahedral coordination.[50] Two emission

---

[*] The other shortest peak wavelength is at 460 to 470 nm, observed in $SrSb_2O_6$,[47] in which Mn^{2+} is considered to be located in an extraordinary weak crystal field (Sr–O distance is as large as 2.5 Å).

Table 10 Crystal Field Parameters for Mn^{2+}

Host	λ (nm)	Dq (cm^{-1})	B (cm^{-1})	C (cm^{-1})	Coordination	Ref.
$MgGa_2O_4$	504	520	624	3468	(4)	48
$LaAl_{11}O_{18}$	517	543	572	3455	4	41
Zn_2SiO_4	525	540	(624)	(3468)	4	48
$Ca_5(PO_4)_3F$	572	760	691	3841	6	68
$Mg_4Ta_2O_9$	659	425	(698)	(3678)	6	55
CaF_2	495	(2375)	770	3449	8	46
hex·ZnS	591[51]	520	630	3040	4	69
$Mn(H_2O)_6^{2+}$	Abs.	1230	860	3850	6	3
Free ion	372.5 (^4G)[25]		860	3850		3

Note: *B* and *C* values in parenthesis, which were obtained from other phosphors, are used for calculating *Dq* values.

bands separated by about 50 nm were recognized long ago in Mn^{2+}-doped alkaline earth silicates.[51]

Even in the case of the same coordination number, different luminescence bands may come from Mn^{2+} ions occupying crystallographically different sites. In $Ca_5(PO_4)_3F$, there are principally Ca(I) and Ca(II) sites having different crystallographic symmetries; several additional sites accompany these two main calcium sites. The correspondence between the luminescence bands and the various sites has been investigated by means of polarized light,[52] ESR,[53] and excitation[52] spectral studies. In the case of the commercially available phosphor $Ca_5(PO_4)_3(F,Cl):Sb^{3+},Mn^{2+}$ (for Cool White fluorescent lamps), the Mn^{2+} band consists of three bands at 585, 584, and 596 nm, originating from Mn^{2+} ions replacing Ca(I), Ca(II), and Cl, respectively.[54]

Figure 21 shows the spectra in $Zn_2SiO_4:Mn^{2+}$, where two zero-phonon lines are observed at very low temperatures (504.6 and 515.3 nm at 4.2K).[55] These lines are assigned to two types of Mn^{2+} differing in their distance to the nearest oxygen; one is 1.90 Å and the other is 1.93 Å. Since the Dq value depends on the fifth power of the distance (Eqs. 9 and 13), a 7% difference in the Dq value is expected between the two types of Mn^{2+} sites; this is consistent with the difference estimated by crystal field theory from the observed line positions (2% difference).[55] The polarization of the luminescence light observed in a single crystal is also related to the site symmetry of Mn^{2+}.[56] The zero-phonon lines are accompanied by broad bands in the longer wavelength side; these originate from lattice-electron interactions and are known as vibronic sidebands (See Section 1.3.) Multi zero-phonon lines resulting from different Mn^{2+} sites are also observed in $Mg_4Ta_2O_9$[55] and $LiAl_5O_8$.[37]

In ZnS doped with high concentrations of Mn^{2+}, although there is only one cation site crystallographically, two zero-phonon lines appear at 558.9 and 562.8 nm at low temperatures. These are ascribed to a single Mn^{2+} ion (τ = 1.65 ms) and a Mn^{2+}-Mn^{2+} pair (τ = 0.33 ms).[57] In this material, the luminescence band shifts to longer wavelength and is accompanied by a decrease in decay time with increasing Mn^{2+} concentration; this is also observed in such hosts as Zn_2SiO_4,[51] $MgGa_2O_4$,[58] $ZnAl_2O_4$,[51] $CdSiO_3$,[51] and ZnF_2.[51] Most of these effects are attributed to Mn^{2+}-Mn^{2+} interactions.

2.2.5.3   UV absorption

Lamp phosphors must absorb the mercury ultraviolet (UV) line at 254 nm. In most cases, Mn^{2+} does not have strong absorption bands in this region. To counter the problem, energy-

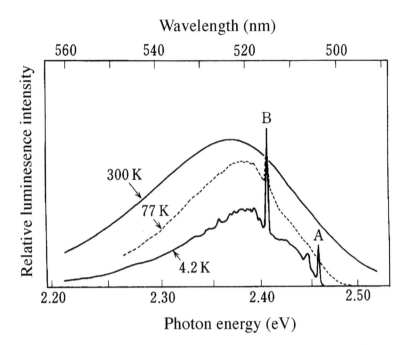

Figure 21 Luminescence spectra of $Zn_2SiO_4:Mn^{2+}$. (From Stevels, A.L.N. and Vink, A.T., *J. Luminesc.*, 8, 443, 1974. With permission.)

transfer mechanisms are utilized to sensitize Mn^{2+}; transfers are effected through the host* or via such ions as Sb^{3+}, Pb^{2+}, Sn^{2+}, Ce^{3+}, and Eu^{2+}, which absorb the UV efficiently through allowed transitions. These ions are called sensitizers for the Mn^{2+} luminescence. In Zn_2SiO_4, a strong absorption band appears at wavelengths shorter than 280 nm when doped with Mn^{2+}.[30] This band is ascribed to $Mn^{2+} \rightarrow Mn^{3+}$ ionization[59] or to a $d^5 \rightarrow d^4s$ transition.[32,60]

2.2.5.4 Luminescence decay time

The decay time of Mn^{2+} luminescence is usually in the millisecond range (Table 9). A shorter decay time is expected in the tetrahedral coordination because it has no center of inversion symmetry. In most practical phosphors, the Mn^{2+} sites are surrounded by six oxygen ions, but the symmetry is lower than octahedral and the sites do not have a center of inversion. (See Table 9.) It follows that the difference in the decay time between the phosphors having Mn^{2+} with four and six coordination numbers is not actually so large. Decay times in the fluoride phosphors are one order of magnitude longer than those in the oxo-acid salt phosphors. This is thought to be due to the fact that Laporte's rule holds more strictly in the fluorides, since odd states do not mix into the d states easily because: (1) Mn^{2+} ions in the fluorides are located at a center of inversion symmetry, and (2) the smaller nephelauxetic effect makes the odd states lie at higher energies than those in oxo-acid salts.

2.2.6 Fe^{3+} Phosphors ($3d^5$)

Luminescence due to Fe^{3+} lies in the wavelength region longer than 680 nm, and only $LiAlO_2:Fe^{3+}$ and $LiGaO_2:Fe^{3+}$ are used for special fluorescent lamp applications.[70] It is easily understood from Figure 13 why the luminescence wavelengths due to Fe^{3+} are so much

---

* The host-absorption wavelength does not always correspond to the bandgap energy of the host material.

Table 11 Crystal Field Parameters for Fe^{3+}

Host	λ (nm)	Dq (cm^{-1})	B (cm^{-1})	C (cm^{-1})	Coordination	Ref.
$LiAl_5O_8$ [a]	680	800	644	2960	4	72
β-$LiAlO_2$	735	883	630	3000	4	73
$Ca(PO_3)_2$	830	1250			6	75
γ-AlF_3	735	1220	895	3000	6	71
$Fe(H_2O)_6^{2+}$	Abs.	1350	820	3878	6	3
Free ion	312 (^4G)^{25}		1015	4800		3

[a] Ordered structure, τ = 7.1 ms.

longer than those due to Mn^{2+}, which has the same electronic configuration of $3d^5$. That is, the larger valency of Fe^{3+} brings about the stronger crystal field, reducing the transition energy of $^4T_1(^4G) \rightarrow {}^6A_1(^6S)$. In fact, as shown in Table 11, Dq/B for Fe^{3+} is ~1.2, even at tetrahedral sites, and larger than that (<1) for Mn^{2+}. The reason for the emission wavelength being as short as 735 nm despite the octahedral coordination in AlF_3 is attributable to the large B value resulting from the small nephelauxetic effect.[71]

The absorption (or excitation) spectrum in the visible region due to Fe^{3+} resembles the shape of that of Mn^{2+}. Zero-phonon lines are also observed in $LiAl_5O_8$ [72] and $LiAlO_2$.[73] In the UV region, contrary to the Mn^{2+} case, however, a strong absorption band supposedly caused by charge transfer appears,[10,72,74] and the Fe^{3+} phosphors can be excited directly by 254-nm light irradiation without the need of sensitization through other ions.

References

1. Kamimura, H., Sugano, S., and Tanabe, Y., *Ligand Field Theory and Its Applications*, Syokabo, Tokyo, 1969 (in Japanese); Sugano, S., Tanabe, Y., and Kamimura, H., *Multiplets of Transition-Metal Ions in Crystals*, Academic Press, 1970.
2. a) Tanabe, Y. and Sugano, S., *J. Phys. Soc. Jpn.*, 9, 753, 1954; b). *ibid.*, 766.
3. Orgel, L.E., *J. Chem. Phys.*, 23, 1004, 1955.
4. McClure, D.S., *Electronic spectra of molecules and ions in crystals. Part II. Spectra of ions in crystals*, in *Solid State Physics*, Seitz, F. and Turnbull, D., Eds., 9, 399, Academic Press, 1959.
5. Griffith, J.S., *The Theory of Transition Metal Ions*, Cambridge Univ. Press, 1964.
6. Ballhausen, C.J., *Introduction to Ligand Field Theory*, McGraw-Hill, 1962.
7. Jørgensen, C.K., *Absorption Specta and Chemical Bonding in Complexes*, Pergamon Press, Elmsford, NY, 1962.
8. Prather, J.L., *National Bureau of Standards Monogr.*, 19, 1, 1961.
9. Di Bartolo, B., *Optical Interactions in Solids*, John Wiley & Sons, 1968.
10. Tippins, H.H., *Phys. Rev.*, B1, 126, 1970.
11. Hoshina, T., Imanaga, S., and Yokono, S., *J. Luminesc.*, 15, 455, 1977.
12. Sluzky, E., Lemoine, M., and Hesse, K., *J. Electrochem. Soc.*, 141, 3172, 1994.
13. Sugano, S. and Tsujikawa, I., *J. Phys. Soc. Jpn.*, 13, 899, 1958; Sugano, S. and Tanabe, Y., *ibid.*, 880.
14. Maiman, T.H., *Nature*, 187, 493, 1960.
15. Tolstoi, N.A., Liu, S., and Lapidus, M.E., *Opt. Spectrosc.*, 13, 133, 1962.
16. Kushida, T., *J. Phys. Soc. Jpn.*, 21, 1331, 1966; Shinada, M., Sugano, S., and Kushida, T., *ibid.*, 1342.
17. Powell, R.C. and Di Bartolo, B., *Phys. Status Solidi (a)*, 10, 315, 1972.
18. Petermann, K. and Huber, G., *J. Luminesc.*, 31&32, 71, 1984.
19. Walling, J.C., Tunable paramagnetic-ion solid-state lasers, in *Tunable Lasers*, Mollenauer, L.F. and White, J.C., Eds., Springer-Verlag, 1987.
20. Moulton, P.F., Tunable paramagnetic-ion lasers, in *Laser Handbook*, Vol. 5, Bass, M. and Stitch, M.L., Eds., Elseviers Science, B. V., 1985.

21. Wood, D.L., Imbusch, G.F., Macfarlane, R.M., Kisliuk, P., and Larkin, D.M., *J. Chem. Phys.*, 48, 5255, 1968.
22. Macfarlane, R.M., *Phys. Rev.*, B1, 989, 1970.
23. Wood, D.L., Ferguson, J., Knox, K., and Dillon, Jr., J.F., *J. Chem. Phys.*, 39, 890, 1963.
24. Pott, G.T. and McNicol, B.D., *J. Solid State Chem.*, 7, 132, 1973.
25. Moore, C.E., *Atomic Energy Levels*, Vol. II, NBS Circular, 1952, 467.
26. Wood, D.L., *J. Chem. Phys.*, 42, 3404, 1965.
27. Imbusch, G.F., Experimental spectroscopic techniques for transition metal ions in solids, in *Luminescence of Inorganic Solids*, Di Bartolo, B., Ed., Plenum Press, 1978, 135.
28. Sevast'yanov, V.P., Sviridov, D.T., Orekhova, V.P., Pasternak, L.B., Sviridova, R.K., and Veremeichik, T.F., *Sov. J. Quant. Electron.*, 2, 339, 1973.
29. Ouweltjes, J.L., Elenbaas, W., and Labberte, K.R., *Philips Tech. Rev.*, 13, 109, 1951.
30. Kröger, F.A., *Some Aspects of Luminescence of Solids*, Elsevier, 1948.
31. Kemeny, G. and Haake, C.H., *J. Chem. Phys.*, 33, 783, 1960.
32. Butler, K.H., *Proc. Int. Conf. Luminesc.*, Budapest, 1966, 1313.
33. Ibuki, S., Awazu, K., and Hata, T., *Proc. Int. Conf. Luminesc.*, Budapest, 1966, 1465.
34. Blasse, G. and Grabmaier, B.C., *Luminescent Materials*, Springer-Verlag, 1994, 128.
35. Geschwind, S., Kisliuk, P., Klein, M.P., Remeika, J.P., and Wood, D.L., *Phys. Rev.*, 126, 1684, 1962.
36. Stade, J., Hahn, D., and Dittmann, R., *J. Luminesc.*, 8, 318, 1974.
37. McNicol, B.D. and Pott, G.T., *J. Luminesc.*, 6, 320, 1973.
38. Dittmann, R. and Hahn, D., *Z. Phys.*, 207, 484, 1967.
39. Travniçek, M. Kröger, F.A., Botden, Th.P.J., and Zahm, P., *Physica*, 18, 33, 1952.
40. Data obtained from Chemical Abstracts in 1948 to 1971, and References 30 and 51.
41. Tamatani, M., *Jpn. J. Appl. Phys.*, 13, 950, 1974.
42. Klasens, H.A., Zahm, P., and Huysman, F.O., *Philips Res. Repts.*, 8, 441, 1953.
43. Narita, K., *J. Phys. Soc. Jpn.*, 16, 99, 1961; *ibid.*, 18, 79, 1963.
44. Antoine, J., Vivien, D., Livage, J., Thery, J., and Collongues, R., *Mat. Res. Bull.*, 10, 865, 1975.
45. Bergstein, A. and White, W.B., *J. Electrochem. Soc.*, 118, 1166, 1971.
46. Alonso, P.J. and Alcalá, *J. Luminesc.*, 22, 321, 1981.
47. Yamada, H., Matsukiyo, H., Suzuki, T., Yamamoto, H., Okamura, T., Imai, T., and Morita, M., *Electrochem. Soc. Fall Meeting*, Abstr. No. 564, 1988.
48. Palumbo, D.T. and Brown, Jr., J.J., *J. Electrochem. Soc.*, 117, 1184, 1970.
49. Palumbo, D.T. and Brown, Jr., J.J., *J. Electrochem. Soc.*, 118, 1159, 1971.
50. Stevels, A.L.N., *J. Luminesc.*, 20, 99, 1979.
51. Leverenz, H.W., *An Introduction to Luminescence of Solids*, John Wiley & Sons, New York, 1950; Recent publications for Zn_2SiO_4:Mn are Barthou, C., Benoit, J., Benalloul, P., and Morell, A., *J. Electrochem. Soc.*, 141, 524, 1994; Ronda, C.R., *Proc. 2nd Int. Display Workshops*, Vol. 1, 1995, 69.
52. Ryan, F.M., Ohlman, R.C., and Murphy, J., *Phys. Rev.*, B2, 2341, 1970.
53. Kasai, P.H., *J. Phys. Chem.*, 66, 674, 1962.
54. Ryan, F.M. and Vodoklys, F.M., *J. Electrochem. Soc.*, 118, 1814, 1971.
55. Stevels, A.L.N. and Vink, A.T., *J. Luminesc.*, 8, 443, 1974.
56. Bhalla, R.J.R.S. and White, E.W., *J. Electrochem. Soc.*, 119, 740, 1972.
57. Busse, W., Gumlich, H.E., Meissmer, B., and Theis, D., *J. Luminesc.*, 12/13, 693, 1976.
58. Brown, Jr., J.J., *J. Electrochem. Soc.*, 114, 245, 1967.
59. Robbins, D.J., Avouris, P., Chang, I.F., Dove, D.B., Giess, E.A., and Mendez, E.E., *Electrochem. Soc. Spring Meeting*, Abstr. No. 513, 1982.
60. Butler, K.H., *Fluorescent Lamp Phosphors*, Pennsylvania State University Press, 1980.
61. Wyckoff, R.W.G., *Crystal Structures*, Interscience Publishers, 1965.
62. Van Noy, B.W. and Mikus, F.F., *Proc. Int. Conf. Luminesc.*, Budapest, 1966, 794.
63. Náray-Szabó, S., *Z. Krist.*, 75, 387, 1930.
64. Tolliday, J., *Nature*, 182, 1012, 1958.
65. Morimoto, N., Appleman, D.E., and Evans, Jr., E.T., *Z. Krist.*, 114, 120, 1960.
66. Soules, T.F., Bateman, R.L., Hewes, R.A., and Kreidler, E.R., *Phys. Rev.*, B7, 1657, 1973.
67. Yamamoto, H., Megumi, K., Kasano, H., Suzuki, T., Ueno, Y., Morita, Y., and Ishigaki, T., *Tech. Digest, Phosphor Res. Soc. 198th Meeting*, 1983 (in Japanese).

68. Uehara, Y., *Toshiba Rev.*, 24, 1090, 1969 (in Japanese).
69. Kushida, T., Tanaka, Y., and Oka, Y., *Solid State Commun.*, 14, 617, 1974.
70. Van Broekhoven, J., *J. Illum. Eng. Soc.*, 3, 234, 1974.
71. Telfer, D.J. and Walker, G., *J. Luminesc.*, 11, 315, 1976.
72. Pott, G.T. and McNicol, B.D., *J. Chem. Phys.*, 56, 5246, 1972.
73. Stork, W.H.J. and Pott, G.T., *J. Phys. Chem.*, 78, 2496, 1974.
74. Tamatani, M. and Tsuda, N., *Tech. Digest, Phosphor Res. Soc. 157th Meeting*, 1970 (in Japanese).
75. Fox, K.E., Furukawa, T., and White, W.B., *J. Am. Cer. Soc.*, 64, C-42, 1981.

chapter two — section three

Principal phosphor materials and their optical properties

Tsuyoshi Kano

Contents

2.3   Luminescence of rare earth ions[1-3]

2.3.1   Electronic configuration

The rare-earth elements usually comprise 17 elements consisting of the 15 lanthanides from La (atomic number 57) to Lu (atomic number 71), of Sc (atomic number 21), and of Y (atomic number 39). The electronic configurations of trivalent rare-earth ions in the ground states are shown in Table 12. As shown in the table, Sc^{3+} is equivalent to Ar, Y^{3+} to Kr, and La^{3+} to Xe in electronic configuration. The lanthanides from Ce^{3+} to Lu^{3+} have one to fourteen 4*f* electrons added to their inner shell configuration, which is equivalent to Xe. Ions with no 4*f* electrons, i.e., Sc^{3+}, Y^{3+}, La^{3+}, and Lu^{3+}, have no electronic energy levels that can induce excitation and luminescence processes in or near the visible region. In contrast, the ions from Ce^{3+} to Yb^{3+}, which have partially filled 4f orbitals, have energy levels characteristic of each ion and show a variety of luminescence properties around the visible region.[1-3] Many of these ions can be used as luminescent ions in phosphors, mostly by replacing Y^{3+}, Gd^{3+}, La^{3+}, and Lu^{3+} in various compound crystals.

The azimutal quantum number (*l*) of 4*f* orbitals is 3, giving rise to 7 (= 2*l* + 1) orbitals, each of which can accommodate two electrons. In the ground state, electrons are distributed so as to provide the maximum combined spin angular momentum (*S*). The spin angular momentum *S* is further combined with the orbital angular momentum (*L*) to give the total angular momentum (*J*) as follows;

$$J = L - S, \text{ when the number of 4}f \text{ electrons is smaller than 7}$$

$$J = L + S, \text{ when the number of 4}f \text{ electrons is lager than 7}$$

An electronic state is indicated by notation $^{2S+1}L_J$, where *L* represents *S*, *P*, *D*, *F*, *G*, *H*, *I*, *K*, *L*, *M*, ..., corresponding to *L* = 0, 1, 2, 3, 4, 5, 6, 7, 8, 9, ..., respectively. More accurately, an actual electronic state is expressed as an intermediate coupling state, which can be described as a mixed state of several $^{2S+1}L_J$ states[2,4] combined by spin-orbit interaction. For qualitative discussions, however, the principal *L* state can be taken to represent the actual

Table 12   Electronic Configurations of Trivalent Rare-Earth Ions in the Ground State

Atomic number	Ions	Corresponding element	4*f* electrons							S Σs	L Σl	J $\Sigma(L+S)$
21	Sc^{3+}	Ar								0	0	0
39	Y^{3+}	Kr								0	0	0
57	La^{3+}									0	0	0
58	Ce^{3+}	Xe	↑							1/2	3	5/2
59	Pr^{3+}	Xe	↑	↑						1	5	4
60	Nd^{3+}	Xe	↑	↑	↑					3/2	6	9/2
61	Pm^{3+}	Xe	↑	↑	↑	↑				2	6	4
62	Sm^{3+}	Xe	↑	↑	↑	↑	↑			5/2	5	5/2
63	Eu^{3+}	Xe	↑	↑	↑	↑	↑	↑		3	3	0
64	Gd^{3+}	Xe	↑	↑	↑	↑	↑	↑	↑	7/2	0	7/2
65	Tb^{3+}	Xe	↑↓	↑	↑	↑	↑	↑	↑	3	3	6
66	Dy^{3+}	Xe	↑↓	↑↓	↑	↑	↑	↑	↑	5/2	5	15/2
67	Ho^{3+}	Xe	↑↓	↑↓	↑↓	↑	↑	↑	↑	2	6	8
68	Er^{3+}	Xe	↑↓	↑↓	↑↓	↑↓	↑	↑	↑	3/2	6	15/2
69	Tm^{3+}	Xe	↑↓	↑↓	↑↓	↑↓	↑↓	↑	↑	1	5	6
70	Yb^{3+}	Xe	↑↓	↑↓	↑↓	↑↓	↑↓	↑↓	↑	1/2	3	7/2
71	Lu^{3+}	Xe	↑↓	↑↓	↑↓	↑↓	↑↓	↑↓	↑↓	0	0	0

state. The mixing due to spin-orbit interaction is small for the levels near ground states, while it is considerable for excited states that have neighboring states with similar J numbers. The effect of mixing is relatively small on the energy of levels, but can be large on their optical transition probabilities.

2.3.2   Electronic processes leading to luminescence

2.3.2.1   4f energy levels and relaxation

The $4f$ electronic energy levels of lanthanide ions are characteristic of each ion. The levels are not affected much by the environment because $4f$ electrons are shielded from external electric fields by the outer $5s^2$ and $5p^6$ electrons. This feature is in strong contrast with transition metal ions, whose $3d$ electrons, located in an outer orbit, are heavily affected by the environmental or crystal electric field. The characteristic energy levels of $4f$ electrons of trivalent lanthanide ions have been precisely investigated by Dieke and co-workers.[5] The results are shown in Figure 22, which is known as a Dieke diagram.[5] The levels were determined experimentally by considering the optical spectra of individual ions incorporated in $LaCl_3$ crystals; this diagram is applicable to ions in almost any environment because the maximum variation of the energy levels is, at most, of the order of several hundred cm^{-1}.

Each level designated by the number J in Figure 22 is split into a number of sublevels by the Stark effect due to the crystal field. The number of split sublevels is, at most, $(2J + 1)$ or $(J + 1/2)$ for J of integer or J of half-integer, respectively. The number of levels is determined by the symmetry of the crystal field surrounding the rare-earth ion. The width of each level shown in Figure 22 indicates the range of splittings within each component.

Light-emitting levels are indicated by semicircles below the energy levels. Most of the emitting levels are separated from the next lower level by at least 2×10^3 cm^{-1} or more. This is because the excited states relax via two competitive paths: one is by light emission and the other by phonon emission. The rate of phonon emission, w, depends on the number of phonons emitted simultaneously to bridge the energy gap and is expressed as:

$$w \propto \exp(-k\Delta E/h\nu_{max}), \tag{24}$$

where ΔE is the energy gap to the nearest lower level and $h\nu_{max}$ is the maximum energy of phonons coupled to the emitting states. The phonon emission rate, w, decreases rapidly with an increase in ΔE, so that the competitive light emission or radiative process becomes dominant.[6] Thus, the well-known high luminescence efficiencies for $^5D_0$ of Eu^{3+} and $^5D_4$ of Tb^{3+} are based on the large energy gap of more than 10^4 cm^{-1} that needs to be bridged to the next lower level of these ions. The above formula implies that large values of $h\nu_{max}$ also quench light emission. This is demonstrated by the fact that luminescence of Eu^{3+} in aqueous solution is almost quenched, but begins to appear if H_2O is replaced by D_2O.[7]

Luminescence originating from electronic transitions between $4f$ levels is predominantly due to electric dipole or magnetic dipole interactions. Electric dipole f-f transitions in free $4f$ ions are parity-forbidden, but become partially allowed by mixing with orbitals having different parity because of an odd crystal field component. The selection rule in this case is $|\Delta J| \leq 6$, (except for 0→0, 0→1, 0→3, 0→5). Typical examples of this mechanism are demonstrated by the luminescence from the ^5D_J states of Eu^{3+}; the intensity of these transitions depends strongly on the site symmetry in a host crystal (See Section 2.3.3.7 Eu^{3+}). Magnetic dipole f-f transitions are not affected much by the site symmetry because they are parity-allowed. The J selection rule in this case is $\Delta J = 0, \pm 1$ (except for 0→0).

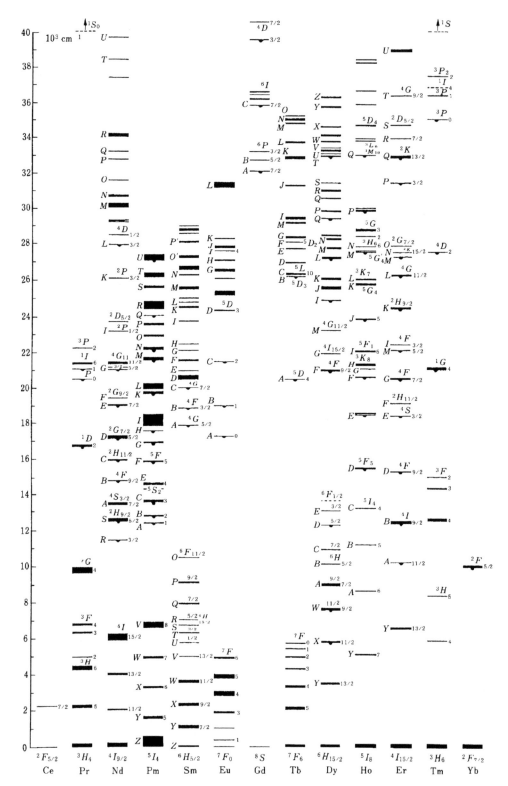

Figure 22 Energy levels of trivalent lanthanide ions. (From Dieke, G.H., *Spectra and Energy Levels of Rare Earth Ions in Crystals*, Interscience, 1968; *American Institute of Physics Handbook*, 3rd edition, McGraw-Hill, 1972, 7-25. With permission.)

Oscillator strengths are of the order of 10^{-5} to 10^{-8} for partially allowed electric dipole transitions, and 10^{-8} for magnetic dipole transitions.

The electric dipole transition probability between $4f$ levels can be calculated using the Judd-Ofelt theory.[8,9] This theory assumes closure of the wavefunctions mixed to a $4f$ orbital and takes an average for the energy separation between the allowed states and the $4f$ levels. In spite of such approximations, the theory gives satisfactory agreement with observed values in many cases. The points of the calculation are sketched out as follows.

The absorption coefficient, $k(\lambda)$, is experimentally determined; here, $k(\lambda) = \ln(I/I_0)/a$, where I_0 is intensity of incident light, I is intensity of transmitted light, and a is sample thickness. By using the value of $k(\lambda)$, the line strength S is given by the following formula.

$$\int k(\lambda)d\lambda = \frac{8\pi^3 e^2 \lambda \rho}{3ch(2J+1)} \cdot \frac{1}{n} \cdot \frac{\left(n^2+2\right)^2}{9} \cdot S \tag{25}$$

where ρ is the density of the lanthanide ion and n is the refractive index. The parameters Ω_2, Ω_4, and Ω_6, giving the light emission probability, are included in S and are determined by a least square fit of:

$$S = \sum_{t=2,4,6} \Omega_t \left| < (S,L)J \left\| U^{(t)} \right\| (S',L')J' > \right|^2 \tag{26}$$

Here, $<(S,L)J \| U^{(t)} \| (S',L')J'>$ ($t = 2, 4, 6$) are reduced matrix elements characteristic of individual ions and available as a table.[8] Using the parameters Ω_2, Ω_4, and Ω_6 for specific host material, the light emission probability, A, between the levels of interest is calculated as follows:

$$A = \frac{64\pi^4 e^2}{3h(2J'+1)\lambda^3} \cdot n \cdot \frac{\left(n^2+2\right)^2}{9} \cdot S' \tag{27}$$

$$S' = \Omega_2 \left[U^{(2)} \right]^2 + \Omega_4 \left[U^{(4)} \right]^2 + \Omega_6 \left[U^{(6)} \right]^2$$

The theory contains some assumptions not strictly valid in actual cases, but still provides useful theoretical explanations for the nature of the luminescence spectra, as well as the excited-state lifetime, of lanthanide ions.[2,10]

Luminescence spectra of various trivalent lanthanide ions in YVO_4 (or YPO_4) are shown in Figure 23.[12] The luminescence and excitation spectra in Y_2O_3 are shown in Figure 24.[13] The luminescence spectra are composed of groups of several sharp lines. Each group corresponds to a transition between an excited and ground state designated by the total angular momentum, J. The assignment of the transition corresponding to each group of lines can be made on the basis of the energy level diagram shown in Figure 22. The excitation spectra generally consist of sharp lines due to the $4f$-$4f$ transition and of broad bands due to the $4f$-$5d$ transition and/or charge-transfer processes. Excited states giving rise to these broad excitation bands will be discussed in the next subsection.

The lifetimes of the luminescence due to $4f \rightarrow 4f$ transitions are mostly in the range of milliseconds because of the forbidden character of the luminescence transition.[11] For luminescence due to a spin-allowed transition between levels having equal spin multiplicity (e.g., $^3P_0 \rightarrow {}^3H_J$ of Pr^{3+}), a relatively short lifetime of ~10^{-5} s is observed.

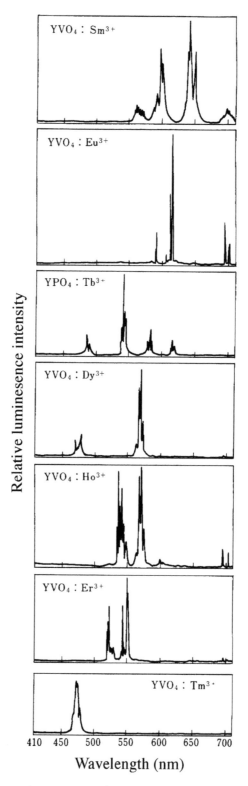

***Figure* 23** Emission spectra of various trivalent rare-earth ions in YVO_4 or YPO_4 hosts under cathode-ray excitation. (From Pallila, F.C., *Electrochem. Technol.*, 6, 39, 1968. With permission.)

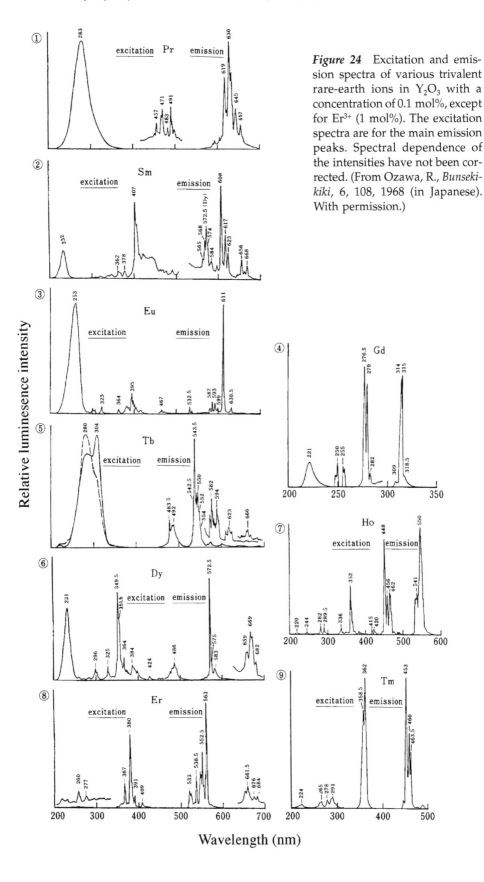

Figure 24 Excitation and emission spectra of various trivalent rare-earth ions in Y_2O_3 with a concentration of 0.1 mol%, except for Er^{3+} (1 mol%). The excitation spectra are for the main emission peaks. Spectral dependence of the intensities have not been corrected. (From Ozawa, R., *Bunseki-kiki*, 6, 108, 1968 (in Japanese). With permission.)

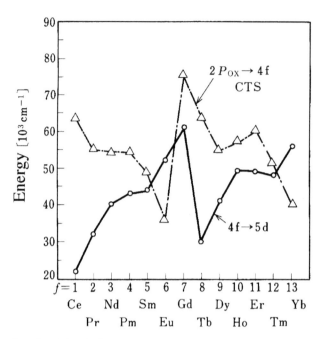

Figure 25 Energies for $4f \rightarrow 5d$ and CTS transitions of trivalent rare-earth ions. (From Hoshina, T., *Luminescence of Rare Earth Ions*, Sony Research Center Rep., 1983 (in Japanese). With permission.)

2.3.2.2 $4f^{n-1} 5d^1$ states and charge-transfer states (CTS)

In the energy region spanned by $4f$ levels, one finds two additional kinds of electronic states with different characters from those levels. They are the $4f^{n-1}5d^1$ states and the charge-transfer states (CTS). In the former, one of the $4f$ electron(s) is transferred to a $5d$ orbital and, in the latter case, electrons in the neighboring anions are transferred to a $4f$ orbital. Both of these processes are allowed and result in strong optical absorptions. They are observed as broadband excitation spectra around 300 nm, as is shown in Figure 24. Optical absorptions due to f-d transitions are found for Pr^{3+} and Tb^{3+}; those due to a charge-transfer transition are found in Eu^{3+}. The broad-band excitation spectra around 230 nm for Sm^{3+}, Dy^{3+}, and Gd^{3+} are due to host absorptions.

The energies of the $4f^{n-1}5d^1$ and CTSs are more dependent on their environments than the energies of $4f$ states, but the relative order of energies of these states are found to be the same for the whole series of rare-earth ions in any host materials. The transition energies from the ground states to these states are shown in Figure 25.[2,14] These energies are obtained by determining the values of parameters so as to agree with absorption spectra of trivalent rare-earth oxides. As shown in the figure, $4f$-$5d$ transitions in Ce^{3+}, Pr^{3+}, Tb^{3+}, and CTS absorptions in Eu^{3+} and Yb^{3+} have energies less than ca. 40×10^3 cm^{-1}. They can, therefore, interact with $4f$ levels, leading to $f \rightarrow f$ emissions. In case the energy levels of these states are lower than those of $4f$ levels, direct luminescence transitions from these levels are found, such as $5d \rightarrow 4f$ transitions in Ce^{3+}, Pr^{3+}, and Eu^{2+}. Spectra of this luminescence vary as a result of crystal field splitting in host crystals (Section 2.3.3). Luminescence due to the transition from CTS has also been reported for Yb^{3+} (See Section 2.3.3.18).

By comparing chemical properties of trivalent rare-earth ions with Figure 25, one can conclude that those ions that are easily oxidized to the tetravalent state have lower $4f \rightarrow 5d$ transition energies, while those that are easily reducible to the divalent state have lower CTS transition energies. It has also been confirmed that $4f^0$, $4f^7$, and $4f^{14}$ electronic configurations are relatively stable.

2.3.2.3   Divalent and tetravalent cations

In appropriate host crystals with divalent constituent ions such as Ca^{2+}, Sr^{2+}, or Ba^{2+}, Sm^{2+}, Eu^{2+}, and Yb^{2+} are stable and can luminesce. The electronic configurations of these ions are the same as those of Eu^{3+}, Gd^{3+}, and Lu^{3+}, respectively. The excited states of the divalent ions, however, are lowered compared with those of the corresponding trivalent ions, because the divalent ions have smaller nuclear charges. The lower $4f$-$5d$ transition energy reflects their chemical property of being easily ionized into the trivalent state. All trivalent ions, from La^{3+} to Yb^{3+}, can be reduced to the divalent state by γ-ray irradiation when doped in CaF_2.[15]

The electronic configurations of the tetravalent cations, Ce^{4+}, Pr^{4+}, and Tb^{4+} are the same as those of trivalent ions La^{3+}, Ce^{3+}, and Gd^{3+}, respectively. Their CTS energy is low, in accordance with the fact that they are easily reduced. When Ce, Pr, or Tb ions are doped in compound oxide crystals of Zr, Ce, Hf, or Th, the resulting powders show a variety of body colors, probably due to the CTS absorption band.[16] Luminescence from these CTSs has not been reported.

2.3.2.4   Energy transfer

The excitation residing in an ion can migrate to another ion of the same species that is in the ground state as a result of resonant energy transfer when they are located close to each other. The ionic separation where the luminescence and energy transfer probabilities become comparable is in the vicinity of several Angstroms. Energy migration processes increase the probability that the optical excitation is trapped at defects or impurity sites, enhancing nonradiative relaxation. This causes concentration quenching, because an increase in the activator concentration encourages such nonradiative processes. As a result, that excitation energy diffuses from ion to ion before it is trapped and leads to emission. On the other hand, a decrease in the activator concentration decreases the energy stored by the ions. Consequently, there is an optimum in the activator concentration, typically 1 to 5 mol% for trivalent rare-earth ions, resulting from the trade-off of the above two factors. In some compounds such as NdP_5O_{14}, the lattice sites occupied by Nd are separated from each other by a relatively large distance (5.6 Å), and a high luminescence efficiency is achieved even when all the sites are occupied by activator ions (Nd). Such phosphors are called *stoichiometric phosphors*.

The energy transfer between different ion species can take place when they have closely matched energy levels. The energy transfer results either in the enhancement (e.g., $Ce^{3+} \rightarrow Tb^{3+}$) or in the quenching (e.g., $Eu^{3+} \rightarrow Nd^{3+}$) of emission. The effects of impurities on the luminescence intensities of lanthanide ions in Y_2O_3 are shown in Figure 26. Energy transfer between $4f$ levels has been shown to originate from the electric dipole-electric quadrupole interaction using glass samples.[17]

The luminescence spectra of Eu^{3+}, as well as that of Tb^{3+}, have strong dependence on the concentration. This is because at higher concentrations, the higher emitting levels, $^5D_1$ of Eu^{3+} and $^5D_3$ of Tb^{3+}, transfer their energies to neighboring ions of the same species by the following cross-relaxations; that is:

$$^5D_1\left(Eu^{3+}\right) + {}^7F_0\left(Eu^{3+}\right) \rightarrow {}^5D_0\left(Eu^{3+}\right) + {}^7F_6\left(Eu^{3+}\right)$$

$$^5D_3\left(Tb^{3+}\right) + {}^7F_6\left(Tb^{3+}\right) \rightarrow {}^5D_4\left(Tb^{3+}\right) + {}^7F_0\left(Tb^{3+}\right)$$

The energy transfer from a host crystal to activators leads to host-excited luminescence. The type of charge carriers to be captured by the doped ions, either electrons or holes, determines the nature of the valence changes in the ions. For a Y_2O_2S host, Tb^{3+} and Pr^{3+}

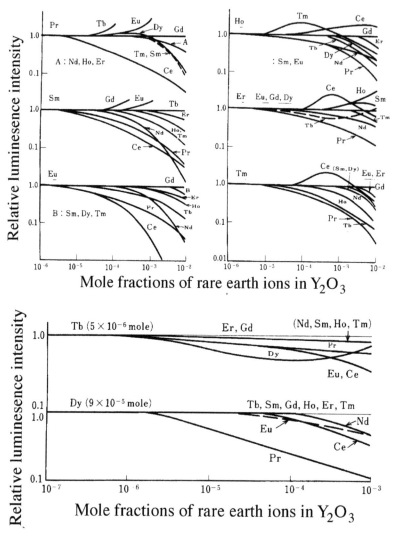

Figure 26 Decrease of luminescence intensities of trivalent rare-earth ions in Y_2O_3 due to the addition of other rare-earth ions. The concentration of ions is 10^{-3} mol% except for Tb and Dy (see the figure). (From Ozawa, R., *Bunseki-kiki*, 6, 108, 1968 (in Japanese). With permission.)

will act as hole traps, while Eu^{3+} will act as an electron trap at the initial stage of host excitation. In the next stage, these ions will capture an opposite charge and produce excitation of $4f$ levels.[18–20] A similar model has also been applied to $Y_3Al_5O_{12}:Ce^{3+},Eu^{3+},Tb^{3+}$.[21] Energy transfer from an excited oxy-anion complex to lanthanide ions is responsible for the luminescence observed in $CaWO_4:Sm^{3+}$,[22] $YVO_4:Eu^{3+}$,[23,24] and $Y_2WO_6:Eu^{3+}$.[25]

2.3.3 Luminescence of specific ions

2.3.3.1 Ce³⁺

Among the lanthanide ions, the $4f \rightarrow 5d$ transition energy is the lowest in Ce^{3+}, but the energy gap from the $5d^1$ states to the nearest level ($^2F_{7/2}$) below is so large that the $5d$ level serves as an efficient light-emitting state. The luminescence photon energy depends strongly on the structure of the host crystal through the crystal-field splitting of the $5d$ state, as shown in Figure 27[26] (see also Reference 27 and the discussion in Section 2.3.3.8

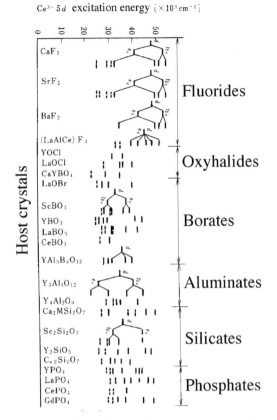

Figure 27 Energies of 5d excited levels of Ce^{3+} in various host crystals. (From Narita, K. and Taya, A., *Tech. Digest, Phosphor Res. Soc. 147th Meeting,* 1979 (in Japanese). With permission.)

on Eu^{2+} described below) and varies from near-ultraviolet to the green region. Typical luminescence spectra of some Ce^{3+}-activated phosphors are shown in Figure 28.[28] The two emission peaks are due to the two terminating levels, $^2F_{5/2}$ and $^2F_{7/2}$, of the $4f$ configuration of Ce^{3+}.

The decay time of the Ce^{3+} emission is 10^{-7} to 10^{-8} s, the shortest in observed lanthanide ions. This is due to two reasons: the $d \rightarrow f$ transition is both parity-allowed and spin-allowed since $5d^1$ and $4f^1$ states are spin doublets.[29] By virtue of the short decay time, $Y_2SiO_5:Ce^{3+}$ and $YAlO_3:Ce^{3+}$ are used for flying spot scanners or beam-index type cathode-ray tubes. Also, Ce^{3+} is often used for the sensitization of Tb^{3+} luminescence in such hosts as $CeMgAl_{11}O_{19}$.[30]

2.3.3.2 Pr^{3+}

Luminescence of Pr^{3+} consists of many multiplets, as follows: ~515 nm ($^3P_0 \rightarrow {}^3H_4$), ~670 nm ($^3P_0 \rightarrow {}^3F_2$), ~770 nm ($^3P_0 \rightarrow {}^3F_4$), ~630 nm ($^1D_2 \rightarrow {}^3H_6$), ~410 nm ($^1S_0 \rightarrow {}^1I_6$), and ultraviolet ($5d \rightarrow 4f$) transitions. The relative intensities of the peaks depend on the host crystals. As an example, the emission spectrum of $Y_2O_2S:Pr^{3+}$ is shown in Figure 29. The radiative decay time of the $^3P_0 \rightarrow {}^3H_J$ or ^3F_J emission is ~10^{-5} s, which is the shortest lifetime observed in $4f \rightarrow 4f$ transitions. For example, in Y_2O_2S host, decay times until 1/10 initial intensity are 6.7 μs for Pr^{3+}, 2.7 ms for Tb^{3+}, and 0.86 ms for Eu^{3+}.[2] The short decay time of Pr^{3+} is ascribed to the spin-allowed character of the transition. Since the short decay time is fit for fast information processing, $Gd_2O_2S(F):Pr^{3+},Ce^{3+}$ ceramic has been developed for an X-ray detector in X-ray computed tomography.[31]

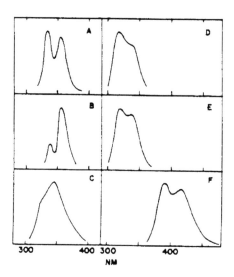

Figure 28 Emission spectra and excitation wavelengths of Ce^{3+} in various hosts. (A) YPO_4, 254-nm excitation; (B) YPO_4, 324-nm excitation; (C) $GdPO_4$, 280-nm excitation; (D) $LaPO_4$, 254-nm excitation; (E) $LaPO_4$, 280-nm excitation; (F) YBO_3, 254-nm excitation. (From Butler, K.H., *Fluorescent Lamp Phosphors, Technology and Theory*, The Pennsylvania State University Press, 1980, 261. With permission.)

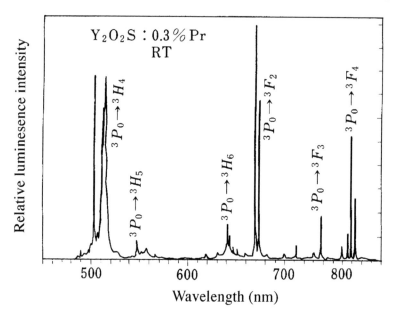

Figure 29 Emission spectrum of $Y_2O_2S:Pr^{3+}$ (0.3%) at room temperature. (From Hoshina, T., *Luminescence of Rare Earth Ions*, Sony Research Center Rep., 1983 (in Japanese). With permission.)

The quantum efficiency of more than 1 was reported for Pr^{3+} luminescence when excited by 185-nm light.[32,33] The excitation-relaxation process takes the following paths: $^3H_4 \rightarrow {}^1S_0$ (excitation by 185 nm), $^1S_0 \rightarrow {}^1I_6$ (405-nm emission), $^1I_6 \rightarrow {}^3P_0$ (phonon emission), $^3P_0 \rightarrow {}^3H_4$ (484.3-nm emission), $^3P_0 \rightarrow {}^3H_5$ (531.9 nm emission), $^3P_0 \rightarrow {}^3H_6$, $^3F_2$ (610.3-nm emission), and $^3P_0 \rightarrow {}^3F_3$, $^3F_4$ (704 nm emission). The sum of the visible light emissions in the above processes was estimated to have a quantum efficiency of 1.4.[33] In some fluoride crystals, the $4f^15d^1$ state was found to be lower than $^1S_0$, resulting in broad-band UV luminescence (see Figure 30).

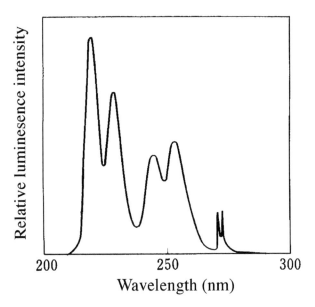

Figure 30 Emission spectrum of $LiYF_4:Pr^{3+}$ (1%) using 185-nm excitation. (From Piper, W.W., Deluca, J.A., and Ham, F.S., *J. Luminesc.*, 8, 344, 1974. With permission.)

2.3.3.3 Nd^{3+}

The four lower-lying levels of Nd^{3+} provide a condition favorable to the formation of population inversion. For this reason, Nd^{3+} is used as the active ion in many high-power, solid-state lasers (at 1.06 μm wavelength); the most common hosts are $Y_3Al_5O_{12}$ single crystals (yttrium aluminum garnet, YAG) or glass. The relative emission intensity of Nd^{3+} in $Y_3Al_5O_{12}$ has been found to be as follows[34];

$$^4F_{3/2} \rightarrow {}^4I_{9/2} \quad (0.87-0.95 \text{ μm}) \qquad : 0.25$$

$$^4F_{3/2} \rightarrow {}^4I_{11/2} \quad (1.05-1.12 \text{ μm}) \qquad : 0.60$$

$$^4F_{3/2} \rightarrow {}^4I_{13/2} \quad (\sim 1.34 \text{ μm}) \qquad : 0.15$$

$$^4F5_{3/2} \rightarrow {}^4I_{9/2} \quad \text{and others} (\tau = 230 \text{ μs}) \quad :\sim 0.010$$

2.3.3.4 Nd^{4+}

Luminescence in the regions ~415, 515, 550, and ~705 nm has been reported in $Cs_3NdF_7:Nd^{4+}$.[35]

2.3.3.5 Sm^{3+}

Red luminescence at ~610 nm ($^4G_{5/2} \rightarrow {}^6H_{7/2}$) and ~650 nm ($^4G_{5/2} \rightarrow {}^6H_{9/2}$) is observed in $Sm.^{3+}$ High luminescence efficiency in Sm^{3+}, however, has not been reported. Sm^{3+} acts as an auxiliary activator in photostimulable $SrS:Eu^{2+}$ (Mn^{2+} or Ce^{3+}) phosphors. Under excitation, Sm^{3+} captures an electron, changing to Sm^{2+}, which in turn produces an excitation band peaking at 1.0 μm.[36,37] (See 2.6.)

2.3.3.6 Sm^{2+}

The $4f^55d^1$ level of Sm^{2+} is located below its $4f$ levels in CaF_2, resulting in band luminescence due to the $5d \rightarrow 4f$ transition (728.6 nm, $\tau \sim$ μs). In SrF_2 and BaF_2, on the other hand, a line spectrum due to the $4f \rightarrow 4f^5 D_0 \rightarrow {}^7F_1$ transition has been observed (696 nm, $\tau \sim$ μs).[38]

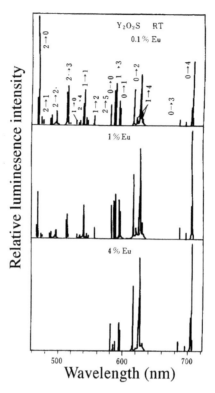

Relative luminescence intensity

Wavelength (nm)

Figure 31 Eu^{3+} concentration dependence of the emission spectrum of $Y_2O_2S:Eu^{3+}$. (From Hoshina, T., *Luminescence of Rare Earth Ions*, Sony Research Center Rep., 1983 (in Japanese). With permission.)

Also, in BaFCl, line emission at 550 to 850 nm due again to $^5D_{0,1} \rightarrow {}^7F_{0\sim4}$ transitions has been reported.[39]

2.3.3.7 Eu^{3+}

A number of luminescence lines due to $^5D_J \rightarrow {}^7F_{J'}$ of Eu^{3+} in Y_2O_2S are shown in Figure 31. As can be seen, the emissions from $^5D_2$ and $^5D_1$ are quenched, with an increase in the Eu^{3+} concentration due to a cross-relaxation process, $(^5D_J \rightarrow {}^5D_0) \rightarrow (^7F_0 \rightarrow {}^7F_{J'})$, as discussed in Section 2.3.2.4. The emission in the vicinity of 600 nm is due to the magnetic dipole transition $^5D_0 \rightarrow {}^7F_1$, which is insensitive to the site symmetry. The emission around 610–630 nm is due to the electric dipole transition of $^5D_0 \rightarrow {}^7F_2$, induced by the lack of inversion symmetry at the Eu^{3+} site, and is much stronger than that of the transition to the $^7F_1$ state. Luminescent Eu^{3+} ions in commercial red phosphors such as YVO_4, Y_2O_3 and Y_2O_2S, occupy the sites that have no inversion synmetry. The strong emission due to the electric dipole transition is utilized for practical applications. If the Eu^{3+} site has inversion symmetry, as in Ba_2GdNbO_5, $NaLuO_2$,[40] and $InBO_3$,[41] the electric dipole emission is weak, and the magnetic dipole transition becomes relatively stronger and dominates, as is shown in Figure 32.

The spectral luminous efficacy as sensed by the eye has its maximum at 555 nm. In the red region, this sensitivity drops rapidly as one moves toward longer wavelengths. Therefore, red luminescence composed of narrow spectra appear brighter to the human eye than various broad red luminescences having the same red chromaticity and emission energy. For the red emission of color TV to be used in the NTSC system, the red chromaticity standard has been fixed at the coordinates x = 0.67, y = 0.33; in 1955, the ideal emission spectra were proposed as a narrow band around 610 nm, before the development of Eu^{3+}

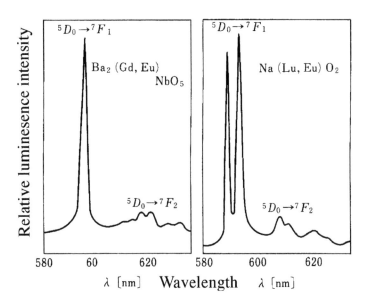

Figure 32 Emission spectra of Eu^{3+} from the sites having the inversion synmetry. (From Blasse, G. and Bril, A., *Philips Tech. Rev.*, 31, 304, 1970. With permission.)

phosphors.[42] This proposal was dramatically fulfilled for the first time in 1964 by newly developed YVO_4:Eu^{3+}.[43] Since then, Eu^{3+} phosphors have completely replaced broad-band emitting Mn^{2+} phosphors or (Zn,Cd)S;Ag, which were predominantly in use at that time. Just after the introduction of YVO_4:Eu^{3+}, another Eu^{3+}-activated phosphor, Y_2O_2S;Eu^{3+}, was developed[44] and is in current use due to its better energy efficiency as well as its stability during recycling in the screening process of CRT production. The possibility of further improvement can occur in materials with single-line emission, as in $Y_2(WO_4)_3$:Eu^{3+}.[45] Use of narrow-band luminescence is also advantageous in three-band fluorescent lamp applications, where both brightness and color reproducibility are required. For high color rendering lamps, Y_2O_3:Eu^{3+} has been used as the red-emitting component.

The sequence of excitation, relaxation, and emission processes in Y_2O_2S;Eu^{3+} is explained by the configurational coordinate model shown in Figure 33.[46] The excitation of Eu^{3+} takes place from the bottom of the $^7F_0$ curve, rising along the straight vertical line, until it crosses the charge-transfer state (CTS). Relaxation occurs along the CTS curve. Near the bottom of the CTS curve, the excitation is transferred to ^5D_J states. Relaxation to the bottom of the ^5D_J states is followed by light emission downward to ^7F_J states. This model can explain the following experimental findings. (1) No luminescence is found from $^5D_3$ in Y_2O_2S:Eu^{3+}. (2) The luminescence efficiency is higher for phosphors with higher CTS energy.[47] (3) The quenching temperature of the luminescence from ^5D_J is higher as J (0,1,2,3) decreases. The excited 4f states may dissociate into an electron-hole pair. This model is supported by the observation that the excitation through the $^7F_0 \rightarrow {}^5D_2$ transition of La_2O_2S:Eu^{3+} causes energy storage that can be converted to luminescence by heating. The luminescence is the result of the recombination of a thermally released hole with an Eu^{2+} ion.[48,49]

By taking a model where CTS is a combination of $4f^7$ electrons plus a hole, one finds that the resulting spin multiplicities should be 7 and 9. It is the former state that affects optical properties related to the ^7F_J state by spin-restricted covalency.[50] The intensity ratio of the luminescence from $^5D_0 \rightarrow {}^7F_2$ and from $^5D_0 \rightarrow {}^7F_1$ decreases with increasing CTS energy sequentially as $ScVO_4$, YVO_4, $ScPO_4$, and YPO_4, all of which have the same type of zircon structure.[51] The above intensity ratio is small in YF_3:Eu^{3+}, even though Eu^{3+} occupies a site without inversion symmetry.[52] It is to be noted that CTSs in fluorides have

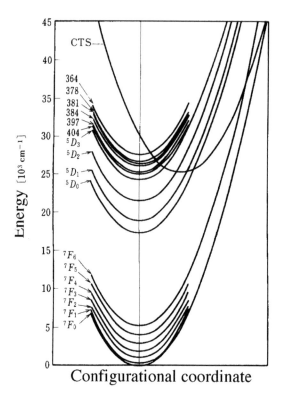

Figure 33 Configurational coordinate model of $Y_2O_2S:Eu^{3+}$. (From Struck, C.W. and Fonger, W.H., *J. Luminesc.*, 1/2, 456, 1970. With permission.)

higher energies than those in oxides. These results suggest that higher CTS energies reduce the strength of the electric dipole transition $^5D_0 \rightarrow {}^7F_2$ in Eu^{3+}.

2.3.3.8 Eu^{2+}

The electronic configuration of Eu^{2+} is $4f^7$ and is identical to that of Gd^{3+}. The lowest excited state of 4f levels is located at about 28×10^3 cm^{-1} and is higher than the $4f^6 5d^1$ level in most crystals, so that Eu^{2+} usually gives broad-band emission due to *f-d* transitions. The wavelength positions of the emission bands depend very much on hosts, changing from the near-UV to the red. This dependence is interpreted as due to the crystal field splitting of the 5d level, as shown schematically in Figure 34.[53] With increasing crystal field strength, the emission bands shift to longer wavelength. The luminescence peak energy of the 5d-4f transitions of Eu^{2+} and Ce^{3+} are affected most by crystal parameters denoting electron-electron repulsion; on this basis, a good fit of the energies can be obtained.[27]

The near-UV luminescence of Eu^{2+} in $(Sr,Mg)_2P_2O_7$ is used for lamps in copying machines using photosensitive diazo dyes. The blue luminescence in $BaMgAl_{10}O_{17}$ is used for three-band fluorescent lamps. (See Figure 35.)[54] Ba(F,Br): Eu^{2+} showing violet luminescence is used for X-ray detection through photostimulation.[55] Red luminescence is observed in Eu^{2+}-activated CaS[36]; the crystal field is stronger in sulfides than in fluorides and oxides.

The lifetime of the Eu^{2+} luminescence is 10^{-5}–10^{-6} s, which is relatively long for an allowed transition. This can be explained as follows. The ground state of $4f^7$ is ^8S, and the multiplicity of the excited state $4f^6 5d^1$ is 6 or 8; the sextet portion of the excited state contributes to the spin-forbidden character of the transition.[29]

Sharp-line luminescence at ~360 nm due to an *f-f* transition and having a lifetime of milliseconds is observed when the crystal field is weak so that the lowest excited state of

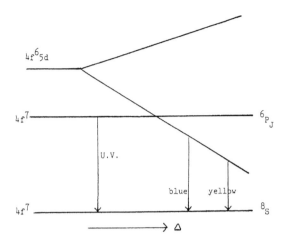

Figure 34 Schematic diagram of the energies of $4f^7$ and $4f^65d^1$ levels in Eu^{2+} influenced by crystal field Δ. (From Blasse, G., Material science of the luminescence of inorganic solids, in *Luminescence of Inorganic Solids*, DiBartolo, B., Plenum Press, 1978, 457. With permission.)

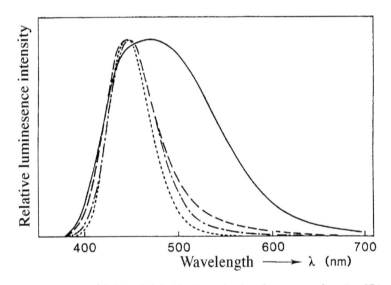

Figure 35 Emission spectra of Eu^{2+} in $BaMgAl_{10}O_{17}$ and related compounds using 254-nm excitation at 300K. ----: $Ba_{0.95}Eu_{0.05}MgAl_{10}O_{17}$, —·—·—: $Ba_{0.825}Eu_{0.05}Mg_{0.5}Al_{10.5}O_{17.125}$, — — —: $Ba_{0.75}Eu_{0.05}Mg_{0.2}Al_{10.8}O_{17.2}$, ————: $Ba_{0.70}Eu_{0.05}Al_{11}O_{17.25}$. (From Smets, B.M.J. and Verlijsdonk, J.G., *Mater. Res. Bull.*, 21, 1305, 1986. With permission.)

$4f^7(^6P_J)$ is lower than the $4f^65d^1$ state, as illustrated in Figure 34. The host crystals reported to produce UV luminescence are $BaAlF_5$, $SrAlF_5$ [56] (see Figure 36), $BaMg(SO_4)_2$,[57] $SrBe_2Si_2O_7$,[58] and $Sr(F,Cl)$.[59]

2.3.3.9 Gd^{3+}

The lowest excited $4f$ level of Gd^{3+} ($^6P_{7/2}$) gives rise to sharp-line luminescence at ~315 nm[60] and can sensitize the luminescence of other rare-earth ions.[61] The energy levels of the CTS and the $4f^65d^1$ states are the highest among rare-earth ions, so that Gd^{3+} causes no quenching in other rare-earth ions. As a consequence, Gd^{3+} serves, as Y^{3+} does, as a good constituent cation in host crystals to be substituted by luminescent rare-earth ions. For X-ray phosphors, Gd^{3+} is better suited as a constituent than Y^{3+} since it has a higher absorption cross-section due to its larger atomic number.

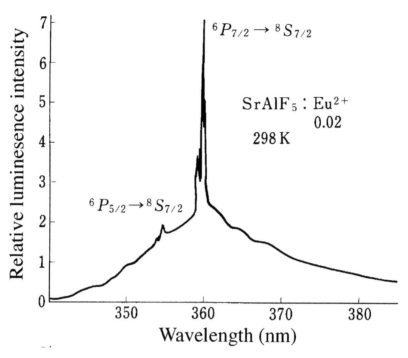

Figure 36 Emission spectrum at 298K of $SrAlF_5:Eu^{2+}$ using 254-nm excitation. (From Hews, R.A. and Hoffman, M.V., *J. Luminesc.*, 3, 261, 1970. With permission.)

2.3.3.10 Tb^{3+}

Luminescence spectra consisting of many lines due to $^5D_J \to {}^7F_{J'}$ are observed for Tb^{3+}. As an example, the spectra of $Y_2O_2S:Tb^{3+}$ are shown in Figure 37, in which the Tb^{3+} concentration varies over a wide range. The intensity of the emissions from $^5D_3$ decreases with increasing Tb^{3+} concentration due to cross-relaxation, as discussed in Section 2.3.2.4. Among the emission lines from the $^5D_4$ state, the $^5D_4 \to {}^7F_5$ emission line at approximately 550 nm is the strongest in nearly all host crystals when the Tb^{3+} concentration is a few mol% or higher. The reason is that this transition has the largest probability for both electric-dipole and magnetic-dipole induced transitions.[2] The Tb^{3+} emission has a broad excitation band in the wavelength region 220 to 300 nm originating from the $4f^8 \to 4f^75d^1$ transition.

The chromaticity due to the Tb^{3+} emission has been estimated by calculation of the various transition probabilities.[62] The spectral region around 550 nm is nearly at the peak in the spectral luminous efficacy; in this region, therefore, the brightness depends only slightly on the wavelength and the spectral width. Thus, the narrow spectral width of the Tb^{3+} emission is not so advantageous in cathode-ray tube applications as compared with the case of red Eu^{3+} emission previously described.

The intensity ratio of the emission from $^5D_3$ to that from $^5D_4$ depends not only on the Tb^{3+} concentration, but also on the host material. In borate hosts such as $ScBO_3$, $InBO_3$, and $LuBO_3$, the relative intensity of $^5D_3$ emission is much weaker than in other hosts, such as phosphates, silicates, and aluminates.[2] Figure 38 shows emission spectra of a series of $Ln_2O_2S:Tb^{3+}(0.1\%)$ (Ln = La, Gd, Y, and Lu) materials having the same crystal structure.[2] It is seen that the relative intensity of the $^5D_3$ emission increases dramatically as one progresses from La $\to$ Gd $\to$ Y $\to$ Lu, with the ionic radii becoming smaller.

In addition to the Tb^{3+} concentration, one needs to consider two additional factors that help determine the ratio of $^5D_3$ to $^5D_4$ intensity. One is the maximum energy of phonons that causes phonon-induced relaxation, as discussed in Section 2.3.2.1; if the maximum phonon

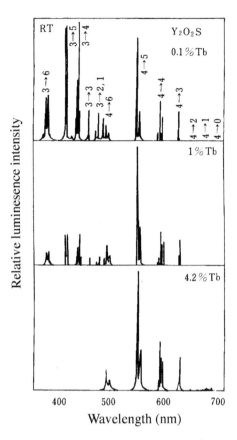

Figure 37 Tb^{3+} concentration dependence of emission spectra of Y$_2$O$_2$S:Tb^{3+} at room temperature. (From Hoshina, T., *Luminescence of Rare Earth Ions*, Sony Research Center Rep., 1983 (in Japanese). With permission.)

energy is large, the ratio of $^5D_3$ to $^5D_4$ intensity becomes small. The luminescence of Tb^{3+} in borate hosts is explained by this factor. The other factor is the energy position of the $4f^75d^1$ level relative to $4f^8$ levels, which can be discussed in terms of the configurational coordinate model. In this model, the potential curve of $4f^75d^1$ can be drawn just like the CTS in Figure 33. If the minimum of the $4f^75d^1$ curve is fairly low in energy and the Frank-Condon shift is fairly large, there is a possibility that an electron excited to the $4f^75d^1$ level can relax directly to the $^5D_4$, bypassing the $^5D_3$ and thus producing only $^5D_4$ luminescence.[2] The net effect of these two factors on the spectra of Ln$_2$O$_2$S:Tb^{3+} in Figure 38 is not known quantitatively.

YVO$_4$ is a good host material for various Ln^{3+} ions, as shown in Figure 23. However, Tb^{3+} does not luminesce in this host. A nonradiative transition via a charge-transfer state of Tb^{4+}-O^{2-}-V^{4+} has been proposed as a cause.[63] The transition energy to this proposed state is considered to be relatively low because both the energies of conversion from Tb^{3+} to Tb^{4+} and that from V^{5+} to V^{4+} are low. The Frank-Condon shift in the transition would be so large that the proposed state would provide a nonradiative relaxation path from excited Tb^{3+} to the ground state.

Tb^{3+}-activated green phosphors are used in practice in three-band fluorescent lamps, projection TV tubes, and X-ray intensifying screens.

2.3.3.11 Dy^{3+} [64,65]

The luminescence lines of Dy^{3+} are in the 470 to 500-nm region due to the $^4F_{9/2} \rightarrow {}^6H_{15/2}$ transition, and in the 570 to 600-nm region due to the $^6F_{15/2} \rightarrow {}^6F_{11/2}$ transition. The color

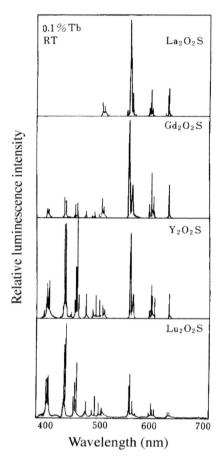

Figure 38 Emission spectra of Ln_2O_2S:Tb^{3+} (0.1%) (Ln = La, Gd, Y, and Lu) at room temperature. (From Hoshina, T., *Luminescence of Rare Earth Ions*, Sony Research Center Rep., 1983 (in Japanese). With permission.)

of the luminescence is close to white. In $Y(P,V)O_4$, the relative intensity of the latter decreases with increasing P concentration. This can be understood if one considers that the $\Delta J = 2$ transition probability decreases with a decrease in the polarity of the neighboring ions as in the case of the $^5D_0 \rightarrow {}^7F_2$ transition of Eu^{3+}. The energy of the CTS and $4f^85d^1$ is relatively large so that direct UV excitation of Dy^{3+} is not effective. The excitation via host complex ions by energy transfer can however be effective. The quantum efficiency of UV-excited (250–270 nm) luminescence of YVO_4:Dy^{3+} has been reported to be as high as 65%.

2.3.3.12 Dy^{2+} [66]

Luminescence of Dy^{2+} has been reported to consist of line spectra at 2.3–2.7 μm at 77K and 4.2K in CaF_2, SrF_2, and BaF_2. Dy^{2+} in these hosts was prepared by the reduction of Dy^{3+} through γ-ray irradiation.

2.3.3.13 Dy^{4+} [67]

Luminescence lines of Cs_3DyF_7:Dy^{4+} at 525 nm due to $^5D_4 \rightarrow {}^7F_5$ transition and at 630 nm due to $^5D_4 \rightarrow {}^7F_3$ transition have been reported.

2.3.3.14 Ho^{3+}

Efficient luminescence of Ho^{3+} has rarely been found due to the crowded energy level diagram of this ion. In $LaCl_3$, cross-relaxation between $(^5S_2 \rightarrow {}^5I_4) \leftrightarrow (^5I_8 \rightarrow {}^5I_7)$ at an interioninc distance of 7.5 Å has been reported.[68] A green luminescence due to the $^5F_4$, $^5S_2 \rightarrow {}^5I_8$ transition has been reported in an infrared-to-visible up-conversion phosphor, $LiYF_4:Yb^{3+},Ho^{3+}$.[69]

2.3.3.15 Ho²⁺

Infrared luminescence of Ho^{2+} in CaF_2 appearing around 1.8 μm at 77K has been reported.[70]

2.3.3.16 Er³⁺

Green luminescence due to the $^4S_{3/2} \rightarrow {}^4I_{15/2}$ transition of Er^{3+} has been reported in infrared-to-visible up-conversion phosphors, such as $LaF_3:Yb^{3+},Er^{3+}$,[71] and $NaYF_4:Yb^{3+},Er^{3+}$.[72] This luminescence was also reported in ZnS,[73] Y_2O_3,[74] and Y_2O_2S.[75] The emission color is a well-saturated green.

Er³⁺ ions embedded in an optical fiber (several hundreds ppm) function as an optical amplifier for 1.55-μm semiconductor laser light. Population inversion is realized between lower sublevels of $^4I_{13/2}$ and upper sublevels of $^4I_{15/2}$. This technology has been developed for optical amplification in the long-distance optical fiber communication systems.[76]

2.3.3.17 Tm³⁺

The blue luminescence of Tm^{3+} due to the $^1G_4 \rightarrow {}^3H_6$ transition has been reported in ZnS,[77] as well as in infrared-to-visible up-conversion phosphors sensitized by Yb^{3+} such as $YF_3:Tm^{3+},Yb^{3+}$.[78] Electroluminescent $ZnS:TmF_3$ has also been investigated as the blue component of multicolor displays.[79] The efficiency of the blue luminescence of Tm^{3+} is low, and is limited by the competitive infrared luminescence, which has a high efficiency.

2.3.3.18 Yb³⁺

The infrared absorption band of Yb^{3+} at about 1 μm due to the $^5F_{5/2} \rightarrow {}^5F_{7/2}$ transition is utilized for Er^{3+}-doped infrared-to-visible up-conversion phosphors as a sensitizer.[71,72] The CTS energy of Yb^{3+} ions is low next to the lowest of Eu^{3+} among the trivalent lanthanide ions (see Figure 25). Yb^{3+} has no 4f energy levels interacting with CTS, so that luminescence due to the direct transition from CTS to the 4f levels can occur. This luminescence has been observed in phosphate[80] and oxysulfide hosts.[81] Figure 39 shows the excitation and emission spectra of $Y_2O_2S:Yb^{3+}$ and $La_2O_2S:Yb^{3+}$.[81] As seen in Figure 33, CTS is characterized by a fairly large Frank-Condon shift. As a result, the emission spectra are composed of two fairly broad bands terminating in $^2F_{5/2}$ and $^2F_{7/2}$, as shown in Figure 39.

2.3.3.19 Yb²⁺

The emission and absorption of Yb^{2+} due to the $4f^{14} \leftrightarrow 4f^{13}5d^1$ transition have been reported.[82] Emission peaks are at 432 nm in $Sr_3(PO_4)_2$ (see Figure 40), 505 nm in Ca_2PO_4Cl, 560 nm in $Sr_5(PO_4)_3Cl$, and 624 nm in $Ba_5(PO_4)_3Cl$. The lifetimes of the emissions are between $1-6 \times 10^{-5}$ s.

References

1. Blasse, G., *Handbook on the Physics and Chemistry of Rare Earths*, ed. by Gschneidner, Jr., K.A. and Eyring, L., Vol. 4, North-Holland Pub. 1979, 237.
2. Hoshina, T., *Luminescence of Rare Earth Ions*, Sony Research Center Rep. (Suppl.) 1983 (in Japanese).
3. Adachi, G., *Rare Earths—Their Properties and Applications*, ed. by Kano, T. and Yanagida, H., Gihodo Pub. 1980, 1 (in Japanese). Kano, T., *ibid*, 173.
4. Ofelt, G.S., *J. Chem. Phys.*, 38, 2171,1963.
5. Dieke, G.H., *Spectra and Energy Levels of Rare Earth Ions in Crystals*, Interscience, 1968; *American Institute of Physics Handbook*, 3rd edition, McGraw-Hill, 1972, 7-25.
6. Riseberg, L.A. and Moos, H.W., *Phys. Rev.*, 174, 429, 1968.

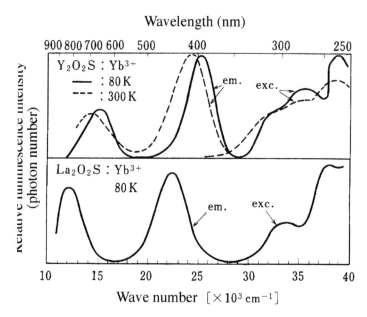

Figure 39 Excitation and emission spectra of $Y_2O_2S:Yb^{3+}$ and $La_2O_2S:Yb^{3+}$. (From Nakazawa, E., *J. Luminesc.*, 18/19, 272, 1979. With permission.)

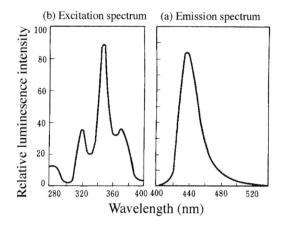

Figure 40 Emission (a) and excitation (b) spectra of $Sr_3(PO_4)_2:Yb^{2+}$ at liquid nitrogen temperature. (From Palilla, F.C., O'Reilly, R.E., and Abbruscato, V.J., *J. Electrochem. Soc.*, 117, 87, 1970. With permission.)

7. Kropp, J.L. and Windsor, M.W., *J. Chem. Phys.*, 42, 1599, 1965.
8. Judd, B.R., *Phys. Rev.*, 127, 750, 1962.
9. Ofelt, G.S., *J. Chem. Phys.*, 37, 511, 1962.
10. Hirao, K., *Rev. Laser Eng.*, 21, 618, 1993 (in Japanese).
11. Barasch, G.E. and Dieke, G.H., *J. Chem. Phys.*, 43, 988, 1965.
12. Pallila, F.C., *Electrochem. Technol.*, 6, 39, 1968.
13. Ozawa, R., *Bunseki-kiki*, 6, 108, 1968 (in Japanese).
14. Joergensen, C.K., Papalardo, R., and Rittershaus, E., *Z. Naturforschg.*, 20-a, 54, 1964.
15. McClure, D.S. and Kiss, Z., *J. Chem. Phys.*, 39, 3251, 1963.
16. Hoefdraad, H.E., *J. Inorg. Nucl. Chem.*, 37, 1917, 1975.
17. Nakazawa, E. and Shionoya, S., *J. Chem. Phys.*, 47, 3211, 1967.
18. McClure, D.S., *The Electrochem. Soc., Extended Abstr.*, 77-1, 365, 1977.

19. Ozawa, L., *The Electrochem. Soc., Extended Abstr.*, 78-1, 850, 1978.
20. Yamamoto, H. and Kano, T., *J. Electrochem. Soc.*, 126, 305, 1979.
21. Robins, D.J., Cockayne, B., Glasper, J.L., and Lent, B., *J. Electrochem. Soc.*, 126, 1221, 1979.
22. Botden, Th.P., *Philips Res. Rpts.*, 6, 425, 1951.
23. Van Uitert, L.G., Soden, R.R., and Linares, R.C., *J. Chem. Phys.*, 36, 1793, 1962.
24. Pallila, F.C., Levin, A.K., and Rinkevics, M., *J. Electrochem. Soc.*, 112, 776, 1965.
25. Blasse, G. and Bril, A., *J. Chem. Phys.*, 51, 3252, 1969.
26. Narita, K. and Taya, A., *Tech. Digest, Phosphor Res. Soc. 147th Meeting*, 1979 (in Japanese).
27. Van Uitert, L.G., *J. Luminesc.*, 29, 1, 1984.
28. Butler, K.H., *Fluorescent Lamp Phosphors, Technology and Theory*, The Pennsylvania State University Press, 1980, 261. Copyright 1996 by The Pennsylvania State University.
29. Blasse, G., Wanmaker, W.L., Tervrugt, J.W., and Bril, A., *Philips Res. Rept.*, 23, 189, 1968.
30. Sommerdijk, J.L. and Verstegen, J.M.P.J., *J. Luminesc.*, 9, 415, 1974.
31. Yamada, H., Suzuki, A., Uchida, Y., Yoshida, M., Yamamoto, H., and Tsukuda, Y., *J. Electrochem. Soc.*, 136, 2713, 1989.
32. Sommerdijk, J.L., Bril, A., and de Jager, A.W., *J. Luminesc.*, 8, 341, 1974.
33. Piper, W.W., Deluca, J.A., and Ham, F.S., *J. Luminesc.*, 8, 344, 1974.
34. Kushida, T., Marcos, H.M., and Geusic, J.E., *Phys. Rev.*, 167, 289, 1968.
35. Vaga, L.P., *J. Chem. Phys.*, 49, 4674, 1968.
36. Urbach, F., Pearlman, D., and Hemmendinger, H., *J. Opt. Soc. Am.*, 36, 372, 1946.
37. Keller, S.P., Mapes, J.E., and Cheroff, G., *Phys. Rev.*, 111, 1533, 1958.
38. Feofilof, P.D. and Kaplyanskii, A.A., *Opt. Spectrosc.*, 12, 272, 1962.
39. Mahbub'ul Alam, A.S. and Baldassare Di Bartolo, B., *J. Chem. Phys.*, 47, 3790, 1967.
40. Blasse, G. and Bril, A., *Philips Tech. Rev.*, 31, 304, 1970.
41. Avella, F.J., Sovers, O.J., and Wiggins, C.S., *J. Electrochem. Soc.*, 114, 613, 1967.
42. Bril, A. and Klassens, H.A., *Philips Res. Rept.*, 10, 305, 1955.
43. Levine, A.K. and Pallila, F.C., *Appl. Phys. Lett.*, 5, 118, 1964.
44. Royce, M.R. and Smith, A.L., *The Electrochem. Soc., Extended Abstr.*, 34, 94, 1968.
45. Kano, T., Kinameri, K., and Seki, S., *J. Electrochem. Soc.*, 129, 2296, 1982.
46. Struck, C.W. and Fonger, W.H., *J. Luminesc.*, 1&2, 456, 1970.
47. Blasse, G., *J. Chem. Phys.*, 45, 2356, 1966.
48. Forest, H., Cocco, A., and Hersh, H., *J. Luminesc.*, 3, 25, 1970.
49. Struck, C.W. and Fonger, W.H., *Phys. Rev.*, B4, 22, 1971.
50. Hoshina, T., Imanaga, S., and Yokono, S., *J. Luminesc.*, 15, 455, 1977.
51. Blasse, G. and Bril, A., *J. Chem. Phys.*, 50, 2974, 1969.
52. Blasse, G. and Bril, A., *Philips Res. Rept.*, 22, 481, 1967.
53. Blasse, G., Material science of the luminescence of inorganic solids, in *Luminescence of Inorganic Solids*, DiBartolo, B., Ed., Plenum Press, 1978, 457.
54. Smets, B.M.J. and Verlijsdonk, J.G., *Mater. Res. Bull.*, 21, 1305, 1986.
55. Takahashi, K., Kohda, K., Miyahara, J., Kanemitsu, Y., Amitani, K., and Shionoya, S., *J. Luminesc.*, 31&32, 266, 1984.
56. Hews, R.A. and Hoffman, M.V., *J. Luminesc.*, 3, 261, 1970.
57. Ryan, F.M., Lehmann, W., Feldman, D.W., and Murphy, J., *J. Electrochem. Soc.*, 121, 1475, 1974.
58. Verstegen, J.M.P.J. and Sommerdijk, J.L., *J. Luminesc.*, 9, 297, 1974.
59. Sommerdijk, J.L., Verstegen, J.M.P.J., and Bril, A., *J. Luminesc.*, 8, 502, 1974.
60. Wickersheim, K.A. and Lefever, R.A., *J. Electrochem. Soc.*, 111, 47, 1964.
61. D'Silva, A.P. and Fassel, V.A., *J. Luminesc.*, 8, 375, 1974.
62. Hoshina, T., *Jpn. J. Appl. Phys.*, 6, 1203, 1967.
63. DeLosh, R.G., Tien, T.Y., Gibbon, F.F., Zacmanidis, P.J., and Stadler, H.L., *J. Chem. Phys.*, 53, 681, 1970.
64. Sommerdijik, J.L. and Bril, A., *J. Electrochem. Soc.*, 122, 952, 1975.
65. Sommerdijik, J.L., Bril, A., and Hoex-Strik, F.M.J.H., *Philips Res. Rept.*, 32, 149, 1977.
66. Kiss, Z.J., *Phys. Rev.*, 137, A1749, 1965.
67. Varga, L.P., *J. Chem. Phys.*, 53, 3552, 1970.
68. Porter, Jr., J.F., *Phys. Rev.*, 152, 300, 1966.

69. Watts, R.K., *J. Chem. Phys.*, 53, 3552, 1970.
70. Weakliem, H.A. and Kiss, Z.J., *Phys. Rev.*, 157, 277, 1967.
71. Hews, R.A. and Sarver, J.F., *Phys. Rev.*, 182, 427, 1969.
72. Kano, T., Yamamoto, H., and Otomo, Y., *J. Electrochem. Soc.*, 119, 1561, 1972.
73. Larach, S., Shrader, R.E., and Yocom, P.N., *J. Electrochem. Soc.*, 116, 47, 1969.
74. Kisliuk, P. and Krupke, W.F., *J. Chem. Phys.*, 40, 3606, 1964.
75. Shrader, R.E. and Yocom, P.N., *J. Luminesc.*, 1&2, 814, 1970.
76. Hagimoto, K. Iwatsuki, K., Takada, A., Nakagawa, M., Saruwatari, M., Aida, K., Hakagawa, K., and Horiguchi, M., *OFC'89 PD-15*, 1989.
77. Shrader, R.E., Larach, S., and Yocom, P.N., *J. Appl. Phys.*, 42, 4529, 1971. (Erratum: *J. Appl. Phys.*, 43, 2021, 1972.)
78. Geusic, J.E., Ostermayer, F.W., Marcos, H.M., Van Uitert, L.G., and Van der Ziel, J.P., *J. Appl. Phys.*, 42, 1958, 1971.
79. Kobayashi, H., Tanaka, S., Shanker, V., Shiiki, M., Kunou, T., Mita, J., and Sasakura, H., *Physi. Stat. Sol. (a)*, 88, 713, 1985.
80. Nakazawa, E., *Chem. Phys. Lett.*, 56, 161, 1978.
81. Nakazawa, E., *J. Luminesc.*, 18/19, 272, 1979.
82. Palilla, F.C., O'Reilly, R.E., and Abbruscato, V.J., *J. Electrochem. Soc.*, 117, 87, 1970.

chapter two — section four

Principal phosphor materials and their optical properties

Makoto Morita

Contents

2.4   Luminescence centers of complex ions

2.4.1   Introduction

Phosphors containing luminescence centers made up of complex ions have been well known since 1900s. The specific electronic structures are reflected in the spectral band shapes and transition energies. These phosphors have been widely used in practical applications. However, in spite of their common usage, it is only in the last three decades that the electronic structures of the complex ions have been explained in terms of the crystal field theory (See 2.2.1); because of this understanding, new phosphors have been prepared with a variety of colors and with high quantum yields.[1] This section will first focus on the luminescence from complex ions with closed-shell electronic structures such as the scheelite compounds and others and describe a variety of applications for these phosphors. Other interesting luminescence centers such as uranyl(II), platinum(II), mixed-valence, and other complexes are discussed subsequently.

2.4.2 Scheelite-type compounds

2.4.2.1 Scheelite compounds and their general properties

Calcium tungstate ($CaWO_4$) has long been known as a practical phosphor, and it is a representative scheelite compound. The luminescence center is the WO_4^{2-} complex ion in which the central W metal ion is coordinated by four O^{2-} ions in tetrahedral symmetry (T_d). Other analogous T_d complexes are *molybdate* (MoO_4^{2-}) and vanadate (VO_4^{3-}). In these three complex ions, the electronic configuration of the outer-shell is [Xe]$4f^{14}$, [Kr], and [Ar] for WO_4^{2-}, MoO_4^{2-}, and VO_4^{3-}, respectively. In general, scheelite phosphors take the form of A_pBO_4 with A standing for a monovalent alkaline, divalent alkaline earth, or trivalent lanthanide metal ion, p for the number of ions, and B for W, Mo, V, or P. Bright luminescence in the blue to green spectral regions was observed in the early 20th century. An introduction to the electronic configurations common to these complex ions is followed by a discussion of the results of investigations, referring to a number of recent review articles on this subject.[1,2]

2.4.2.2 Electronic structures of closed-shell molecular complex centers

As a typical complex ion center, consider the electronic structure of the MnO_4^- ion. In this case, the Mn^{7+} ion has a closed-shell structure with no d electrons. Using a one-electron transition scheme, consider a one-electron charge transfer process from the oxygen $2p$ orbital (t_1 symmetry in T_d) to the $3d$ orbital (e and t_2 symmetry) of the Mn^{7+} ion. A molecular orbital calculation[3] leads to e^* and a t_1 states for the lowest unoccupied molecular orbital (LUMO) and the highest occupied molecular orbital (HOMO), respectively. By taking the $e \rightarrow t_1$ transition into account, the excited electronic states of t_1^5e electronic configuration in T_d symmetry are found to consist of $^3T_1 \lesssim {}^3T_2 < {}^1T_1 < {}^1T_2$ in order of increasing energies, the ground state being a $^1A_1$ state. The orbital triplets ($^3T_1$, $^3T_2$) have degenerate levels in the spectral region of 250 to 500 nm. By employing more advanced calculations (the Xα method), similar results for the $e \leftarrow t_1$ transition have been calculated for the VO_4^{3-} ion.[4]

Electronic structures of the scheelite compounds have in common closed-shell electronic configurations as explained for the MnO_4^- ion, and their luminescence and absorption processes are exemplified in the model scheme of the MO_4^{n-} complex. Generally speaking, the luminescence of MO_4^{n-} ion is due to the spin-forbidden $^3T_1 \rightarrow {}^1A_1$ transition that is made allowed by the spin-orbit interaction. The corresponding $^3T_1 \leftarrow {}^1A_1$ absorption transition is not easily observed in the excitation spectrum due to the strong spin selection rule, and the first strong absorption band is assigned to the spin-allowed $^1T_1 \leftarrow {}^1A_1$ transition. Electronic levels and their assignments are given schematically in Figure 41 for the MO_4^{n-} ion.[2] In this model, assume an energy level scheme for the MO_4^{n-} complex in a tetrahedral environment. The energy separation between $^3T_1$ and $^3T_2$ has been estimated to be about 500 cm^{-1} for the VO_4^{3-} complex from luminescence experiments.[2] The splitting of $^3T_1$, shown in the figure, amounts to several tens of cm^{-1} and is due to the lowering of the crystal field symmetry from T_d and to the inclusion of the spin-orbit interaction. We want to understand changes of spectral properties and decay times of the luminescence from these complexes at temperatures between room temperature and 77K. Then, the

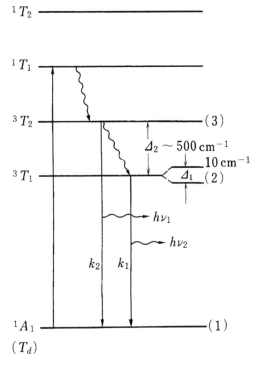

$^1 T_2$

$^1 T_1$

$^3 T_2$ — (3)

$\Delta_2 \sim 500 \, \text{cm}^{-1}$

$10 \, \text{cm}^{-1}$

$^3 T_1$ — Δ_1 (2)

$h\nu_1$

$h\nu_2$

k_2 k_1

$^1 A_1$ — (1)

(T_d)

Figure 41 Three-level energy scheme for luminescence processes of MO_4^{n-} ion in scheelite compounds. It is necessary to take into account of the splittings of the $^3T_1$ state to analyze changes of emission decay times at very low temperatures. (From Blasse, G., *Structure and Bonding*, 42, 1, 1980. With permission.)

simplified three-level model based on the two excited states ($^3T_1$, $^3T_2$) and the ground state $^1A_1$ is quite satisfactory.

Figure 41 illustrates a simple but useful model for the energy levels of ions in scheelite compounds. If the species of the central metal ion M are changed, the position of the higher excited states and the splitting of these levels will change considerably .[4] However, the ordering of the states is rigorously observed. Higher excited states due to the $t_1^5 t_2$ configuration have also been examined theoretically.[5] Excited-state absorption from the $t_1^5 e$ to the $t_1^5 t_2$ have been investigated in $CaWO_4$ crystals.[6]

2.4.2.3 Luminescence centers of VO_4^{3-} ion type

Yttrium vanadate (YVO_4) is a very useful phosphor in use for a long time. This compound does not show luminescence at room temperature; but at temperatures below 200K, it shows blue emission centered at 420 nm, as shown in Figure 42.[7] The broad band has a full width at half maximum (FWHM) of about 5000 cm^{-1}, with a decay time of several milliseconds. Even at 4K, no vibronic structure is seen. The first excitation band is located at about 330 nm, separated by 6000 cm^{-1} from the emission band. The emission and excitation are due to the $^3T_1 \leftrightarrow {}^1A_1$ transition, and the large Stokes' shift is due to the displacement between the excited- and the ground-state potential minima in the configurational coordinate model. In YVO_4, energy migration tends to favor nonradiative transition processes; because of this thermal quenching, luminescence is not observed at room temperature. However, room-temperature luminescence is observed in $YPO_4:VO_4^{3-}$ mixed crystals. Bright luminescence from VO_4^{3-} ions is commonly observed in other vanadate complexes such as $Mg_3(VO_4)_2$, $LiZnVO_4$, $LiMgPO_4:VO_4^{3-}$, and $NaCaVO_4$. If trivalent rare-earth ions such as Eu^{3+} and Dy^{3+} are incorporated into the YVO_4 host, bright luminescence

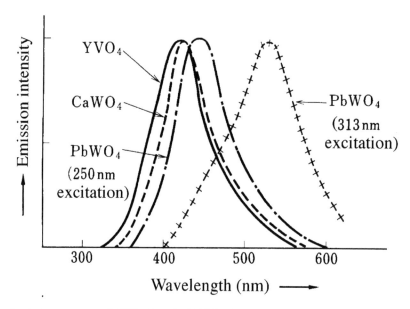

Figure 42 Emission spectra of YVO$_4$ (——), CaWO$_4$ (------), and PbWO$_4$ (—·—·—, +++++) under 250-nm excitation at 77K. Two emission bands of blue and green colors are seen in PbWO$_4$ under 313-nm excitation. (From Blasse, G., *Radiationless Processes*, DiBartolo, B., Ed., Plenum Press, New York, 1980, 287. With permission.)

due to the dopant ions is observed because of efficient energy transfer processes from the vanadate ions.

2.4.2.4 Luminescence centers of MoO$_4^{2-}$ ion type

Many molybdate phosphors containing MoO$_4^{2-}$ centers are known with a general chemical formula MMoO$_4$ (where M^{2+} = Ca^{2+}, Sr^{2+}, Cd^{2+}, Zn^{2+}, Ba^{2+}, Pb^{2+}, etc.). Luminescence properties do not depend significantly on the ion M. In PbMoO$_4$, a green emission band due to the $^3T_1 \rightarrow {}^1A_1$ transition is observed at around 520 nm at low temperatures (77K), as shown in Figure 43.[8] The FWHM of this broad band is about 3300 cm^{-1}. The lifetime is 0.1 ms, shorter than that of VO$_4^{3-}$ compounds. The degree of polarization in luminescence has been measured in some molybdate single crystals as a function of temperatures in the low-temperature region.[9] From these studies, the upper triplet state $^3T_2$ separation has been determined to be $\Delta_2 \approx 550$ cm^{-1}, with the triplet $^3T_1$ being lowest. The decay time from $^3T_2$ to $^1A_1$ is in the 1 to 0.1 µs range.

Orange-to-red luminescence is also observed in some molybdate complexes in addition to the green luminescence. In CaMoO$_4$,[10] for example, green emission appears under UV-light excitation (250–310 nm), but the orange emission at 580 nm is observable only if the excitation light of wavelengths longer than 320 nm is used. Orange emission was thus observed under excitation just below the optical bandgap. The intensity of the orange emission decreases or increases when CaMoO$_4$ is doped with Y^{3+} or Na$^+$ ions.[2] Therefore, this orange emission is ascribed to lattice defects. In the case of PbMoO$_4$,[8] red emission (centered at 620 nm) is also observed under photoexcitation at 360 nm at room temperature, as shown in Figure 43. Deep-red emission can be seen under 410-nm excitation at 77K. These bands are thought to be due to defect centers of MoO$_4^{2-}$ ions coupled to O^{2-} ion vacancies.

Thermoluminescence of MoO$_4^{2-}$ salts[11] has been investigated to clarify the electronic structure of the defect centers and impurities in these materials. Studies of the luminescence of molybdate compounds containing trivalent rare-earth ions as activators, such as Gd$_2$(MoO$_4$)$_3$:Er^{3+} (abbreviated as GMO:Er^{3+}),[12] Na$_5$Eu(MoO$_4$)$_4$, and KLa(MoO$_4$)$_2$:Er^{3+}, have been

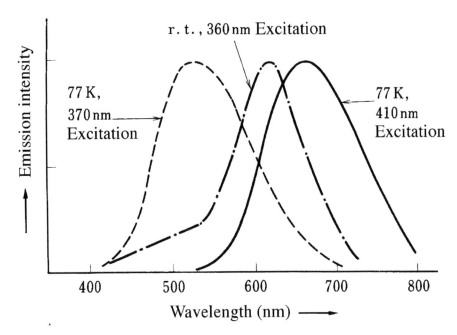

Figure 43 Spectral features of emission from PbMoO$_4$. Orange-to-red emission (—·—·—) is observed at room temperature under photoexcitation at 360 nm. This emission is compared with the deep-red one (——) at 77K under 410-nm excitation and the green one (------) under 370-nm excitation also at 77K. (From Bernhardt, H.J., *Phys. Stat. Sol. (a)*, 91, 643, 1985. With permission.)

reported. Strong, sharp luminescence due to rare-earth ions has been reported in the visible and the near-infrared spectral regions due to efficient energy transfer from the MoO$_4^{2-}$ ion.

2.4.2.5 Luminescence centers of WO$_4^{2-}$ ion type

There are many blue phosphors of interest in the metal tungstate series of complexes having the chemical formula MWO$_4$ (M^{2+} = alkaline earth metal ion). The splitting Δ_1 of the $^3T_1$ state, shown in Figure 41, is about 20 cm^{-1} for the WO$_4^{2-}$ ion center. The spin-orbit interaction in the MO$_4^{n-}$ ion becomes stronger with increasing atomic numbers of the metal; thus, VO$_4^{3-}$ < MoO$_4^{2-}$ < WO$_4^{2-}$. In order of increasing L-S coupling, the spin-forbidden $^3T_1 \leftrightarrow {}^1A_1$ transition probability is enhanced and the emission lifetime decreases correspondingly. The lifetime of the blue emission from the WO$_4^{2-}$ ion is as short as 10 µs; this is 100 times shorter than that of the VO$_4^{3-}$ ion.

A representative tungstate phosphor is CaWO$_4$; this material emits a bright blue emission in a broad band (centered at 420 nm) with FWHM of about 5000 cm^{-1}. The mixed crystal (Ca,Pb)WO$_4$ produces a very strong green emission with high quantum yields reaching 75%.[7] The blue emission spectra of CaWO$_4$ and PbWO$_4$ under 250-nm excitation are shown in Figure 42. In CaWO$_4$, there is a weak emission band at Ý530 nm superimposed on the longer wavelength tail of the blue emission. PbWO$_4$ manifests the presence of the orange band under 313-nm excitation. The orange luminescence was interpreted as being due to impurity ions or to Schottky defects. In decay time measurements of CaWO$_4$,[13] the fast decay component of about 30 µs was found at temperatures between 1.5 and 5.0K, which cannot be explained as being due to the crystal field splitting of the emitting level $^3T_1$. It has also been confirmed by studies of the emission and excitation spectra that only a single, broad blue emission band exists in pure single crystals of CdWO$_4$ and ZnWO$_4$.[14]

Ba$_2$WO$_3$F$_4$ has a crystal structure similar to MgWO$_4$ and this structure is considered to be most favorable to realize a high quantum efficiency. This is because a substitution of the F$^-$ ion for O^{2-} seems to reduce the magnitude of the phonon energy and this in turn quenches nonradiative transition processes in the [WO$_3$F]$^-$ tetrahedron. The emission process was analyzed using the configurational coordinate diagram,[15] and quantum yields of 75% have been reported in this material.[16]

2.4.2.6 Other closed-shell transition metal complex centers

There are other interesting emission centers with closed-shell configurations besides VO$_4^{3-}$, MoO$_4^{2-}$, and WO$_4^{2-}$ ions.[2] They form a series of phosphors of the [MO$_4^{n-}$] type, where M = Ti^{4+}, Cr^{6+}, Zr^{4+}, Nb^{5+}, Hf^{4+}, and Ta^{5+}. These complexes have been investigated extensively as possible new media for solid-state lasers.[17] The luminescence spectra from KVOF$_4$, K$_2$NbOF$_5$,[18] and SiO$_2$ glass:Cr^{6+} [19,20] have been reported recently as new complex centers possessing this electronic configuration.

2.4.3 Uranyl complex centers

2.4.3.1 Electronic structure

The uranyl ion is a linear triatomic ion with a chemical formula [O=U=O]$^{2+}$ ($D_{\infty h}$ symmetry). The strong, sharp line luminescence from this center has been known for more than half a century. Jørgensen and Reisfeld[21] have thoroughly discussed the historical background and theoretical aspects of the luminescence of these centers. The electronic structure of uranyl ions is particularly interesting.

As for the excited states of uranyl ions, first consider the charge-transfer process of an electron from O$_2^{4-}$ to U^{6+}. The resulting U^{5+} (5f^1) ion has the following atomic orbitals: σ_u (5f_0), π_u (5$f_{\pm 1}$), δ_u (5$f_{\pm 2}$), ϕ_u (5$f_{\pm 3}$). The electronic levels, $^2F_{7/2}$ and $^2F_{5/2}$, consist of several states having total angular momentum Ω_1 = 1/2, 3/2, 5/2, 7/2 in $D_{\infty h}$ symmetry. On the other hand, O$_2^{3-}$ has molecular orbital configurations, ($\pi_u^4 \sigma_u$) and ($\pi_u^3 \sigma_u^2$). A combination of these states gives total angular momentum Ω_2 = 1/2, 3/2. From vector coupling of Ω_1 and Ω_2,[21,22] the UO$_2^{2+}$ ion can be expressed as possessing total angular momentum of Ω = 0, 1, 2, 3, 4, 5.[21,22] On the basis of investigations of the polarized absorption and the isotope effects, Denning et al.[23] have determined that the lowest excited state is Ω = 1 ($^1\Pi_g$, $\sigma_u \delta_u$) (σ_u and δ_u stand for the electronic states of O$_2^{3-}$ and the 5f^1 ion, respectively), as shown in Figure 44. The luminescence of UO$_2^{2+}$ corresponds to a $^1\Pi_g \rightarrow {}^1\Sigma_g^+$ ($D_{\infty h}$) magnetic dipole-allowed transition. More precise molecular orbital calculations[24] and absorption experiments in Cs$_2$UO$_2$Cl$_{4-x}$Br$_x$ mixed crystals[25] confirm the ($\sigma_u \delta_u$) state as the lowest excited state. The states arising from the ($\pi_u^3 \delta_u$) configuration must be taken into account to consider the higher electronic excited states. Until the nature of the excited electronic state of Ω = 1 ($^1\Pi_g$, $\sigma_u \delta_u$) was finally clarified in 1976, the odd parity state $^1\Gamma_u$ was thought to be the lowest excited state. Therefore, reports on uranyl ions published before 1976 must be read with this reservation in mind. Figure 44 shows assignments and positions (in units of cm^{-1}) of electronic levels of uranyl ions as determined from the absorption spectra of Cs$_2$UO$_2$Cl$_4$.[23]

2.4.3.2 Luminescence spectra

A luminescence spectrum from a Cs$_2$UO$_2$Cl$_4$ single crystal at 13K, accompanied by vibronic structure due to Morita and Shoki,[26] is shown in Figure 45. The Frank-Condon pattern shows vibronic progressions of the fundamental vibrations, ν_s = 837 cm^{-1} and ν_{as} = 916 cm^{-1}, of the UO$_2^{2+}$ ion. By applying the configurational coordinate model to Cs$_2$UO$_2$Cl$_4$, the nuclear displacement ΔQ is estimated to be 0.094 Å for the two potential minima of the ^1E_g($^1\Pi_g$) excited state and the $^1A_{1g}$($^1\Sigma_g^+$) ground state in D_{4h} ($D_{\infty h}$) symmetry.[26] Emission peaks with symbol * in the figure are due to traps, and these peaks disappear above 20K. The fine structures seen in the vibronic progressions are electric dipole-allowed transitions due to coupling with odd-parity lattice vibrations.

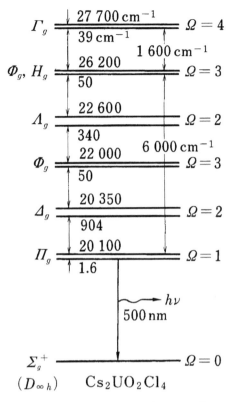

Figure 44   Energy levels and their assignments of UO_2^{2+} ion in $D_{\infty h}$ symmetry. Emission is due to the magnetic dipole-allowed $^1\Pi_g \rightarrow {}^1\Sigma_g^+$ ($D_{\infty h}$) transition. (From Denning, R.G., Snellgrove, T.R., and Woodwark, D.R., *Molec. Phys.*, 32, 419, 1976. With permission.)

Flint and Tanner[27] have investigated the luminescence of various other uranyl complexes, of the series $A_2UO_2Cl_4 \cdot nH_2O$ (A = Rb^+, Cs^+, K^+, $(CH_3)_4N^+$). They found good agreement between the molecular vibrations observed in the luminescence spectra and those reported in infrared and Raman spectra. Dynamic aspects of luminescence of $[UO_2Cl_4]^{2-}$ phosphors have also proved to be of interest. Krol[28] has investigated the decay of the luminescence of $Cs_2UO_2Cl_4$ at 1.5K under strong laser irradiation and obtained nonexponential decays; these decays are thought to be due to the presence of biexcitons associated with interionic interactions. Localization of excitons has also been reported in $CsUO_2(NO_3)_3$.[29] Excitation energy transfer to traps has been studied in $Cs_2UO_2Br_4$[30] in the temperature range between 1.5 and 25K and compared with a diffusion-limited transfer model. There are additional spectral features in uranyl compounds. For example, optically active single crystals of $NaUO_2(CH_3COO)_3$ exhibit[31] a series of complicated vibronic lines due to the presence of two emission centers, which are resolved by the difference of the degree of circular polarization in luminescence. Decay times of the luminescence of uranyl β-diketonato complexes[32] in liquid solvents have been found to be in the 1 to 500-ns range; the drastic variations are understood in terms of changes in the nonradiative rate constants correlated to the energy position of the zero-phonon emission line.

2.4.4   Platinum complex ion centers

Platinum(II) and mixed-valence platinum(II, IV) complex ions have also been investigated extensively. The best known platinum(II) complex is a yellow-green compound, barium tetracynoplatinate (II) $Ba[Pt(CN)_4] \cdot 4H_2O$ (abbreviated BCP), which possesses a linear chain

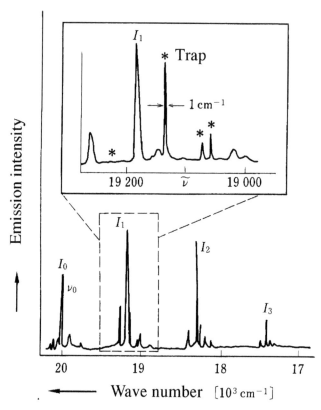

Figure 45 Emission spectra of $Cs_2UO_2Cl_4$ at 13K, showing the vibronic progressions. Inserted figure shows details of vibronic structures and trap centers are denoted by the symbol *. (From Morita, M. and Shoki, T., *J. Luminesc.*, 38/39, 678, 1987 and unpublished results. With permission.)

structure. Mixed-valence complexes such as the bromide-doped potassium tetracyano-platinate $K_2[Pt(CN)_4]$ $Br_{0.3} \cdot 3H_2O$ (KCP:Br) and Wolfram's red salt (WRS salt), i.e., $[Pt(II)L_4]$ $[Pt(IV)L_4X_2]X_4 \cdot 2H_2O$ (L = ethylamine $C_2H_5NH_2$; X = Cl, Br) have also been studied comprehensively. With reference to earlier review articles,[1,33] the electronic structure of one-dimensional platinum(II) complexes is described below and along with the unique spectroscopic character of these complexes.

2.4.4.1 $[Pt(CN)_4]^{2-}$ Complex ions

The very strong green luminescence of the anisotropic platinum(II) complex BCP has been known for more than 65 years. The $[Pt(CN)_4]^{2-}$ ion forms a flat tetragonal plane, with the Pt^{2+} being located in the center; in BCP, the Pt^{2+} forms a linear chain as shown in Figure 46. X-ray diffraction analysis[34] confirms a linear chain structure of planar $[Pt(CN)_4]^{2-}$ along the *c*-axis. Since the Pt^{2+}-to-Pt^{2+} distance in BCP is as short as 0.327 nm, the direct overlap of the $5d_z{}^2$ orbitals is possible. Monreau-Colin[35] has tabulated optical properties of these materials obtained from studies of the reflection, absorption, and emission spectra for a series of platinum complexes of the general form: $M[Pt(CN)_4] \cdot nH_2O$ (where M = Mg^{2+}, Ba^{2+}, Ca^{2+}, $Li_2{}^{2+}$ and $K_2{}^{2+}$). Large spectral shifts of the emission bands are observed with changes in the M ion. It has been established that these shifts are correlated to the Pt^{2+}-to-Pt^{2+} interaction along the one-dimensional platinum(II) chain.

From molecular orbital calculations,[36] $5d$ and $6p$ orbitals of Pt(II) can couple with the π^* orbital of the CN^- ion to form a_{1g} ($5d_z{}^2$) HOMO and $a_{2u}{}^*(6p_z)$ LUMO orbitals. The emission in BCP is due to the $a_{2u}{}^* \rightarrow a_{1g}$ transition in D_{4h} symmetry and shows a strong polarization

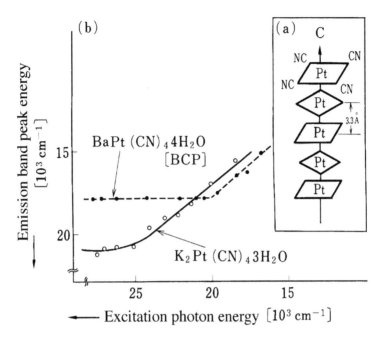

Figure 46   (a) Schematic structure of the one-dimensional platinum(II) complex Ba[Pt(CN)$_4$]·4H$_2$O (BCP). (b) Abnormal shifts of emission band peeks of platinum(II) complexes, BCP (dotted line) and K$_2$[Pt(CN)$_4$]·3H$_2$O (solid line), with changing excitation photon energy at 4.2K. (From Murata, K. and Morita, M., *Tech. Rep. Seikei Univ.*, 18, 1383, 1974 and unpublished results. With permission.)

dependence along the *c*-axis (Pt^{2+}-Pt^{2+} chain). The room-temperature emission of BCP is centered at 520 nm; the emission shows a large blue shift to 440 nm when Ba^{2+} is replaced by K$^+$. The emission process is interpreted as being due to Frenkel excitons.[37] This is because the position of the emission band shows a red shift proportional to R^{-3}, where R is the Pt-Pt distance and is a function of ionic radius of the M ion. The emission band position of BCP at 4.2K shifts continuously to lower energies as the wavelength of the excitation light is varied from the ultraviolet to visible spectral region.[38] Figure 46 shows this phenomenon, one that cannot be interpreted by present theoretical understanding. This effect is likely related to the dynamic relaxation of excitons in one-dimensional platinum(II) chains.

2.4.4.2   Other platinum complex ions

There are many additional luminescent materials containing [Pt(CN)$_4$]$^{2-}$ complexes besides those mentioned above. Luminescence of [Pt(CN)$_4$]$^{2-}$ is observed in mixed crystals[39] of K$_2$[Pt(CN)]$_6$:K$_2$[Pt(CN)$_4$] (1:1) under nitrogen laser excitation arising from intervalence transitions between Pt(IV) and Pt(II) ions. When the Ba ion in BCP is replaced by rare-earth(III) ions, new complexes Ln$_2$[Pt(CN)$_4$]$_3$·nH$_2$O (Ln = Eu^{3+}, Sm^{3+}, Er^{3+}, etc.) can be formed.[33] Sharp line luminescence of Ln(III) ions is normally observed in these materials due to energy transfer from the [Pt(CN)$_4$]$^{2-}$ complex. Extensive investigations have been conducted at low temperature using high pressure and strong magnetic fields.[33,40]

A mixed-valence platinum complex is KCP:Br, with a valency equal to 2.3. This compound is metallic in appearance and does not fluoresce. A typical example of a luminescent mixed complex is the WRS salt mentioned previously; it contains Pt(II) and Pt(IV) ions. The structure of the WRS salt consists of an alternative stack of a tetragonal plane unit [Pt(II)L$_4$] and an octahedral unit with six coordination atoms [Pt(IV)L$_4$X$_2$]. These units form a quasi-one-dimensional linear chain of [X-Pt(IV)-X--Pt(II)--X] bridged by halogens. The

WRS complex has been known since the late 19th century; it was determined by Day[37] to be a typical low-dimensional compound. Tanino and Kobayashi[41] first reported resonance Raman scattering and NIR luminescence at about 1 μm in WRS salts at 4.2K. A similar mixed-valence complex, $[Pt(en)_2][Pt(en)_2I_2](ClO_4)_4$ (en = $NH_2CH_2CH_2NH_2$), was found to show a luminescence band at 1 μm, with a lifetime of about 200 ps at 2K.[42] The origin of this luminescence band is ascribed to self-trapped-excitonic (STE) states.[42]

2.4.5   Other complex ion centers

2.4.5.1 WO_6^{6-} Ion

In Bi_2WO_6,[43] the emission center is identifiable as WO_6^{6-} in a cubic crystal field. Red emission is observed at 4.2K, with a peak at 600 nm and a FWHM of about 1500 cm^{-1}. The excitation spectrum consists of a band at 390 nm with an apparent Stokes' shift of 9000 cm^{-1}. According to self-consistent field molecular orbital calculations,[44] the emitting levels in WO_6^{6-} are due to two $^3T_{1u}$ states. By employing Figure 41, the emission and absorption can be assigned to $^3T_{1u} \rightarrow A_{1g}$ and $^1T_{1u} \leftrightarrow {}^1A_{1g}$ transitions, respectively. Emission centers of the WO_6^{6-} ion were also reported in many compounds with the perovskite structure A_2BWO_6 (A = Ca^{2+}, Sr^{2+}, Ba^{2+}; B = Mg^{2+}, Ca^{2+}, Sr^{2+}, Ba^{2+}). Wolf and Kammler-Sack[45] reported infrared emission of rare-earth ions incorporated into a very complicated compound $18R-Ba_6Bi_2W_3O_{18}$. In this case, there are three WO_6^{6-} ion sites in the compound with an hexagonal closed-packed polymorphic structure. The emission spectra consist of two bands at 21700 and 17000 cm^{-1} due to two 6c sites and one 3a site, respectively. The corresponding excitation bands are at 36000 cm^{-1} (6c) and 29000 cm^{-1} (3a), respectively. The luminescence of WO_6^{6-} ions can also be seen in other materials such as Li_6WO_6, 12R-$Ba_2La_2MgW_2O_{12}$, and $Ca_3La_2W_2O_{12}$.

2.4.5.2   Perspective of other interesting centers

The above-mentioned WO_6^{6-} luminescence center is one of the closed-shell transition metal complex ions, generally expressed as $[MO_6]^{n-}$ (where M = Ti, Mo, Nb, Zr, Ta, and W). Two papers[2,46] on the luminescence properties of MoO_4^{2-} and MoO_6^{6-} complexes have been published. Recently, luminescence from a europium octamolybdate polymer, $Eu_2(H_2O)_{12}[Mo_8O_{27}]$ $6H_2O$[47] and the picosecond decay of the transient absorbance of $[W_{10}O_{32}]^{4-}$ in acetonitrile[48] have been reported. The luminescence of uranate (UO_6^{6-}) centers in solids have been reviewed by Bleijenberg.[49]

Thus far, this discussion of luminescence centers of complex ions focused on practical phosphors. However, under the category of complex ions, a more general survey is possible. Complex compounds consist of a central metal ion and surrounding anions or organic ligands. In these compounds, there are—in principle—four possible luminescence processes that originate from the central metal ion, from the ligand, from ligand-to-metal charge-transfer (LMCT), and from metal-to-ligand charge-transfer (MLCT) transitions. Due to these different transition processes, the luminescence from complex ions can either be sharp or broad, and can occur in a broad spectral region. Uranyl complexes luminescing of green-yellow color are examples of central metal ion transitions. Eu(III) β-diketonato complex, a typical NMR shift reagent, also shows bright and sharp red luminescence due to the central Eu(III) ion. For more than half a century, the luminescence of the Zinc(II) 8-hydroxyquinolinato complex has been shown to be due to the aromatic organic ligands. Emission transitions due to the LMCT scheme is found in scheelite compounds. Phosphorescence due to MLCT transitions is predominant in complexes such as ruthenium(II) tris-bipyridyl ([$Ru(bpy)_3$]$^{2+}$), metal-phthalocyanines (e.g., Cu-Pc, a famous pigment), and met-alloporphyrins (e.g., Mg-TPP). The latter two complexes are usually considered as organic phosphors because of 16-membered π-ring structures.

In the future, one will be able to design new phosphors of complex ion types that can be excited by various excitation sources such as high electron beams, X-ray lasers, and NIR-laser diodes. Phosphors of complex ions will continue to play a useful role in luminescence applications.

References

1. Morita, M., MoO_4^{2-}, WO_4^{2-} compounds, and one-dimensional compounds, in *Hikaribussei Handbook (Handbook of Optical Properties of Solids)*, Shionoya, S., Toyozawa, Y., Koda, T., and Kukimoto, H., Eds., Asakura Shoten, Tokyo, 1984, chap. 2. 12. 6 and 2. 19. 2. (in Japanese).
2. Blasse, G., *Structure and Bonding*, 42, 1, 1980.
3. Ballhausen, C.J. and Liehr, A.D., *J. Mol. Spectrosc.*, 4, 190, 1960.
4. Ziegler, T., Rank, A., and Baerends, E.J., *Chem. Phys.*, 16, 209, 1976.
5. Kebabcioglu, R. and Mueller, A., *Chem. Phys. Lett.*, 8, 59, 1971.
6. Koepke, C., Wojtowica, A.J., and Lempicki, A., *J. Luminesc.*, 54, 345, 1993.
7. Blasse, G., Radiationless processes in luminescent materials, in *Radiationless Processes*, DiBartolo, B., Ed., Plenum Press, New York, 1980, 287.
8. Bernhardt, H.J., *Phys. Stat. Sol.(a)*, 91, 643, 1985.
9. Rent, E.G., *Opt. Spectrosc. (USSR)*, 57, 90, 1985.
10. Groenink, J.A., Hakfoort, C., and Blasse, G., *Phys. Stat. Sol.(a)*, 54, 329, 1979.
11. Böhm, M., Erb, O., and Scharman, A., *J. Luminesc.*, 33, 315, 1985.
12. Herren, M. and Morita, M., *J. Luminesc.*, 66/67, 268, 1996.
13. Blasse, G. and Bokkers, G., *J. Solid. State. Chem.*, 49, 126, 1983.
14. Shirakawa, Y., Takahara, T., and Nishimura, T., *Tech. Digest, Phosphor Res. Soc. Meeting*, 206, 15, 1985.
15. Tews, W., Herzog, G., and Roth, I., *Z. Phys. Chem. Leipzig*, 266, 989, 1985.
16. Blasse, G., Verhaar, H.C.G., Lammers, M.J.J., Wingelfeld, G., Hoppe, R., and De Maayer, P., *J. Luminesc.*, 29, 497, 1984.
17. Koepke, C., Wojtowicz, A.J., and Lempicki, A., *IEEE J. Quant. Elec.*, 31, 1554, 1995.
18. Hazenkamp, M.F., Strijbosch, A.W.P.M., and Blasse, G., *J. Solid State Chem.*, 97, 115, 1992.
19. Herren, M., Nishiuchi, H., and Morita, M., *J. Chem. Phys.*, 101, 4461, 1994.
20. Herren, M., Yamanaka, K., and Morita, M., *Tech. Rep. Seikei Univ.*, 32, 61, 1995.
21. Jørgensen, C.K. and Reisfeld, R., *Structure and Bonding*, 50, 122, 1982.
22. Denning, R.G., Foster, D.N.P., Snellgrove, T.R., and Woodwark, D.R., *Molec. Phys.*, 37, 1089 and 1109, 1979.
23. Denning, R.G., Snellgrove, T.R., and Woodwark, D.R., *Molec. Phys.*, 32, 419, 1976.
24. Dekock, R.L., Baerends, E.J., Boerrigter, P.M., and Snijders, J.G., *Chem. Phys. Lett.*, 105, 308, 1984.
25. Denning, R.G., Norris, J.O.W., and Laing, P.J., *Molec. Phys.*, 54, 713, 1985.
26. Morita, M. and Shoki, T., *J. Luminesc.*, 38/39, 678, 1987 and unpublished results.
27. Flint, S.D. and Tanner, P.A., *Molec. Phys.*, 44, 411, 1981.
28. Krol, D.M., *Chem. Phys. Lett.*, 74, 515, 1980.
29. Thorne, J.R.G. and Denning, R.G., *Molec. Phys.*, 54, 701, 1985.
30. Krol, D.M. and Roos, A., *Phys. Rev.*, 23, 2135, 1981.
31. Murata, K. and Morita, M., *J. Luminesc.*, 29, 381, 1984.
32. Yayamura, T., Iwata, S., Iwamura, S., and Tomiyasu, H., *J. Chem. Soc. Faraday Trans.*, 90, 3253, 1994.
33. Gliemann, G. and Yersin, H., *Structure and Bonding*, 62, 89, 1985.
34. Krogmann, K., *Angew. Chem.*, 81, 10, 1969.
35. Monreau-Colin, M.L., *Structure and Bonding*, 10, 167, 1972.
36. Moncuit, S. and Poulet, H., *J. Phys. Radium*, 23, 353, 1962.
37. Day, P., Collective states in single and mixed valence metal chain compounds, in *Chemistry and Physics of One-Dimensional Metals*, Keller, H.J., Ed., NATO-ASI Series B25, Plenum Press, New York, 1976, 197.
38. Murata, K. and Morita, M., *Tech. Rep. Seikei Univ.*, 18, 1383, 1974 and unpublished results.

39. Wiswarath, A.K., Smith, W.L., and Patterson, H.H., *Chem. Phys. Lett.*, 87, 612, 1982.
40. Yersin, H. and Stock, M., *J. Chem. Phys.*, 76, 2136, 1982.
41. Tanino, H. and Kobayashi, K., *J. Phys. Soc. Japan*, 52, 1446, 1983.
42. Wada, Y., Lemmer, U., Göbel, E.O., Yamashita, N., and Toriumi, K., *Phys. Rev.*, 55, 8276, 1995.
43. Blasse, G. and Dirkson, G.J., *Chem. Phys. Lett.*, 85, 150, 1982.
44. Van Oosternhout, A.B., *J. Chem. Phys.*, 67, 2412, 1977.
45. Wolf, D. and Kemmler-Sack, S., *Phys. Stat. Sol. (a)*, 86, 685, 1984.
46. Wiegel, K. and Blasse, G., *J. Solid State Chem.*, 99, 388, 1992.
47. Yamase, T. and Naruke, H., *J. Chem. Soc. Dalton Trans.*, 1991, 285.
48. Duncan, D.C., Netzel, T.L., and Hill, C.L., *Inorg. Chem.*, 34, 4640, 1995.
49. Bleijenberg, K.C., *Structure and Bonding*, 50, 97, 1983.

Principal phosphor materials and their optical properties

Shigeo Shionoya

Contents

2.5   Ia-VIIb compounds

2.5.1   Introduction

Ia-VIIb compounds—that is, alkali halides—are prototypical colorless ionic crystals. Their crystal structure is of the rock-salt type except for three compounds (CsCl, CsBr, and CsI) that possess the cesium chloride structure. Melting points are generally in the 620–990°C range, which is relatively low.

Research on the luminescence of alkali halides has a long history. Since the 1920s, luminescence studies on ns^2-type (Tl$^+$-type) ions incorporated in alkali halides have been actively pursued (See 2.1). In the basic studies of those early days, alkali halides were used as hosts for various luminescence centers, because as nearly ideal ionic crystals theoretical treatments of the observations were possible. Since the 1940s, the optical spectroscopy, including luminescence of color centers, has become an active area of study. It was discovered in the late 1950s that excitons in pure alkali halides are self-trapped and can themselves produce luminescence.

Although alkali halides are important from the point of view of basic research as mentioned above, their luminescence is rarely utilized in practical applications. This is because alkali halides are water soluble and have low melting points, so that they are unsuitable as hosts of practical phosphors. Alkali halide phosphors presently in common use include $NaI:Tl^+$ and $CsI:Na^+$ only as discussed below.

2.5.2 Intrinsic optical properties

2.5.2.1 Band structure and exciton

All alkali halides have band structures of the direct transition type. Both the bottom of the conduction band and the top of the valence band are located at the k = 0 point (Γ point) in **k**-space. The bandgap energies E_g are as follows. The largest is 13.6 eV for LiF, and the smallest is 6.30 eV for KI, with RbI and CsI having almost the same value. For NaCl, a representative alkali halide, E_g is 8.77 eV. E_g decreases with increasing atomic number of the cations or anions making up the alkali halides.

A little below the fundamental absorption edge, sharp absorption lines due to excitons are observed. The valence band is composed of *p* electrons of the halogen ions, and it is split into two compounds, with the inner quantum numbers representing total angular momentum j = 3/2 and 1/2. Corresponding to this splitting, two sharp exciton absorption lines are observed. The binding energy of excitons is 1.5 eV (NaF) at its largest and 0.28 eV (NaI) at its smallest; it is 0.81 eV in NaCl.

2.5.2.2 Self-trapping of excitons and intrinsic luminescence[1]

Excitons in all alkali halides except iodides do not move around in the crystal, unlike excitons in IIb-VIb and IIIb-Vb compounds; these excitons are self-trapped immediately after their creation as a result of very strong electron-lattice coupling. Self-trapped excitons emit luminescence called *intrinsic luminescence*. In iodides, excitons can move freely for some distance before becoming self-trapped and emitting their intrinsic luminescence.

Before discussing the self-trapping of excitons, let us consider the self-trapping of positive holes. In alkali halides, holes do not move freely, but are self-trapped and form V_K centers. As shown in Figure 47, the V_K center is a state in which two nearest-neighbor anions are attracted to each other by trapping a positive hole between them, so that it takes the form of a molecular ion denoted as X_2^-. The self-trapped exciton is a state in which an additional electron is trapped by the V_K center.

The spectral position of the intrinsic luminescence is shifted considerably toward lower energies from the exciton absorption. This is because an exciton undergoes a large lattice relaxation, emitting phonons to reach a self-trapped state. Emission spectra are composed of two broad bands in most cases. The luminescence of the short-wavelength band, called σ-luminescence, is polarized parallel, while that of the long-wavelength band, called π-luminescence, is polarized perpendicular to the molecular axis of V_K centers. In NaCl, the peaks of these two types of luminescence are σ: 5.47 eV and π: 3.47 eV. σ-luminescence is due to an allowed transition from the singlet excited state, while π-luminescence is due to a forbidden transition from the triplet excited state. As a result, the decay time of the former is short (about 10 ns), while that of the latter is relatively long (about 100 μs).

2.5.3 Color centers[2]

In alkali halides, lattice defects that trap an electron or a hole have absorption bands in the visible region, and hence color the host crystals. Therefore, such defects are called *color centers*. Figure 47(a) illustrates the principal electron-trapping centers F, F_A, F′, M, and M′. Principal hole-trapping centers V_K, V_{KA}, H, and H_A are illustrated in Figure 47(b). Most of color centers emit luminescence. The F center has an electronic energy structure analogous to the hydrogen atom. It shows strong absorption and emission due to the $s \leftrightarrow p$ transition.

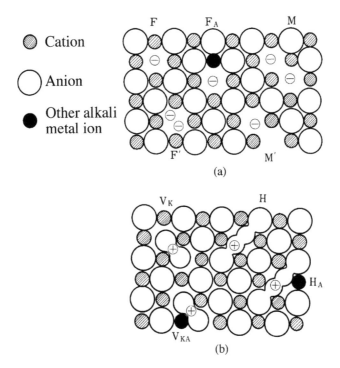

Figure 47 (a) Principal electron trapping centers and (b) principal positive hole trapping centers in alkali halides.

In NaCl, the absorption is at 2.75 eV and the emission at 0.98 eV. Alkali halide crystals containing some kind of color centers, typically an F_A center, are used as materials for tunable solid-state lasers operating in the near-infrared region.

2.5.4 Luminescence centers of ns^2-type ions

For some time, ns^2-type (Tl$^+$-type) ion centers in alkali halides have been investigated in detail from both experimental and theoretical points of view as being a typical example of an impurity center in ionic crystals (See 2.1). Almost all ns^2-type ions, i.e., Ga$^+$, In$^+$, Sn^{2+}, Pb^{2+}, Sb^{3+}, Bi^{3+}, Cu$^-$, Ag$^-$, and Au$^-$, have been studied; Tl$^+$ has been studied in significant detail. For these ion centers, the absorption and emission due to the $s^2 \leftrightarrow sp$ transition, their spectral shapes, the polarization correlation between the absorption and the emission, and the dynamical Jahn-Teller effect in excited states due to electron-lattice interactions have been investigated thoroughly. The range of phenomena have been well elucidated in the literature and are an example of the remarkable contributions that optical spectroscopic studies have made to our understanding of impurity centers.

However, the luminescence in alkali halides is almost worthless from a practical point of view of application, so that further detailed description will not be provided here. The only example of practical use of these materials is NaI:Tl$^+$ and CsI:Tl$^+$ single-crystal phosphors (near-ultraviolet to blue emitting), which have been used as scintillators. For alkali halides such as NaI and CsI, it is easy to grow large single crystals so that they are suitable for these applications in particle and high-energy radiation detection.

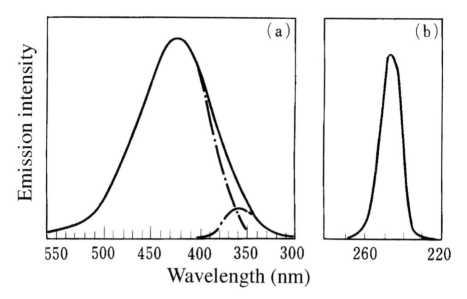

Figure 48 (a) Emission spectrum (300K) and (b) excitation spectrum of CsI:Na⁺. (From Hsu, O.L. and Bates, C.W., *Phys. Rev.*, B15, 5821, 1977. With permission.)

2.5.5 Luminescence of isoelectronic traps

An example of an isoelectronic trap (See 1.4.4) showing luminescence is CsI:Na⁺. It emits blue luminescence with high efficiency when excited by high-energy radiation.[3] Presently, CsI:Na⁺ films prepared by a vapor deposition method are used for X-ray image intensifiers. The concentration of Na⁺ is very low, 6 ppm being the optimum value.

Emission and excitation spectra of CsI:Na⁺ are shown in Figure 48.[4] The peak of the excitation spectrum agrees well with the calculated value of an exciton bound to an isoelectronic trap Na⁺. The luminescence is considered to arise from the relaxed exciton state of this bound exciton, which is assumed to be a V_{KA} center trapping an electron. If so, the luminescence should be polarized parallel or perpendicular to the molecular axis of the V_{KA} center. However, no polarization was observed in experiments, leaving the structure of the emitting state undetermined as yet.

References

1. Review articles:
   (a) Song, K.S. and Williams, R.T., *Self-Trapped Excitons*, (Springer Series in Solid-State Sciences 105), Springer-Verlag, Berlin, 1993.
   (b) Kan'no K., Tanaka, K., and Hayashi, T., *Rev. Solid State Sci.*, 4(2/3), 383, 1990.
2. Review articles:
   (a) Schulman, J.H. and Compton, W.D., *Color Centers in Solids*, Pergamon Press, Oxford, 1963.
   (b) Fowler, W.B., Ed., *Physics of Color Centers*, Academic Press, New York, 1968.
3. Brinckmann, P., *Phys. Lett.*, 15, 305, 1965.
4. Hsu, O.L. and Bates, C.W., *Phys. Rev.*, B15, 5821, 1977.

chapter two — section six

Principal phosphor materials and their optical properties

Hajime Yamamoto

Contents

2.6   IIa-VIb compounds

2.6.1   Introduction

Phosphors based on alkaline earth chalcogenides, mostly sulfides or selenides, are one of the oldest classes of phosphors. Many investigations were made on these phosphors from the end of 19th century to the beginning of 20th century, particularly by Lenard and co-workers as can be seen in Reference 1. For this reason, these phosphors are still called *Lenard phosphors*.

    Even with their long history, progress in understanding the fundamental physical properties for these phosphors has been quite slow. There are good reasons for this: these materials are hygroscopic and produce toxic H_2S or H_2Se when placed in contact

Table 13   The Lattice Constant, Dielectric Constants, and Phonon Frequencies
of IIa-VIb Compounds[17]

Compounds	Lattice constant (nm)	Dielectric constants		Phonon frequency (cm^{-1})	
		ε_0	ε_∞	ω_{TO}	ω_{LO}
MgO[a]	0.4204	9.64	2.94[d]	401	725
CaO	0.4812	11.1[b], 11.6[c]	3.33[b], 3.27[c]	295[b], 311[c]	577[b], 585[c]
CaS	0.5697	9.3	4.15[d]	229	342
CaSe	0.5927	7.8	4.52[d]	168	220
SrO	0.5160	13.1[b], 14.7[c]	3.46[d]	231[b], 229[c]	487[b], 472[c]
SrS	0.6019	9.4	4.06[d]	185	282
SrSe	0.6237	8.5	4.24[d]	141	201
BaO	0.5524	32.8[c]	3.61[b], 3.56[d]	146[c]	440[c]
BaS	0.6384	11.3	4.21[d]	150	246
BaSe	0.6600	10.7	4.41[d]	100	156

Note: The notation ε_0 and ε_∞ indicate static and optical dielectric constant, and ω_{TO} and ω_{LO} the
frequency of the transverse and longitudinal optical phonon, respectively.

[a] From Reference 18.

[b] From Reference 19.

[c] From Reference 20.

[d] From Reference 21.

with moisture. Further, their luminescence properties are sensitive to impurities and nonstoichiometry. Such problems make it difficult to obtain controlled reproducibility in performance when technologies for ambiance control and material purification were insufficient.

In the 1930s and 1940s, research on this family of materials was carried out actively to meet demands for military uses, mostly for the detection of infrared light by the photostimulation effect. After this period, these materials were ignored for many years, until around 1970 when Lehmann[2-6] demonstrated that the alkaline earth chalcogenides could be synthesized reproducibly. He also showed that an exceptionally large number of activators can be introduced into CaS, many of which exhibit high luminescence efficiency. He showed that these features of CaS phosphors were attractive to applications and seem to compensate for the drawback caused by their hygroscopic nature. Accordingly, Lehmann's work revived research interest in these materials and this activity has continued through the years.

In the 1980s, some attempts were made to apply CaS phosphors to CRTs.[7-10] Also in this period, single crystals were grown for many of the IIa-VIb compounds.[11-12] The band structure was investigated and the basic optical parameters were obtained in these single crystals.[11-14,17]

2.6.2   Fundamental physical properties

2.6.2.1   Crystal structures

Most of the IIa-VIb family have NaCl-type structures, except for MgTe, which crystallizes in the zinc-blende structure, and BeO, which favors the wurtzite lattice.[15] The lattice parameters are shown in Table 13.

The compounds in the NaCl-type structure can form solid solutions in a wide range of composition. As a consequence, the emission color can be varied by changing the host composition as well as the activator species and concentration. Such diversity is one of the advantageous features of the IIa-VIb compounds.

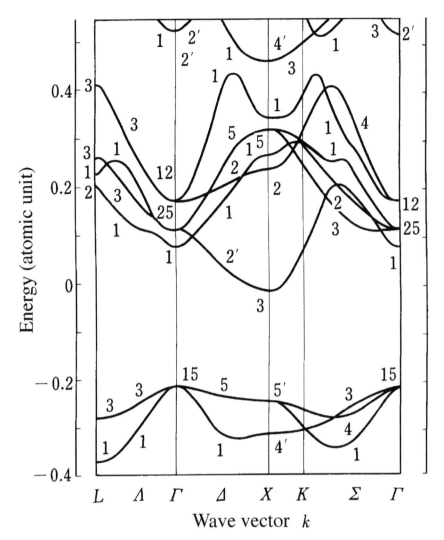

Figure 49 The band structure of SrS calculated by the self-consistent APW (augmented plane wave) method. The energy scale is shown by the atomic unit (= Rydberg constant × 2 = 13.6 × 2 eV). Note that energy values obtained in the figure (e.g., bandgap energies) do not necessarily agree with experimentally obtained values. (From Hasegawa, A. and Yanase, A., *J. Phys. C*, 13, 1995, 1980. With permission.)

2.6.2.2   Band structure

The band structure calculated for SrS is shown by Figure 49.[16] The first thing one should notice is that the absorption edge is of the indirect transition type,[13,16] with the valence band maximum at the Γ point ($\mathbf{k}$ = 000), and the conduction band minimum at the X point ($\mathbf{k}$ = 100). However, the optical transition corresponding to this edge is forbidden because phonons having the momentum and parity required to induce a phonon-assisted direct transition are not available in the NaCl-type structure. The lowest direct transition occurs at the X point of the valence band and not at the Γ point.

In SrS, the conduction band is composed mainly of 5s and 4d orbitals of the Sr atom, and the Γ point has mainly 5s character while the X point possesses 4d character. The 5s and 4d orbitals are close to each other in energy, allowing the X point to be located at lower energy than the Γ point when the SrS crystal is formed. The same feature is also

Table 14 The Observed Bandgap Energies of IIa-VIb Compounds (in eV)

| Compound | Bandgap energies at 2K[a] | | | Absorption edge at 300K[c] |
	$E_g(X_c - \Gamma_v)$[b]	Gap at X point	Gap at Γ point	
CaO	—	6.875	—	—
CaS	4.434	5.343	5.80	4.20
CaSe	3.85	4.898	—	—
SrO	—	5.793	6.08	—
SrS	4.32	4.831	5.387	4.12
SrSe	3.813	4.475	4.570	3.73
BaO	—	3.985	8.3	—
BaS	3.806	3.941	5.229	3.49
BaSe	3.421	3.658	4.556	3.20

[a] Data from Kaneko, Y. and Koda, T., *J. Crystal Growth*, 86, 72, 1988.

[b] The gap between X point in the conduction band and Γ point in the valence band.

[c] Data from Morimoto, K., Masters Thesis, The University of Tokyo, 1982 (in Japanese).

found in Ca and Ba chalcogenides. In contrast to this, Mg atoms have $3d$ orbitals that lie some 40 eV higher than $3s$; as a consequence, MgO has a direct bandgap. The bandgap energies measured at 2K[13] and the absorption edge energies at 300K[17] for IIa-VIb compounds are shown in Table 14. The exciton binding energies of about 40 to 70 meV are obtained at Γ point.[13]

The spectral shape of optical absorption near the edge follows Urbach's rule quite well.[13,22] That is, the absorption coefficient α is expressed as a function of photon energy E by the following formula.

$$\alpha(E) = \alpha_0 \exp\{-\sigma(E_0 - E)/kT\} \tag{28}$$

where α_0, σ, and E_0 are material constants. For SrS, $\alpha_0 = 4 \times 10^7$ cm^{-1}, $\sigma = 1.07$, and $E_0 = 4.6$ eV, which is nearly equal to the lowest exciton energy at the X point.[12] This fact indicates that the absorption tail at room temperature appears as a result of interaction between excitons and phonons at the X point. The absorption edge due to the forbidden indirect transition is masked by the absorption tail of the direct transition. At sufficiently low temperatures, however, the indirect absorption appears and reveals a spectral shape characterized by $\alpha \propto (E_0 - E)^2$.

2.6.2.3 Phonon energies and dielectric constants

The dielectric constants and optical phonon energies obtained from infrared reflection spectra of single crystals[11] are given in Table 13.

2.6.3 Overview of activators

Activators that can be introduced into CaS and their main luminescence properties are summarized in Tables 15 and 16.[4] Elements not appearing in these tables were found to be nonluminescent.[4] However, radioactive elements except for U and Th, the platinum-group elements, and Hg and Tl have not been examined. As for the description of luminescence properties, it is noted that the tables present only typical examples since the luminescence spectra and decay characteristics depend on the activator concentration.

As an example of the activators listed in Table 15, the luminescence and absorption spectra of Bi^{3+} in CaS due to $6s^2 \rightarrow 6s6p$ transitions are shown in Figure 50.[6]

Table 15 Activators and Coactivators in CaS and Luminescence Properties

Activators	Coactivators[a]	Luminescence color	Luminescence spectrum[b]	Peak (eV)	Type of decay curve	Decay time constant[c]
O	Nothing	Bluish-green	Band	2.53	Exponential	6.5 μs
P	**Cl**, Br	Yellow	Band	2.13	Hyperbolic	~500 μs
Sc	**Cl**, Br, Li	Yellowish-green	Band	2.18	—	—
Mn	Nothing	Yellow	Narrow band	2.10	Exponential	4 ms
Ni	Cu, Ag	Red to IR	Broad band	—	—	—
Cu	**F, Li, Na**, Rb, P, Y, As	Violet to blue	Two bands	2.10	Hyperbolic	50 μs
Ga	Nothing, or Cu, Ag	Orange, red, and yellow	Broad band	—	—	—
As	F, Cl, Br	Yellowish-orange	Band	2.00	—	—
Y	**F, Cl**, Br	Bluish-white	Broad band	2.8	Hyperbolic	~200 μs
Ag	Cl, Br, Li, Na	Violet	Band	—	Hyperbolic	~1 ms
Cd	Nothing	UV to IR	Very broad band	—	—	—
In	Na, K	Orange	Broad band	—	—	—
Sn	**F**, Cl, Br	Green	Band	2.3	Hyperbolic	~500 μs
Sb	Nothing, or **Li, Na**, K	Yellowish-green	Band	2.27	Exponential	0.8 μs
La	Cl, Br, I	Bluish-white	Broad band	2.55	Hyperbolic	~200 μs
Au	Li, K, Cl, I	Blue to bluish-green	Two bands	—	Hyperbolic	~10 μs
Pb	**F, Cl, Br**, I, P, As, Li	UV	Narrow band	3.40	Hyperbolic	~1 μs
Bi	**Li, Na, K**, Rb	Blue	Narrow band	2.77	Hyperbolic	~1 μs

Note: The rare-earth elements are listed in Table 16.

[a] Efficient co-activators are shown in bold letters.

[b] A spectrum changes depending on a co-activator.

[c] The period when the luminescence intensity falls to $1/e$ times the initial value.

From Lehmann, W., *J. Luminesc.*, 5, 87, 1972. With permission.

Table 16 Rare Earth Activators in CaS and Luminescence Properties

Ion	Luminescence color	Luminescence spectrum	Type of decay curve	Decay time constant[b]
Ce^{3+}	Green	Two bands, peaked at 2.10, 2.37 eV	Hyperbolic	~1 μs
Pr^{3+}	Pink to green	Lines, green, red, and IR	Green: exponential	260 μs
(Nd^{3+})[a]	—	—	—	—
Sm^{3+}	Yellow	Lines, yellow, red, and IR	Yellow: exponential	5 μs
Sm^{2+}	Deep red (low temperature)	Lines, green, red, and IR	—	—
Eu^{2+}	Red	Narrow band, peaked at 1.90 eV	Hyperbolic	~1 μs
Gd^{3+}	—	Lines, UV	Exponential	1.5 ms
Tb^{3+}	Green	Lines, UV to red	Green: exponential	1.8 ms
Dy^{3+}	Yellow & bluish-green	Lines, yellow, bluish-green, and IR	Yellow: $(1+t/\tau)^{-1}$	150 μs
Ho^{3+}	Greenish-white	Lines, blue to IR	Green: $(1+t/\tau)^{-1}$	150 μs
Er^{3+}	Green	Lines, UV, green, and IR	Green: $(1+t/\tau)^{-1}$	370 μs
Tm^{3+}	Blue with some red	Lines, blue and red	Blue: exponential	1.05 ms
Yb^{3+}	—	Lines, IR	—	—
Yb^{2+}	Deep red	Band, peaked at 1.66 eV	Hyperbolic	~10 μs

[a] Luminescence of Nd^{3+} is not identified.
[b] The period when the luminescence intensity falls to 1/e times the initial value. For the type expressed by $(1+t/\tau)^{-1}$, the time constant means τ.
From Lehmann, W., *J. Luminesc.*, 5, 87, 1972. With permission.

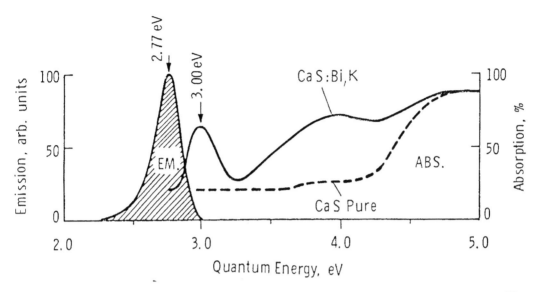

Figure 50 Luminescence and absorption spectra of CaS:Bi^{3+} (0.01%),K$^+$ at room temperature. The hatched zone indicates the luminescence spectrum. The solid line shows the absorption spectrum of CaS:Bi^{3+},K$^+$, and the broken line that of pure CaS. (From Lehmann, W., *Gordon Research Conference Report*, July 1971. With permission.)

One significant difference between IIa-VIb compounds and IIb-VIb compounds is that the concept of the donor or acceptor in the latter is not applicable in the former. For example, donor-acceptor pair luminescence is not observed for IIa-VIb compounds doped with (Cu$^+$, Cl$^-$) or (Cu$^+$, Al^{3+}) pairs. Another example is that alkali ions act as "co-activators" of Ag or Au activators, just as halogen ions do. As this example indicates, many activator/co-activator combinations violate the charge compensation rule in an ionic crystal. In fact, the co-activators do not play a donor role similar to that of halogen ions in IIb-VIb compounds; instead, these ions help the activator diffuse into the host lattice by creating lattice defects. These observations are presumably related to the stronger ionicity of IIa-VIb compounds compared to IIb-VIb compounds. As is the case with alkali halides, luminescence centers are localized in IIa-VIb compounds.

As described previously, many materials of this group have high luminescence efficiency. Cathodoluminescence efficiencies for various phosphors are shown in Table 17.[5,7] Above all, CaS:Ce^{3+} shows an efficiency nearly as high as ZnS:Ag,Cl or ZnS:Cu,Al, which are the most efficient cathode-ray phosphors. The efficiencies of CaS:Eu^{2+},Ce^{3+} and MgS:Eu^{2+} are much higher than the efficiency of Y$_2$O$_2$S:Eu^{3+} (about 13% in energy efficiency) and very good red-emitting phosphors. (Here, Ce^{3+} acts as a sensitizer of Eu^{2+} luminescence.) Luminance of CaS:Eu^{2+},Ce^{3+} is, however, only 80% that of Y$_2$O$_2$S:Eu^{3+}, because the emission peak of CaS:Eu^{2+},Ce^{3+} is at 650 nm, while that of Y$_2$O$_2$S:Eu^{3+} is at 627 nm.

It should be noted that the data listed in Table 17 were measured at low electron-beam current density. It was found that for the Eu^{2+}, Mn^{2+}, and Ce^{3+} emissions, the efficiency decreases with increasing current density. This saturation in luminescence intensity can be reduced by increasing the activator concentration, but not eliminated; the reason for the saturation is not clear as yet.

As shown in Table 18, the luminescence peak shifts to shorter wavelengths in Eu^{2+} and Ce^{3+} (*f-d* transitions) and in Mn^{2+} (*d-d* transition) when the host lattice is varied from CaS to SrS to BaS. This shift is reasonable from theoretical points of view because the

Table 17   Cathodoluminescence Efficiency of Alkali Earth Chalcogenide Phosphors

Phosphors	Energy efficiency (%)	Luminescence color
$MgS:Eu^{2+}$	16	Orange-red
$CaS:Mn^{2+}$	16	Yellow
$CaS:Cu$	18	Blue-violet
$CaS:Sb$	18	Yellow-green
$CaS:Ce^{3+}$	22	Green
$CaS:Eu^{2+}$	10	Red
$CaS:Eu^{2+},Ce^{3+}$	16	Red
$CaS:Sm^{3+}$	12 (+IR)	Yellow
$CaS:Pb^{2+}$	17	UV
$CaO:Mn^{2+}$	5	Yellow
$CaO:Pb^{2+}$	10	UV

Note: The efficiency was measured at room temperature relative to a standard material. For $MgS:Eu^{2+}$, excitation was made at an accelerating voltage of 18 kV and a current density of 10^{-7} A/cm². The standard was $Y_2O_2S:Eu^{3+}$. For other phosphors, excitation was made at 8 kV and 10^{-6} A/cm² or less. The standard was P-1, P-22, or $MgWO_4$. The measurement error is ±10%.

From Lehmann, W., *J. Electrochem. Soc.*, 118, 1164, 1971; Lehmann, W., *Gordon Research Conference Report*, July 1971; Kasano, H., Megumi, K., and Yamamoto, H., *Abstr. Jpn. Soc. Appl. Phys. 42nd Meeting*, No. 8P-Q-11, 1981. With permission.

Table 18   Spectral Peak Shift by a Host Material for Eu^{2+}, Ce^{3+}, and Mn^{2+} Activation

Host	Peak wavelength (nm)			Distance between the nearest neighbor ions (nm)
	Eu^{2+} 0.1%	Ce^{3+} 0.04%	Mn^{2+} 0.2%	
CaS	651	520	585	0.285
SrS	616	503	~550	0.301
BaS	572	482	~541	0.319

From Kasano, H., Megumi, K., and Yamamoto, H., *Abstr. Jpn. Soc. Appl. Phys. 42nd Meeting*, No. 8P-Q-11, 1981. With permission.

crystal field parameter $(10Dq)$ (see 2.2.1) decreases in the above order. See also 2.2.5 for Mn^{2+} luminescence and 2.3.3 for Eu^{2+} and Ce^{3+} luminescence.

2.6.4   Typical examples of applications

2.6.4.1   Storage and stimulation

It is another remarkable feature of the IIa-VIb compounds that they show various phenomena related to traps, e.g., storage, photostimulation (infrared stimulation), and photoquenching (see 1.7).

$Ca_{0.7}Sr_{0.3}S:Bi^{3+},Cu$ was developed as a particularly efficient storage material.[24] This material doped with Bi^{3+} shows bluish-violet emission due to the $6s6p \rightarrow 6p^2$ transition of Bi^{3+}; the addition of Cu shifts the emission toward longer wavelengths and improves its luminance (see 12.2).

$CaS:Bi^{3+}$, reported as early as in 1928 by Lenard,[1] is one of the best known of all CaS phosphors. This phosphor requires the Bi^{3+} luminescent centers to be co-activated by an

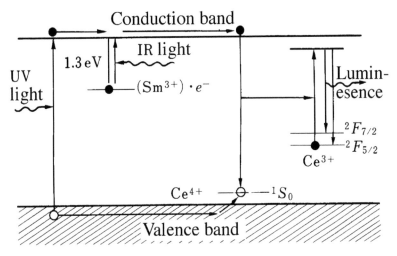

Figure 51 Schematic diagram of infrared stimulation mechanism of $SrS:Ce^{3+},Sm^{3+}$. The original figure has been simplified. (From Keller, S.P. and Pettit, G.D., *Phys. Rev.*, 111, 1533, 1958. With permission.)

alkali metal ion. The luminescence and absorption spectra of $CaS:Bi^{3+},K^{+}$ are shown in Figure 50.[4]

Photostimulation attracted attention as a means to detect infrared light. During World War II, an enormous volume of research was carried out on this phenomenon in Japan[23] and the U.S. for military purposes. Some of the materials developed in this period includes $(Ca,Sr)S:Ce^{3+},Bi^{3+}$,[23] $SrS:Eu^{2+},Sm^{3+}$ and $SrS:Ce^{3+},Sm^{3+}$.[24] In these compositions, the primary activator (i.e., Ce^{3+} or Eu^{2+}) determines the luminescence spectrum, while the auxiliary activator (i.e., Bi^{3+} or Sm^{3+}) forms the necessary traps that determine the stimulable wavelength in the infrared. These wavelengths range from 0.8 to 1.4 μm for Sm^{3+} and 0.5 to 1.0 μm for Bi^{3+}.

The stimulation mechanism proposed for $SrS:Ce^{3+},Sm^{3+}$ is schematically shown in Figure 51.[25] Here, one assumes that Sm^{3+} forms electron traps and Ce^{3+} forms hole traps. Electrons trapped by Sm^{3+} ions are released to the conduction band by absorption of infrared light of energy corresponding to the trap depth. After migration through the lattice, some of the electrons are retrapped by Ce^{3+}, which already has trapped holes. The electron and the hole recombine in Ce^{3+}, releasing energy characteristic of the Ce^{3+} luminescence. In $SrS:Eu^{2+},Sm^{3+}$, it is thought that Eu^{2+} plays a similar role as Ce^{3+}. The concept of an activator ion working either as an electron or a hole trap is also applicable to other materials with localized luminescence centers.

2.6.4.2 Cathode-ray tubes

Application of CaS phosphors applied to CRTs have attracted attention because of their high efficiency and diversity in emission colors.[26] Green-emitting $CaS:Ce^{3+}$ with weak temperature quenching has been tested for application in heavily loaded projection tubes.[8,9] However, $CaS:Ce^{3+}$ was found to be not satisfactory for projection tube uses because it shows serious luminance saturation at high current density. Amber-emitting $(Ca,Mg)S:Mn^{2+}$ has been tested in terminal display tubes.[10] The commercial application of this phosphor was also abandoned because its efficiency and persistence were found to be unsatisfactory. Screening technology for this family of phosphors has improved, but evolution of toxic H_2S gas in the manufacturing and reclaiming processes remains a serious impediment for the widespread application of these phosphors.

Table 19 (a) Color and Luminance of DC Electroluminescence of
CaS and SrS Powder Phosphors

Phosphors	Luminescence color	Luminance (cd/m²) (Applied voltage is shown in parenthesis)	
		Continuous drive	Pulse drive[a]
CaS:Ce³⁺	Green	1700 (70 V)	600 (110 V)
CaS:Er³⁺	Green	300 (80 V)	85 (120 V)
CaS:Tb³⁺	Green	17 (80 V)	50 (120 V)
CaS:Eu³⁺	Red	100 (50 V)	17 (120 V)
SrS:Ce³⁺	Bluish-green	400 (70 V)	200 (110 V)
SrS:Mn²⁺	Green	270 (120 V)	—
SrS:Cu,Na	Green	270 (80 V)	17 (120 V)

[a] Pulse width is 10–20 μs. Duty is 1–1¹/₄%.

From Vecht, A., *J. Crystal Growth*, 59, 81, 1982. With permission.

Table 19 (b) Properties of Thin-Film AC Electroluminescence Devices Using
CaS or SrS Phosphors

Chemical composition	Maximum luminance (cd/m²)	Maximum efficiency (lm/W)	Luminescence color
CaS:Eu²⁺	200	0.05	Red
CaS:Ce³⁺	150	0.1	Green
SrS:Ce³⁺	900	0.44	Greenish-blue

Note: The driving voltage has a frequency of 1 kHz.

From Ono, Y.A., *Electroluminescence Displays*, Series for Information Display, Vol. 1, World
Scientific Publishing, Singapore, p. 84, 1995. With permission.

2.6.4.3 Electroluminescence (EL)

Phosphors based on CaS and other IIa-VIb compounds are important electroluminescent materials because they can provide colors other than the orange color provided by ZnS:Mn²⁺ (See 1.10). Table 19(a) shows the properties of the EL cells made of fine phosphor particles manufactured by Phosphor Products Co.[27] (See also [Sections 2.6.6.1). Among the materials listed in this table, CaS:Ce³⁺ is nearly as bright as ZnS:Mn²⁺. Although the degradation has been improved, the lifetime of this material is still at an impractical level.

In thin-film electroluminescent devices, CaS and SrS phosphors provide luminances that are higher than ZnS phosphors in the red and green-blue regions. The luminance and luminous efficiency of the three primary colors obtained by IIa-VIb compounds are shown in Table 19(b).[28]

As SrS:Ce EL was developed in 1984[29], formation process of SrS:Ce thin-films has been investigated by various techniques, e.g., electron-beam deposition, sputtering, hot-wall deposition, and molecular beam epitaxy. A recent study has shown that thin-film formation in excess sulfur promotes introduction of Ce³⁺ ions in SrS lattice by creating Sr vacancies[30]. By this optimization, luminous efficiency higher than 1 lm/W was achieved at 1 KHz driving frequency[30]. However, the emission color of SrS:Ce is not saturated blue, but greenish-blue with color coordinates $x=0.30$ and $y=0.52$ in the above case[30]. Luminescence of more saturated blue with $x=0.18$ and $y=0.34$ can be achieved either by codoping of Rb as a charge compensator[31] or by supplying H_2O vapor during SrS:Ce film deposition[32].

A thin-film of SrS:Cu shows better performance of blue EL[33]. Moderately high luminous efficiency of 0.22 lm/W was obtained at 60 Hz driving with saturated blue of color coordinates, $x=0.15$ and $y=0.23$.

Addition of Ag to Cu further improves color coordinates to $x=0.17$ and $y=0.13$, which is close to the primary blue in CRTs. Luminous efficiency of SrS:Cu, Ag was reported to be 0.15 lm/W at 60 Hz driving.

Photoluminescence studies have shown that a SrS:Cu thin film has the emission peak at around 480 nm, while CaS:Cu shows the peak at around 420 nm. Accordingly, the emission peak can be tuned by the formation of the solid solution (Sr, Ca):Cu[34]. However, EL by the solid solution has not been obtained yet, though EL of CaS:Cu thin films was reported recently[35].

2.6.5 Host excitation process of luminescence

Figure 52 shows a luminescence spectrum of SrSe containing trace amounts of Ba^{2+}.[13] The line at 3.74 eV can be assigned to a free exciton at the indirect bandgap (i.e., an indirect exciton), while the broad band at lower energy arises from the recombination of a localized indirect exciton trapped by the short-range potential of Ba^{2+} acting as an isoelectronic impurity. Presumably, the localized indirect excitons also produce luminescence by other types of activators.

It was observed that the excitation spectrum of CaS:Eu^{2+} in the vacuum-UV region shows that the luminescence efficiency increases with a step-like shape at the energy position twice that of the direct exciton transition (Figure 53).[22] This fact shows that an excited electron can efficiently create two direct excitons through an Auger process. Direct excitons thus created are scattered and transformed to indirect excitons, which eventually transfer energy to an impurity producing luminescence. These experimental results provide evidence that excitation energy given by the band-to-band transition is efficiently transferred to activators via excitons in alkali earth sulfides and selenides.

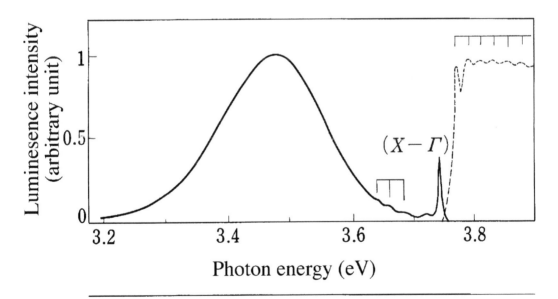

Figure 52 A luminescence (a solid line) and an excitation spectrum (a broken line) of SrSe containing a trace of Ba^{2+} at 2K. The notation $(X-\Gamma)$ indicates a recombination transition of an exciton from the X point of the conduction band to the Γ point of the valence band. The vertical lines above the spectra show phonon structures. The broad emission band is due to localized excitons at the isoelectronic Ba center. (From Kaneko, Y. and Koda, T., *J. Crystal Growth*, 86, 72, 1988. With permission.)

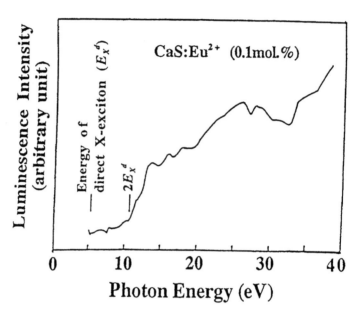

Figure 53 An excitation spectrum of CaS:Eu²⁺ (0.1 mol%) in the vacuum UV region at 77K. (From Kaneko, Y., Ph.D. Thesis, The University of Tokyo, 1984 (in Japanese). With permission.

2.6.6   *Preparation methods of phosphors*

2.6.6.1   *Sulfides*

The preparation of sulfide phosphors can be classified into two methods; one entails the sulfurization of alkaline earth oxides or carbonates and the other involves the reduction of sulfates. The following agents are known to sulfurize or reduce the starting materials; the sulfurizing agents are H_2S, CS_2, S+C (in many cases, starch or sucrose are used as the source of carbon), and Na_2CO_3+S (or Na_2S), and the reducing agents are H_2 and C.

In addition to these agents, fluxes are often added to the starting materials at the level of several to 10 wt%. Typical fluxes are alkali carbonates, alkali sulfates, and NH_4Cl. Lithium compounds are particularly effective in promoting crystal growth and diffusion of activator ions into the sulfide lattice. This is probably because Li^+, which has a small ionic radius, enters interstitial sites and generates cation vacancies; ionic diffusion is accelerated through these means. Fluxes that promote crystal growth and ion diffusion effects remarkably, however, may have a side effect to degrade luminescence efficiency because the constituent ions of the fluxes (e.g., Li^+ F^-) are likely to remain in the phosphor lattice as impurities. The material and the quantity of a flux are selected by considering these two kinds of effect.

Typical preparation methods are shown below for CaS:Ce³⁺.[36]

$$CaSO_4 \xrightarrow{\ H_2+H_2S\ } CaS$$

$$CaS + Ce_2S_3 + NH_4Cl + S \xrightarrow[\text{2 hr}]{\ 1200°C\ } CaS:Ce^{3+}$$

(29)[3]

$$CaSO_4 + Ce(NO_3)_3 \cdot 6H_2O + Na_2SO_4 + C \xrightarrow[2\,hr]{900°C} CaS:Ce^{3+} \qquad (30)^{37}$$

$$CaCO_3 + Ce_2(SO_4)_3 \cdot 8H_2O + Na_2CO_3 + S \xrightarrow[1-72\,hr]{800-1100°C} CaS:Ce^{3+} \quad (31)^{38,39}$$

$$CaCO_3 + CeO_2 \xrightarrow[2\,hr]{1300°C} CaO:Ce^{3+}$$

$$(32)^{40}$$

$$CaO:Ce^{3+} \xrightarrow[1200°C,\,2\,hr]{H_2S+PCl_3} CaS:Ce^{3+}$$

$$Ca(CH_2O)_2 + 2H_2S \xrightarrow[2\,hr]{} Ca(HS)_2 + 2CH_3OH$$

$$(33)$$

$$Ca(HS)_2 + Ce_2S_3 \xrightarrow[1000°C,\,2\,hr]{N_2} CaS:Ce^{3+} + H_2S$$

When the sulfurizing or reducing agents are in the solid or liquid state, the reaction can be performed in an encapsulated crucible. When the agents are in the gas phase, however, the reaction must be done in a quartz tube that allows a gas flow. In this review, the former will be called the *crucible method* and the latter the *gas-flow method*. These two methods are described below.

The crucible method.   Examples of this method are given in the second reaction of Eq. 29, and Eqs. 30 and 31. By selecting an appropriate flux, this method provides particles of fairly large size and good dispersion characteristics. On the other hand, contact with the flux can introduce impurities into the phosphor, resulting in degraded efficiency. Insufficient sulfurization or partial oxidation may also occur by exposure to oxygen, since quantities of a flux generally used are insufficient to cover all the particle surfaces. When firing must be carried out for many hours at high temperature, a double-crucible config-uration is used; one crucible nestles in the other with carbon between, thus preventing phosphors from oxidation.

The gas-flow method.   Examples are given in the first reaction of Eq. 29, and Eqs. 32 and 33. Alkali compounds, which are used as fluxes in the crucible method, cannot be used in this case because they can vaporize and react with the quartz tube during the firing. As a result, this method provides smaller particles with poor dispersion characteristics. Improvement is obtained in some cases if a small amount of PCl_3 gas is supplied for a period of time, as in the method for Eq. 32.

On the other hand, the gas-flow method can give high luminescence efficiency because contamination of phosphors with impurities are less probable than in the crucible method and also because stoichiometry may be controlled through adjustment of the compositions and flow rates of the gases. In this case, firing may be repeated and different preparation methods can be experimented with and/or combined, if necessary. The method in Eq. 33 is used to obtain fine particles for electroluminescent powder phosphors.

Preparation methods can have considerable effects on luminescence properties. The luminescence peak positions of CaS:Ce^{3+} prepared by the flux method show a peak that is blue-shifted by about 600 cm^{-1} relative to the positions obtained in phosphors prepared by the gas-flow method. Differences are also observed in the excitation spectra and temperature dependencies of the luminescence intensity.[41]

Sulfides other than CaS can essentially be obtained by the same process. However, polysulfides are formed more easily from sulfides of the heavier cation elements. In other words, the sulfurizing reaction proceeds more slowly for the sulfides of lighter elements. It has been reported that the reduction of sulfates is a better way for the synthesis of SrS and BaS. For example, the following reactions have been employed:[23]

$$SrSO_4 + S + \text{starch (or sucrose)} \longrightarrow SrS + (CO_2, CO, H_2O, SO_2) \qquad (34)$$

The synthesis of MgS by reduction of MgO or MgSO$_4$ requires repeated reactions to complete sulfurization.[7] Another method for MgS synthesis starting with Mg metal and employing CS$_2$, a more powerful sufurizing agent, has also been used. This latter method is reported to be particularly effective in producing MgS.[23]

$$Mg \text{ powder} + S \longrightarrow MgS \text{ (an explosive reaction)}$$
$$2MgO + CS_2 \longrightarrow 2MgS + CO_2 \qquad (35)$$

2.6.6.2 Selenides[23]

Selenides are synthesized by methods similar to those to form sulfides using elemental Se or H$_2$Se instead. The following reaction can be used to prepare CaSe:

$$CaO + \text{amidol (a weak organic base)} \longrightarrow CaSe + (CO_2, CO, NH_3, H_2O, SeO_2) \quad (36)$$

SrSe and BaSe are obtained by firing Sr or Ba nitrates with Se and starch. The fired products include Se and polyselenides, which are then vaporized by annealing in vacuum at about 600°C. After annealing, a single phase of SrSe or BaSe is obtained.

References

1. Lenard, P., Schmidt, F., and Tomascheck, R., *Handb. Exp. Phys.*, Vol. 23, Akadem. Verlagsges, Leipzig, 1928.
2. Lehmann, W., *J. Electrochem. Soc.*, 117, 1389, 1970.
3. Lehmann, W. and Ryan, F.M., *J. Electrochem. Soc.*, 118, 477, 1971.
4. Lehmann, W., *J. Luminesc.*, 5, 87, 1972.
5. Lehmann, W., *J. Electrochem. Soc.*, 118, 1164, 1971.
6. Lehmann, W., *Gordon Research Conference Report*, July 1971.
7. Kasano, H., Megumi, K., and Yamamoto, H., *J. Electrochem. Soc.*, 131, 1953, 1984.
8. Kanehisa, O., Megumi, K., Kasano, H., and Yamamoto, H., *Abstr. Jpn. Soc. Appl. Phys. 42nd Meeting*, No. 8P-Q-11, 1981.
9. Tsuda, N., Tamatani, M., and Sato, T., *Tech. Digest, Phosphor Res. Soc. 199th Meeting*, 1984 (in Japanese).
10. Yamamoto, H., Megumi, K., Kasano, H., Kanehisa, O., Uehara, Y., and Morita, Y., *J. Electrochem. Soc.*, 134, 2620, 1987.

11. Kaneko, Y., Morimoto, K., and Koda, T., *J. Phys. Soc. Japan*, 51, 2247, 1982.
12. Kaneko, Y., Morimoto, K., and Koda, T., *J. Phys. Soc. Japan*, 52, 4385, 1983.
13. Kaneko, Y. and Koda, T., *J. Crystal Growth*, 86, 72, 1988.
14. Kaneko, Y., Morimoto, K., and Koda, T., *Oyo Buturi*, 50, 289, 1981 (in Japanese).
15. Krebs, H., *Fundamentals of Inorganic Crystal Chemistry*, McGraw-Hill, London, 1968, 159 and 163.
16. Hasegawa, A. and Yanase, A., *J. Phys. C*, 13, 1995, 1980.
17. Morimoto, K., Masters Thesis, The University of Tokyo, 1982 (in Japanese).
18. Jasperse, J.R., Kahan, A., Plendel, J.N., and Mitra, S.S., *Phys. Rev.*, 146, 526, 1966.
19. Jacobsen, J.L. and Nixon, E.R., *J. Phys. Chem. Solids*, 29, 967, 1968.
20. Galtier, M., Montaner, A., and Vidal, G., *J. Phys. Chem. Solids*, 33, 2295, 1972.
21. Boswarva, I.M., *Phys. Rev. B1*, 1698, 1970.
22. Kaneko, Y., Ph.D. Thesis, The University of Tokyo, 1984 (in Japanese).
23. Kameyama, N., *Theory and Applications of Phosphors*, Maruzen, Tokyo, 1960 (in Japanese).
24. Keller, S.P., Mapes, J.F., and Cheroff, G., *Phys. Rev.*, 108, 663, 1958.
25. Keller, S.P. and Pettit, G.D., *Phys. Rev.*, 111, 1533, 1958.
26. *Japanese Patent Publication (Kokoku) 47-38747*, 1972.
27. Vecht, A., *J. Crystal Growth*, 59, 81, 1982.
28. Ono, Y.A., *Electroluminescent Displays*, Series for Information Displays, Vol. 1, World Scientific Publishing, Singapore, 1995, 84.
29. Barrow W.A., Coovert R.E., and King, C.N., *Digest of Technical Papers, 1984 SID Intl. Symp.* 249, 1984.
30. Ohmi, K., Fukuda, H., Tokuda, N., Sakurai, D., Kimura, T., Tanaka, S., and Kobayashi, H., *Proc. 21st Intl. Display Research Conf., (Nogoya)*, 1131, 2001.
31. Fukada, H., Sasakura, A., Sugio, Y., Kimura, T., Ohmi, K., Tanaka, S. and Kobayashi, H., *Jpn. J. Appl. Phys.*, 41 L941, 2002.
32. Takasu, K., Usui, S., Oka, H., Ohmi, K., Tanaka, S., and Kobayashi, H, *Proc. 10th Intl. Display Workshop, (Hiroshima)*, 1117, 2003.
33. Sun, S.S., Dickey, E., Kane, J., and Yocom, P.N., *Proc. 17th Intl. Display Research Conf., (Toronto)*, 301, 1997.
34. Ehara, M., Hakamata, S., Fukada, H., Ohmi, K., Kominami, H., Nakanish, Y., and Hatanaka, Y, *Jpn. Appl. J. Phys.* 43, 7120-7124, 2004.
35. Hakamata, S., Ehara, M., Fukada, H., Kominami, H., Nakanishi, Y., and Hatanaka, Y., *Appl. Phys. Lett.*, 85, 3729-3730, 2004.
36. Okamoto, F. and Kato, K., *Tech. Digest, Phosphor Res. Soc. 196th Meeting*, 1983 (in Japanese).
37. Vij, D.R. and Mathur, V.K., *Ind. J. Pure Appl. Phys.*, 6, 67, 1968a.
38. Okamoto, F. and Kato, K., *J. Electrochem. Soc.*, 130, 432, 1983.
39. Kato, K. and Okamoto, F., *Jpn. J. Appl. Phys.*, 22, 76, 1983.
40. Yamamoto, H., Manabe, T., Kasano, H., Suzuki, T., Kanehisa, O., Uehara, Y., Morita, Y., and Watanabe, N., *Electrochem. Soc. Meeting, Extended Abstracts*, No. 496, 1982.
41. Kanehisa, O., Yamamoto, H., Okamura, T., and Morita, M., *J. Electrochem. Soc.*, 141, 3188, 1994.

chapter two — section seven

Principal phosphor materials and their optical properties

Shigeo Shionoya

Contents

2.7   IIb-VIb compounds

2.7.1   Introduction

IIb-VIb compounds include the oxides, sulfides, selenides, and tellurides of zinc, cadmium, and mercury. Among these compounds, those with bandgap energies (E_g) larger than 2 eV (i.e., ZnO, ZnS, ZnSe, ZnTe, and CdS) are candidate materials for phosphors that emit visible luminescence; ZnS is the most important in this sense. In this section, the fundamental optical properties and luminescence characteristics and mechanisms of this class of phosphors will be explained.[1] The term IIb-VIb compounds used below will be limited to the above-mentioned compounds.

ZnS-type phosphors are presently very important as cathode-ray tube (CRT) phosphors. These phosphors have a long history, dating back about 130 years. At the

International Conference on Luminescence held in 1966 in Budapest, a presentation titled "The Century of the Discovery of Luminescent Zinc Sulfide" was given,[2] in which the history of luminescent ZnS was discussed. In 1866, a young French chemist, Théodore Sidot, succeeded in growing tiny ZnS crystals by a sublimation method. Although his original purpose was to study crystal growth, the crystals grown exhibited phosphorescence in the dark. The experiments were repeated, the observations confirmed, and a note to the Academy of Sciences of Paris was presented. This note was published by Becquerel.[3] These phosphorescent ZnS (zinc-blende) crystals were thereafter called Sidot's blende. From present knowledge, one can conclude that Sidot's blende contained a small quantity of copper as an impurity responsible for the phosphorescence. The historical processes of the evolution of Sidot's blende to the present ZnS phosphors are described in 2.7.4.

2.7.2   Fundamental intrinsic properties

Important physical properties of IIb-VIb compounds related to luminescence are shown in Table 20.

2.7.2.1   Crystal structure

IIb-VIb compounds crystallize either in the cubic zinc blende (ZB) structure or in the hexagonal wurtzite (W) structure; ZnO, CdS, and CdSe crystallize in the W structure, while ZnSe, ZnTe, and CdTe in the ZB structure. ZnS crystallizes into both the W type (traditionally called α-ZnS) and the ZB type (β-ZnS). The ZB structure corresponds to the low-temperature phase; the ZB $\rightarrow$ W transition temperature is known to be about 1020°C.

2.7.2.2   Melting point and crystal growth

In IIb-VI compounds, the sublimation pressure is very high. As a result, the compounds, with the exception of the tellurides, do not melt at atmospheric pressure. They do melt at pressures of several tens of atmospheres of argon, but the melting points are pretty high: 1975°C for ZnO, 1830 ± 20°C for ZnS, and about 1600°C for ZnSe. ZnS powder phosphors are prepared by firing ZnS powders at 900 to 1200°C. Phosphor particles fired at relatively low temperature (below about 1000°C) are of the ZB structure, while those fired at temperatures above 1000°C are of the W structure.

In the past, single crystals of IIb-VIb compounds with high sublimation pressure were grown by the sublimation-recrystallization method, the vapor phase reaction method, the vapor phase chemical transport method, or by the high-pressure melt growth method. Recently, various epitaxial growth methods, such as molecular beam epitaxy (MBE), metalorganic chemical vapor deposition (MOCVD), and atomic layer epitaxy (ALE), have been actively developed, especially for ZnSe and ZnS (see 3.7.6). As a result, thin single-crystal films with very high purity and high crystallinity are presently available for these two compounds.

2.7.2.3   Band structure

In the IIb-VIb compounds treated in this section, the conduction band has the character of the *s* orbital of the cations, while the valence band has the character of the *p* orbital of the anions. These compounds are all direct-transition type semiconductors, and both the bottom of the conduction band and the top of the valence band are located at the Γ point [k = (000)] in **k**-space. This is simply shown in Figures 7(a) and (b) in 1.2 for the ZB and W structure. The band structure of ZnSe of the ZB structure obtained by nonlocal pseudo-potential calculations is shown in Figure 54.[4]

Table 20 Important Physical Properties of IIb-VIb Compounds Related to Luminescence

	Crystal structure	Lattice constant (Å)	Static dielectric constant	Bandgap energy (eV) 4K	RT	Exciton energy, 4K (eV)	Exciton binding energy (meV)	Effective mass m^*/m_0 Electron	Hole
ZnO	W	a = 3.2403 c = 5.1955	c ∥ 8.8 c ⊥ 8.5	3.436	3.2	3.375	59	0.28	0.59
ZnS	W	a = 3.820 c = 6.260	8.6	3.911	3.8	3.871	40	0.28	c ∥ 1.4 c ⊥ 0.49
	ZB	5.4093	8.3	3.84	3.7	3.799	36	0.39	
ZnSe	ZB	5.6687	8.1	2.819	2.72	2.802	17	0.16	0.75
ZnTe	ZB	6.1037	10.1	2.391	2.25	2.381	11	0.09	Heavy: 0.6 Light: 0.16
CdS	W	a = 4.1368 c = 6.7163	c ∥ 10.3 c ⊥ 9.35	2.582	2.53	2.552	28	0.2	c ∥ 5.0 c ⊥ 0.7
CdSe	W	a = 4.30 c = 7.02	c ∥ 10.65 c ⊥ 9.70	1.840	1.74	1.823	15	0.112	c ∥ 2.5 c ⊥ 0.45
CdTe	ZB	6.4818	10.2	1.606	1.53	1.596	10	0.096	Heavy: 1.0 Light: 0.1

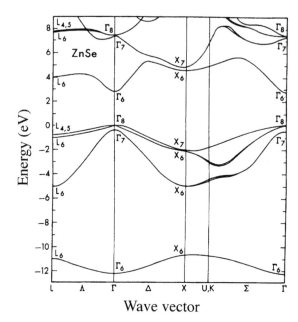

Figure 54 Band structure of ZnSe. (From Chelikowsky, J.R. and Cohen, M.L., *Phys. Rev.*, B14, 556, 1976. With permission.)

As shown in Figure 7(a) of 1.2, the valence band in the ZB structure is split by the spin-orbit interaction into a higher lying Γ_8(A) state (in which the orbital state is doubly degenerate) and a nondegenerate Γ_7(B) state. In the W structure as shown in Figure 7(b), on the other hand, all the orbital degeneracy is lifted by the spin-orbit interaction and the aniosotropy of crystal field, and the split states are Γ_9(A), Γ_7(B), and Γ_7(C) in descending order of energy. The case of ZnO is an exception: Γ_9(A) and Γ_7(B) are reversed, so that the order is Γ_7(A), Γ_9(B), and Γ_7(C) instead. This originates from the fact that in ZnO the splitting by the spin-orbit interaction is negative and smaller than that due to the crystal field anisotropy, unlike other IIb-VIb compounds. The negative spin-orbit splitting arises because of mixing of the d orbitals of Zn with the valence band.

In MX-type compound semiconductors, the bandgap energy E_g usually increases if M or X is replaced by a heavier element. Looking at E_g values in Table 20, it can be noted that this general rule is usually observed, except in the case of ZnO, where the E_g value is smaller than that of ZnS. This is also caused by the mixing of the Zn d orbital with the valence band.

It is seen in the band structure of ZnSe, shown in Figure 54, that in the conduction band there are two minima in upper energy regions at the L [k = (111)] and the X [k = (100)] points with energies of 1.2 and 1.8 eV above the bottom of the conduction band, respectively. The conduction band structure of ZnS is very similar, having the two upper minima at the same points. The existence of these two upper minima plays an important role in the excitation process of high-field, thin-film electroluminescence in ZnS (See 1.10).

The fact that IIb-VIb compounds are direct-gap semiconductors means that they are appropriate host materials for phosphors. If one compares the radiative recombination coefficient of electrons and holes for direct and indirect transitions, the value for the former is four orders of magnitude larger. In practical phosphors, the radiative emission is not caused by direct recombination, but by transitions taking place via energy levels of activators introduced as impurities. For impurities as donors or acceptors, their energy levels

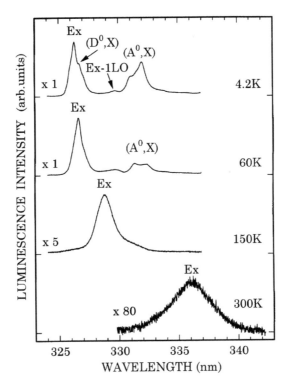

Figure 55 Exciton luminescence spectra of ZnS (ZB type) at various temperatures. (From Nakamura, S., Sakashita, T., Yoshimura, K., Yamada, Y., and Taguchi, *Jpn. J. Appl. Phys.*, 36, L491, 1997. With permission.)

are generated by perturbation on the conduction or valence band. Therefore, the impurity energy levels take on the same character as their parent bands, and the radiative recombination processes and rates in these levels are similar to those in the pure host material. ZnS:Cu,Al and ZnS:Ag,Cl phosphors, which are very important as phosphors for cathode ray tubes (CRT), are typical examples of this type of phosphor, as will be explained in 2.7.4. This is the reason why direct-gap type materials are most favorable as phosphor hosts.

2.7.2.4 Exciton

In IIb-VIb compounds, the exciton structure is clearly observed at low temperature in absorption and reflection spectra near the fundamental absorption edge. Absorption spectra of CdS shown in Figure 11 of 1.2 are a typical example. The exciton energy and its binding energy are shown in Table 20.

An exciton is annihilated, emitting a photon by the recombination of the constituent electron and hole pair. Figure 55 shows exciton luminescence spectra from a high-quality epitaxial layer of ZB-type ZnS grown by MOCVD at various temperatures.[5] At 4.2K, the Ex line from intrinsic free excitons at 326.27 nm is the strongest. The line Ex-1LO is the free exciton line accompanied by the simultaneous emission of one longitudinal optical (LO) phonon.

Even in very pure crystals of IIb-VIb compounds, a trace amount of impurities, at concentration levels of 10^{14} to 10^{15} cm^{-3}, are found. Excitons are bound to these impurities, and luminescence from these bound excitons is also observed at low temperature. Lines (D^0, X) and (A^0, X) are from excitons bound to neutral donors (D^0) and neutral acceptors (A^0). These bound exciton lines are customarily denoted as I_2 and I_1, respectively. Frequently,

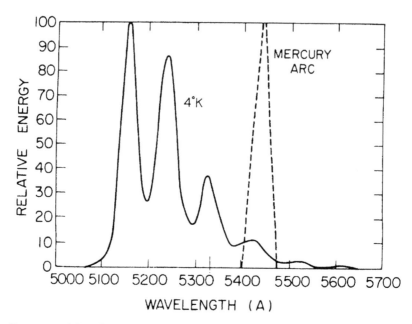

Figure 56 Spectrum of the edge emission of CdS at 4K; dashed line represents spectrometer window. (From Klick, C.C., *J. Opt. Soc. Am.*, 37, 939, 1947. With permission.)

an I_3 line due to excitons bound to ionized donors is observed at wavelength a little shorter than I_2. With increasing temperature from 4.2K, bound excitons are released from impurities, so that only the luminescence line due to free excitons is observed as can be seen in the figure. The binding energy of the exciton in ZB-type ZnS is 36 meV, so that exciton luminescence is observed up to room temperature.

The exciton binding energy in ZnO is as large as 59 meV and is the largest among IIb-VIb compounds. Luminescence of free excitons is observed at 385 nm at room temperature in pure ZnO. This ultraviolet luminescence was found as early as the 1940s,[6] but it was not recognized at that time that this luminescence originates from excitons. This luminescence persists up to fairly high temperatures; it is still observed at temperatures as high as 770K.[7]

2.7.2.5   Type of conductivity and its control

As-grown single crystals of ZnO, ZnS, ZnSe, and CdS are usually *n*-type in conductivity, while those of ZnTe are *p*-type. The conductivity control of IIb-VIb compounds, especially for ZnSe, has made remarkable progress recently. This progress is due to the demand to develop blue and blue-green light-emitting diodes and semiconductor lasers. The preparation of *p*-type ZnSe with high conductivity has been a fundamental problem, which was solved recently by introducing nitrogen acceptors using nitrogen plasma (See 2.7.6). Presently, it is possible to control the type of conductivity in most IIb-VIb compounds.

2.7.3   Luminescence of shallow donors and acceptors

Since the 1940s, it has been known that pure CdS crystals show luminescence at low room temperature with a characteristic spectral structure on the low-energy side of the fundamental absorption edge. This luminescence was called *edge emission*. Its spectrum is shown in Figure 56.[8] It has been established that the characteristic edge emission is observed in all IIb-VIb compounds except for ZnO.

The lines in Figure 56 are equally spaced, with an interval of about 40 meV, which is equal to the energy of longitudinal optical (LO) phonons in CdS. The halfwidth of the lines is approximately 5 meV. The relative intensities of the lines in the figure (numbered n = 0, 1, 2, ... from the short wavelength side) decrease toward longer wavelengths with ratios of 1.00:0.87:0.38:0.12:0.030:0.015. This ratio exactly obeys the Poisson distribution $I_n = e^{-s}S^n/n!$ with S = 0.87. The n = 0 line is known as the zero-phonon line, while lines of n = 1, 2, ... are caused by simultaneous emission of 1, 2, ... LO phonons.

It has been established that the characteristics of the edge emission are satisfactorily interpreted in terms of donor(D)-acceptor(A) pair luminescence (see 1.4.4). The transition energy E of this luminescence is a function of the distance r between D and A in a pair, and is given by:

$$E(r) = E_g - (E_D + E_A) + e^2/4\pi\varepsilon r \qquad (37)$$

where E_D and E_A are the ionization energies of a neutral donor and acceptor, respectively, and ε is the static dielectric constant. The transition probability W also depends on r and is expressed by

$$W(r) = W_0 \exp(-2r/r_B) \qquad (38)$$

where r_B is Bohr radius of the donor electron and W_0 is a constant related to the D-A pair.

The mechanism for donor-acceptor pair luminescence was first verified in the edge emission in GaP doped with S donors and Si acceptors (see 1.4.4 and 2.8).[9] The intra-pair distance r is distributed discretely, so that a spectrum consisting of discrete lines is expected. In GaP:Si,S, a great number of sharp lines were observed adjacent to the high-energy tail of the n = 0 line, and the value of r for each line was determined. On this basis, the main part of the n = 0 line is thought to be composed of a large number of unresolved pair lines for pairs with relatively large r values.

A great number of sharp lines were also observed in the edge emission of CdS[10,11] and ZnSe.[12,13] These facts present clear evidence as to the origin of the edge emission. In ZnSe, the identification of each line has been made in analogy to the GaP case; in CdS, the analysis is not easy to make since the spectra are much more complicated because of the W structure.

Eqs. 37 and 38 indicate that the pair emission energy shifts to lower energies and the decay time becomes longer with increasing r values. Then one expects that in the time-resolved spectra of the edge emission, the peaks of the lines composed of unresolved pair lines should shift to lower energies as a function of time after pulse excitation. This has been observed in CdS,[14] and presents further evidence for the pair emission mechanism in the edge emission.

The fact that the relative intensity ratio of the edge emission lines obeys a Poisson distribution indicates that the configurational coordinate model (see 1.3.2) is applicable to each pair center with a different r value. The donors and acceptors participating in the edge emission are shallow. In these cases, the constant S appearing in the Poisson distribution, which is called the Huang-Rhys-Pekar factor and a measure of the strength of the electron-phonon interaction, is small, of the order of 1 or less; the phonon coupled to the center is the LO phonon of the entire lattice, but not a local mode phonon.

The depths of donor and acceptor levels (E_D and E_A) in IIb-VIb compounds are determined from bound exciton emission lines, edge emission spectra, or absorption spectra between a donor or acceptor level and the band. The available data on levels are

Table 21 Depths of Donor and Acceptor Levels, E_D and E_A (meV) in IIb-VIb Compounds

(a) Donor	$E_{D, calc}$	B	Al	Ga	In	F	Cl	Br	I	Li[a]
ZnS	110		100							
ZnSe	29±2	25.6	25.6	27.2	28.2	28.2	26.2			21
ZnTe			18.5				20.1			
CdS	33.9			33.1	33.8	35.1	32.7	32.5	32.1	28
CdSe	20±2									

[a] Interstitial Li.

(b) Acceptor	$E_{A, calc}$	Li	Na	Cu	Ag	Au	N	P	As
ZnS		150	190	1250	720	1200			
ZnSe	108	114	102	650	430	550	110	85,500	110
ZnTe	62	60.5	62.8	148	121	277		63.5	79
CdS		165	169	1100	260			120,600	750
CdSe		109							

Note: Calculated values by the effective mass approximation[15], $E_{D, calc}$ and $E_{A, calc}$, are also shown.

shown in Table 21. Calculated values of E_D E_A by the effective mass approximation are also shown.[15]

2.7.4 ZnS-type phosphors

2.7.4.1 Luminescence of deep donors and acceptors

ZnS type phosphors such as the green-emitting ZnS:Cu,Al and the blue-emitting ZnS:Ag,Cl are very important from a practical point of view, especially as phosphors for cathode-ray tubes. Luminescence centers in these phosphors are formed from deep donors or deep acceptors, or by their association at the nearest-neighbor sites. In this subsection, a brief history of the development of these phosphors will be given first, and then the characteristics and the mechanisms of their luminescence will be explained.

(a) History. After the research by Sidot described in 2.7.1, it became gradually clear that when ZnS powders are fired with the addition of a small amount of metallic salt, luminescence characteristic of that metal is produced. In the 1920s, it was established that a small amount of copper produces green luminescence, while silver produces blue luminescence. In this sense, copper and silver were called *activators* of luminescence. The firing is made at 900 to 1200°C with the addition of halides (such as NaCl) with low melting points as fluxes. It was found that if the firing is made without the addition of activators but with a halide flux, blue luminescence is produced. Thus, this type of blue luminescence was called *self-activated luminescence*.

In the 1930s and 1940s, research on ZnS-type phosphors was very active. Results of the research are described in detail in a book by Leverenz[16] published in 1950. In this book, the emission spectra of a great number of phosphors in ZnS, (Zn,Cd)S, or Zn(S,Se) hosts activated with Cu or Ag are shown. The spectral data shown in the book are still very useful. It should be noted that this book was written before the concept of the co-activator was conceived, so that chemical formulas of some phosphors given in the book are always not appropriate, and care must be exercised. For example, a phosphor written as ZnS:Ag[NaCl] should be written as ZnS:Ag,Cl according to the rule in use today. The ZnS self-activate phosphor is shown as ZnS:[Zn] in the book, but should be ZnS:Cl(orAl) instead, as will be explained.

In the late 1940s, Kröger and co-workers[17,18] demonstrated that halide flux added in the firing process to ZnS phosphors not only promotes crystal growth, but introduces halide ions (VIIb group anions) into the ZnS lattice, and that these halide ions participate in the formation of luminescence centers. Kröger et al. assumed that the copper or silver activators are in the monovalent state and substitute for Zn^{2+} ions, and that charge compensation for the monovalent activators is accomplished by introducing VIIb group anions substituting for S^{2-} ions. It was supposed that charge compensation should occur not only with VII group anions, but also with IIIb group cations, such as Al^{3+}, substituting for Zn^{2+} ions. Kröger's group[19] clearly showed that if Al^{3+} ions are introduced without using halide fluxes, similar kinds of luminescence are produced, and thus evidenced the above assumption. The VIIb or IIIb ions were called co-activators. These ions are indispensable for the formation of luminescence centers, but the luminescence spectrum is determined only by the kind of Ib ion activators and is almost independent of the kind of co-activators. This is the reason for the naming of co-activators.

In those days, the nature of the electronic transitions responsible for the luminescence in ZnS phosphors was actively discussed. The so-called Schön-Klasens model, first proposed by Schön and then discussed in detail by Klasens,[20] gained general acceptance. This model assumes that the luminescence is caused by the recombination of an electron in the conduction band, with a hole located in a level a little above the valence band.

Prencer and Williams[21] pointed out that Ib group activators and VIIb or IIIb group co-activators should be recognized, respectively, as the acceptors and the donors. It was assumed that donors and acceptors are spatially associated in some way; then it was proposed that the luminescence takes place in centers of pairs of donors (co-activators) and acceptors (activators) associated at the second and third nearest-neighbor site, and that the luminescence transition occurs from the excited state of donors to the ground state of acceptors. This was the first proposal for the donor-acceptor pair luminescence concept, which was later recognized as a basis for understanding semiconductor luminescence as mentioned in 2.7.3.

The above narration touches upon the essential points of the progress in research in this area up until the 1950s. This research was actively pursued in the 1960s. As a result, the luminescence mechanism of ZnS-type phosphors using activators of Ib elements has been elucidated quite thoroughly. This will be described below.

(b) Classification and emission spectra. The luminescence of ZnS-type phosphors using Ib group activators (Cu, Ag) and IIIb (Al, Ga, In) or VIIb (Cl, Br, I) group co-activators can be classified into five kinds, depending on the relative ratio of the concentrations of activators (X) and co-activators (Y). This condition is shown in Figure 57.[22] The range of concentrations for X and Y is 10^{-6} to 10^{-4} mol/mol. The labels of the luminescence in the figure originate from the emission color in the case that the activator is Cu; that is, G = green, B = blue, and R = red. R-Cu,In appears only when the co-activator is a IIIb group element. SA means the self-activated blue luminescence.

Figure 58 depicts the emission spectra of these five kinds of phosphors at room temperature and at 4.2K.[23] As shown in Table 20, the bandgap energy E_g of ZnS is 0.08 eV larger for the W structure than for the ZB structure. Corresponding to this, the emission peaks of phosphors with the W structure are shifted by almost this amount toward shorter wavelength. In general, phosphors prepared by firing above 1000°C have the W structure, while those below this have the ZB structure. Emission peaks at room temperature are located at longer wavelengths than those at 4.2K, except in the case of the SA luminescence. The long wavelength shift is almost proportional to that of E_g. The SA luminescence shows the inverse behavior; that is, the peak at room temperature is located at shorter

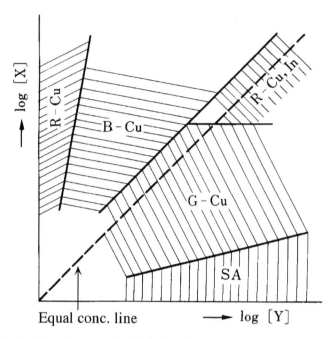

Figure 57 Five kinds of luminescence in ZnS phosphors classified from the point of view of the relative ratio of the concentrations of activators (X) and co-activators (Y). G-Cu: green Cu, B-Cu: blue Cu, R-Cu: red Cu, R-Cu, In: red Cu, In, SA: self-activated blue. (From van Gool, W., *Philips Res. Rept. Suppl.*, 3, 1, 1961. With permission.)

wavelengths. These emission spectra are almost independent of the kind of co-activators, except for the case of the SA luminescence. The G-Cu emission spectra of ZnS:Cu,Al shown in Figure 58 are almost the same as those of ZnS:Cu,Cl. In the case of the SA luminescence, the spectra of ZnS:Cl and ZnS:Al are a little different. The spectrum of ZnS:Al is slightly shifted to longer wavelengths.

If the activator is changed from Cu to Ag, the emission peaks are shifted by 0.4 to 0.5 eV to shorter wavelengths. The blue luminescence of ZnS:Ag,Cl (peak at 45 nm, W type) corresponds to the G-Cu luminescence. Au, a Ib group element, also acts as an activator. The luminescence of ZnS:Au,Al corresponding to the G-Cu luminescence has its peak at 550 nm in the ZB structure, which is shifted slightly to longer wavelengths relative to ZnS:Cu,Al.

ZnS and CdS, and also ZnS and ZnSe, form binary alloys (solid solutions) with relatively simple properties. E_g changes almost in proportion to the composition. For example, E_g in (Zn,Cd)S (W) changes from 3.91 eV for ZnS to 2.58 eV in CdS, almost in proportion to the ratio of Cd. The five kinds of luminescence discussed above also appear in the alloyed materials and have similar properties. In $Zn_xCd_{1-x}S$:Ag,Cl (W), the emission peak changes almost proportionally to E_g, i.e., changes from 435 nm for x = 1 to 635 nm for x = 0.4 continuously. It is possible to obtain a desired luminescence color from blue to red by simply adjusting the composition. In ZnS_xSe_{1-x}:Ag,Cl, the situation is similar, but the change of the emission peak is not always proportional to E_g and sometimes a weak subband appears.

Among the five kinds of luminescence discussed above, the important one for practical use is the G-Cu luminescence, which is produced when the concentrations of the activator and co-activator are nearly equal; in this case, charge compensation is readily and simply attained. ZnS:Cu,Al (green-emitting) and ZnS:Ag,Cl (blue-emitting) phosphors are

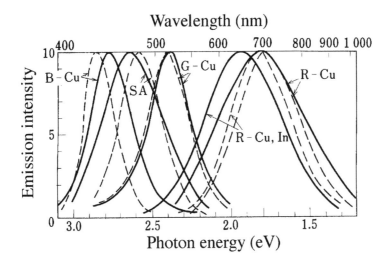

Figure 58 Spectra of the five kinds of luminescence in ZnS phosphors at room temperature (solid line) and 4.2K (dotted line). Activators and co-activators of these phosphors and the crystal structure are shown below. (Shionoya, S., Koda, T., Era, K., and Jujiwara, H., *J. Phys. Soc. Japan*, 19, 1157, 1964. With permission.)

Luminescence	Activator	Co-activator	Crystal structure
G-Cu	Cu	Al	W
B-Cu	Cu	I	ZB
SA	—	Cl	ZB
R-Cu	Cu	—	W
R-Cu,In	Cu	In	ZB

extremely important in CRT applications. ZnS:Cu,Au,Al (green-emitting) phosphors in which both Cu and Au are used as activators also find usage in this area.

The excitation spectra for these five kinds of luminescence consist of two bands in all cases. The first one, having the peak at 325 to 340 nm, corresponds to the fundamental absorption edge (or the exciton position) of the ZnS host crystal, and is called the host excitation band. The second, having the peak at 360 to 400 nm in the longer wavelength region, is characteristic of the luminescence center, and is called the characteristic excitation band. This band is produced by the transition from the ground state of the center (corresponding to the acceptor level) to the excited state of the center (corresponding to the donor level) or to the conduction band. As an example, the excitation spectra for the SA luminescence in a ZnS:Cl single crystal (ZB) are shown in Figure 59(a).[1a,24] An absorption spectrum of the crystal and the absorption band of the SA center are shown in Figure 59(b) for comparison.

(c) Atomic structure of luminescence centers and luminescence transitions. The atomic structure of the luminescence centers and the nature of luminescence transitions for the five kinds of luminescence mentioned above were elucidated in 1960s, mostly by the research of Shionoya and co-workers, as will be described below. Experimental tools that played important roles in clarifying these subjects included measurements of the polarization of luminescence light using phosphor single crystals grown by melting powder phosphors under high argon pressure[25] and measurements of time-resolved emission spectra. The essential characteristics derived from the results of these measurements, as well as the

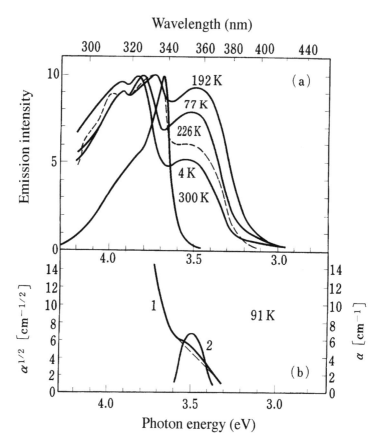

Figure 59 Excitation spectra (a) for the SA luminescence at various temperatures and an absorption spectrum (b) at 91K in a ZnS:Cl single crystal. In (b), curve 1 is the absorption spectrum plotted as $\alpha^{1/2}$ vs. E (α = absorption coefficient, E = photon energy), and curve 2 is the absorption band of the SA center obtained from curve 1 and plotted as α vs. E. (From Koda, T. and Shionoya, S., *Phys. Rev. Lett.*, 11, 77, 1963; Koda, T. and Shionoya, S., *Phys. Rev.*, 136, A541, 1964; Shionoya, S., in *Luminescence of Inorganic Solids*, Goldberg, P., Ed., Academic Press, New York, 1966, chap. 4. With permission.)

atomic structure of the luminescence centers and the nature of the luminescence transitions obtained from these results, are summarized in Table 22.

(i) Polarization of luminescence.   Luminescence light from centers located within the host crystals with uniaxial symmetry, like the wurtzite structure crystals, shows polarization due to the anisotropic crystal fields. In the case of isotropic crystals, if the luminescence center is a spatially associated center, including an activator and/or a co-activator with the site symmetry which is characteristic of the center and is lower than the symmetry of the host crystal, then the polarization of luminescence light results when polarized excitation light is used. Therefore, one can determine the nature of the site symmetry of the luminescence center by observing the difference in the polarization of luminescence light when polarized and unpolarized light are used for excitation.

Among the five kinds of luminescence of ZnS phosphors, polarization measurements were first conducted for the SA luminescence of a ZnS:Cl single crystal.[24] This crystal had the cubic ZB structure, but contained a considerable amount of stacking disorder, leading to small volumes in which hexagonal structure occurred. The hexagonal regions have the *c*-axis defined by the $[111]_c$ axis of the ZB structure. In polarization

Table 22 Characteristics of the Five Kinds of Luminescence in ZnS-type Phosphors

Luminescence	Phosphor	Polarization of luminescence	Symmetry of center	Shift of emission peak	Type of luminescence transition	Structure of center
G-Cu	ZnS:Cu,Al(Cl)	ZB:no pol. W :pol. $\perp$ c	No charact. symm.	Observed	D-A pair	Cu^+_{sub} and $Al^+_{sub}(Cl^-_{sub})$ (random distrib.)
SA	ZnS:Cl(Al)	Charact. pol.	Charact. symm. ZnS:Cl, C_{3v} ZnS:Al, C_s	Observed (ZnS:Cl)	D-A pair	$V(Zn^{2+})$-Cl^-_{sub} (Al^{3+}_{sub}) (associated)
B-Cu	ZnS:Cu,I(Cl)	Charact. pol.	Charact. symm. C_{3v}	No shift	Intra-center	Cu^+_{sub}-Cl^+_{int} (associated)
R-Cu	ZnS:Cu	Charact. pol.	Charact. symm. C_{3v}	No shift	Intra-center	$V(S^{2-})$-Cu^+_{sub} (associated)
R-Cu,In	ZnS:Cu,In	Charact. pol.	Charact. symm. C_s	No shift	Intra-center	Cu^+_{sub}-In^{3+}_{sub} (associated)

Note: The polarization of luminescence, the symmetry of the luminescence center obtained from the characteristics of the polarization, the spectral shift of emission peak in time-resolved emission spectra, the type of luminescence transition inferred from the spectral shift, and the atomic structure of the center. (V = vacancy; sub = substitutional; int = interstitial).

measurements, one of the crystal surfaces was irradiated perpendicularly by polarized excitation light, and the polarization was measured for the luminescence light emitted from the opposite surfaces.

The measurements were made for (110), (112), and (111) planes. Excitation was made with 340-nm light belonging to the host excitation band (H) and with 365-nm light belonging to the characteristic excitation band (C). Both polarized and unpolarized light was used. The results were expressed in terms of the degree of polarization observed, i.e., $P(\theta) = [(I_{\parallel} - I_{\perp})/(I_{\parallel} + I_{\perp})]_{\theta}$, where $I_{\parallel}$ and $I_{\perp}$ are the emission intensities measured with the analyzer parallel and perpendicular, respectively, to the polarizer, and θ is the angle between the optical axis of the polarizer and a particular crystal axis. In the case of unpolarized excitation, θ is the angle between the optical axis of the analyzer and a particular crystal axis.

The $P(\theta)$ curves measured at 77K are shown in Figure 60.[24] In the case of the characteristic (C) excitation, the $P(\theta)$ curves depend critically on whether the excitation light is polarized or unpolarized; under polarized excitation, $P(\theta)$ shows specific angular dependencies that vary for different crystal planes. In the case of host (H) excitation, on the other hand, the $P(\theta)$ curves are quite independent of the polarization of excitation light, and are the same as those obtained under the unpolarized C excitation.

The results observed under the polarized characteristic excitation clearly indicate that the center has the characteristic symmetry, which is lower than that of the host lattice. Prener and Williams[26] proposed a model for the structure of the SA center, which assumes that the center consists of a Zn^{2+} vacancy $[V(Zn^{2+})]$ and one of the charge-compensating co-activators associated at one of the nearest substitutional sites, i.e., a Cl^- co-activator at the nearest S^{2-} site or an Al^{3+} co-activator at the nearest Zn^{2+} site. Figure 61 shows a model of the SA center in ZnS:Cl. The results of polarization measurements were analyzed assuming this model.

According to this model, the SA center in ZnS:Cl has C_{3v} symmetry (in the case of ZnS:Al, C_s symmetry). Dipole transitions that are allowed in a C_{3v} center are those due to a σ-dipole perpendicular to and a π-dipole parallel to the symmetry axis. The angular dependence of the polarization of luminescence due to a σ- or a π-dipole was calculated assuming that the symmetry axes of the centers are distributed uniformly along the various directions of the Zn-S bonding axes.

The results of the calculation are represented in Figure 60 by the thin solid lines. Comparing the experimentally observed $P(\theta)$ curves under polarized C excitation with those calculated, it can be concluded that the observed anisotropy of the luminescence results from the σ-dipole. Thus, the Prener and Williams model for the atomic structure of the SA center was confirmed by these observations. The polarization of the luminescence perpendicular to the c-axis observed under unpolarized C excitation and under polarized and unpolarized H excitation is ascribed to the crystal structure inclusive of the hexagonal domains arising from the stacking disorder.

Measurements of the polarization of luminescence were also made for the SA luminescence in a ZnS:Al crystal.[27] The observed $P(\theta)$ curves showed characteristic angular dependencies under polarized C excitation similar to the SA luminescence in ZnS:Cl. The results were analyzed assuming C_s symmetry for the center as in the Prener and Williams model, and were well explained by ascribing the luminescence to a dipole lying in the mirror plane of the center (∥ dipole).

Results of measurements of the polarization for the five kinds of luminescence are summarized in Table 22. B-Cu,[28] R-Cu,[29] and R-Cu,In[30] luminescence showed characteristic polarizations under C excitation, indicating that the luminescence centers are some kinds of spatially associated centers. The symmetries of the centers determined by analysis are also shown in the table.

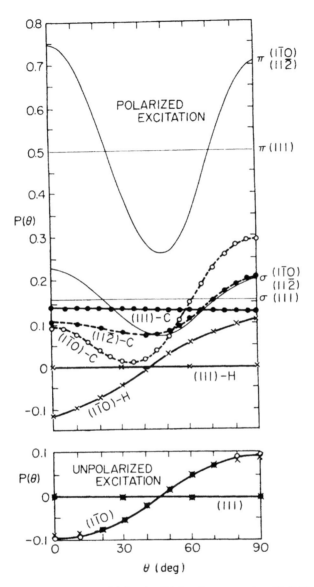

Figure 60 The degree of polarization P(θ) for the SA luminescence in a ZnS:Cl single crystal at 77K under polarized and unpolarized excitation. C and H denote, respectively, the characteristic and host excitation. θ is the angle between the optical axis of the polarizer (or the analyzer in the case of unpolarized excitation) and particular crystal axis [111]c for the surface (110) and (112), and [112] for the surface (111)$_c$. Open circles, closed circles, and crosses show experimental results. The curves drawn with thin solid lines are P(θ) curves calculated assuming σ- or π-dipole. (From Koda, T. and Shionoya, S., *Phys. Rev. Lett.*, 11, 77, 1963; *Phys. Rev.*, 136, A541, 1964. With permission.)

The G-Cu luminescence does not show characteristic polarization differently from the other four kinds of luminescence.[31] In a ZnS:Cu,Al crystal of the W structure, the polarization of luminescence perpendicular to the *c*-axis was observed independently of whether the excitation was due to the C or H band and was also independent of whether the excitation was polarized or unpolarized. In the case of a crystal of the ZB structure, no polarization was observed. These facts indicate that activators and co-activators forming the G-Cu centers are not associated with each other spatially, but are randomly distributed in a crystal occupying respective lattice sites.

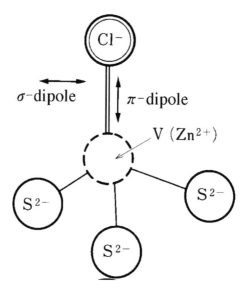

Figure 61   Model of the SA luminescence center in ZnS:Cl phosphors.

(ii) Time-resolved emission spectra.   In D-A pair luminescence, spectral lines or bands are composed of a large number of unresolved pair lines. In time resolved emission spectra, as explained in 2.7.3, the peak of the lines or bands should shift to lower energies with the lapse of time. Figure 62 shows time-resolved spectra of the green luminescence of ZnS:Cu,Al.[32] It is clearly seen that the peak shifts dramatically to lower energies as a function of time.

   If the excitation intensity is increased over a wide range, the D-A pair luminescence peak should shift to higher energies under certain conditions. With sufficient intensity, the lines that originate from pairs with larger intra-pair distances, and hence have long lifetimes, can be saturated, thus leading to the shift. Figure 63 shows changes in the green luminescence spectra of ZnS:Cu,Al observed with changing excitation intensity over a range of five orders of magnitude.[32] The peak is seen to shift to higher energies with excitation intensity. Also for the blue luminescence of ZnS:Ag,Al, very similar spectral peak shifts were observed, both as a function of time and with increasing excitation intensity.[32] In Table 22, shifts of emission peaks observed in the time-resolved spectra are listed for the five kinds of luminescence.

(iii) Atomic structure of various centers and luminescence transitions.   The atomic structure of luminescence centers and the nature of luminescence transitions deduced from above-mentioned experimental observations and their analysis are summarized in Table 22 for the five kinds of luminescence. The luminescence for which peak shifts are observed is thought to be caused by D-A pair type transitions, while the luminescence not showing peak shifts is surely due to intra-center transitions.

   G-Cu center—This center is formed by activators (A) (Cu, Ag, or Au) and co-activators (D) (Al or Cl, Br) introduced with nearly equal concentrations and distributed randomly in the lattice occupying the appropriate lattice sites. Luminescence transition takes place from the D level to the A level in various D-A pairs having different intra-pair distances. Emission spectra consist of broad bands that are quite different from those of the edge emission. In the G-Cu luminescence, the A level is much deeper than those levels involved in the edge emission. The electron-phonon interaction for the acceptors is much stronger, resulting in a Huang-Rhys-Pekar factor

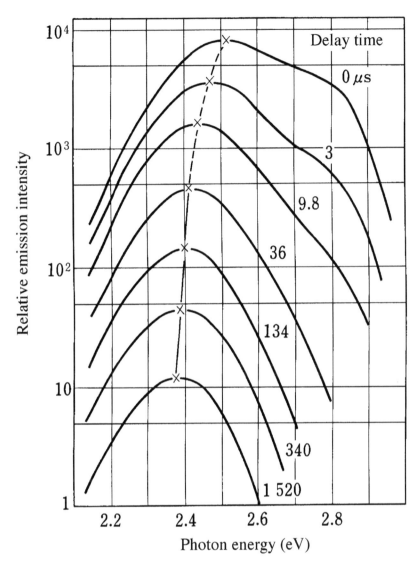

Figure 62 Time-resolved spectra of the green luminescence of ZnS:Cu,Al phosphor at 4.2K. (From Era, K., Shionoya, S., and Washizawa, T., *J. Phys. Chem. Solids*, 29, 1827, 1968; Era, K., Shionoya, S., Washizawa, Y., and Ohmatsu, H., *J. Phys. Chem. Solids*, 29, 1843, 1968. With permission.)

S much larger than 1 in the configurational coordinate model. Spectra for S = 20 are broad Gaussians. The spectra of the G-Cu luminescence can be interpreted in this way.

The energy levels of ZnS:Cu,Al are shown in Figure 64.[33] Before excitation, Cu is monovalent (1+), while Al is trivalent (3+), so that charge compensation is realized in the lattice. Absorption A of the figure located at about 400 nm gives the characteristic excitation band of the center. When excited, Cu and Al become divalent (2+). The levels of Cu^{2+} ($3d^9$ configuration) are split by the crystal field into $^2T_2$ and ^2E states, with $^2T_2$ lying higher in the ZB structure. Under excitation, three induced absorption bands B, C, and D are observed, as shown in Figure 65.[34] The C absorption taking place inside Cu^{2+} states has a sharp zero-phonon line at 1.44 μm. Luminescence of Cu^{2+} due to the downward transition of C is observed.

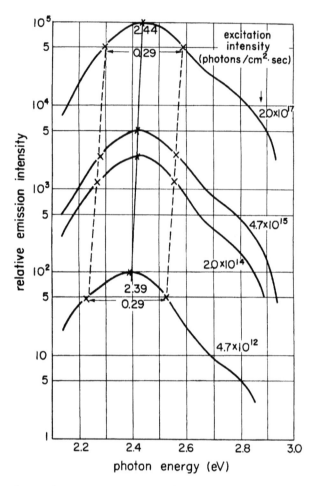

Figure 63 Changes of green luminescence spectra of ZnS:Cu,Al phosphor observed with changing excitation intensity in a wide range at 4.2K. (From Era, K., Shionoya, S., and Washizawa, T., *J. Phys. Chem. Solids*, 29, 1827, 1968; Era, K., Shionoya, S., Washizawa, Y., and Ohmatsu, H., *J. Phys. Chem. Solids*, 29, 1843, 1968. With permission.)

Utilizing the induced absorption bands due to Cu^{2+} (C band) and to Al^{2+} (B band), direct evidence for the D-A pair emission mechanism of the G-Cu luminescence can be obtained.[33] If the green luminescence of ZnS:Cu,Al originates from Cu-Al pairs, the decay rate of the luminescence R_{lum} must be correlated with the decay rates of the intensities of the Cu- and Al-induced absorption, R_{Cu} and R_{Al}. Results of studies of the luminescence decay and the decays of Cu and Al absorption intensities are shown in Figure 66.[33] It is seen in the figure that R_{Cu} and R_{Al} are equal to each other, and both are always equal to the half value of R_{lum} during decay, namely $R_{lum} = R_{Cu} + R_{Al}$. This experimentally observed relation presents very clear and direct evidence for the Cu-Al pair mechanism.

SA center—This center is formed by the spatial association of a Zn^{2+} vacancy with a co-activator Cl^- (or Al^{3+}) at the nearest-neighbor site. An emission peak shift is observed as shown in Tables 22, so that the initial state of this emission is not a level in the associated center, but is considered to be the level of an isolated Cl (Al) donor. A Zn^{2+} vacancy needs two Cl^- ions for charge compensation; one of the two Cl^- ions forms the associated center, while the other is isolated and is responsible for the initial state of the emission. The polarization of luminescence is determined by the symmetry of the surroundings of the hole at a Zn^{2+} vacancy.

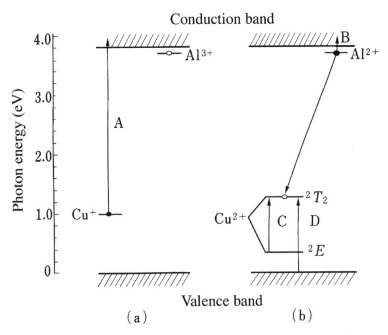

Figure 64 Energy levels and absorption transitions of ZnS:Cu,Al phosphor before excitation (a) and during excitation (b). (From Suzuki, A. and Shionoya, S., *J. Phys. Soc. Japan*, 31, 1455, 1971. With permission.)

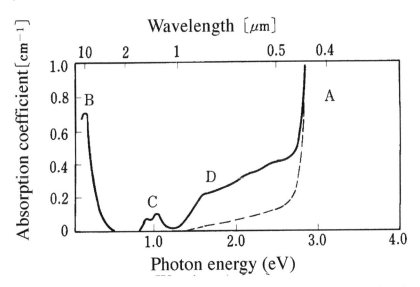

Figure 65 Spectrum of induced absorption (solid line) of a ZnS:Cu,Al single crystal under excitation at 77K, and absorption spectrum before excitation (dashed line). (From Suzuki, A. and Shionoya, S., *J. Phys. Soc. Japan*, 31, 1455, 1971. With permission.)

R-Cu center—The polarization characteristics of this luminescence can be interpreted by assuming C_{3v} symmetry and by taking into account the crystal field splitting of the Cu^{2+} $3d$ orbital in this symmetry.[29] It has been concluded that this center is formed by the spatial association of a substitutional Cu^+ and a S^{2-} vacancy at the nearest-neighbor sites. Therefore, this center is one in which the relation between activator and co-activator is just reversed from that of the SA center. In this sense, this center can be also called the self-coactivated

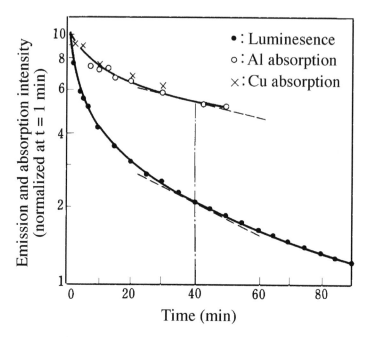

Figure 66 Decay curves of green luminescence and induced absorption intensities due to Cu and Al in a ZnS:Cu,Al single crystal at about 10K. (From Suzuki, A. and Shionoya, S., *J. Phys. Soc. Japan*, 31, 1455, 1971. With permission.)

center. No spectral shift is observed, so that this luminescence is due to intra-center transitions and the initial state is the level of the S^{2-} vacancy.

R-Cu,In center—The polarization characteristics of this luminescence can be interpreted by assuming C_s symmetry and by taking into account the splitting of the Cu^{2+} $3d$ orbital.[30] There is no spectral shift. Therefore, this center is formed by the spatial association of a substitutional Cu^+ and In^{3+} at the nearest neighbor sites, and the luminescence is due to intra-center transitions from the In donor level to the Cu acceptor level.

B-Cu center—The polarization characteristics of this luminescence can be explained if one assumes either C_{3v} or C_s symmetry for the center.[28] A model of this center has been proposed, which suggests that the center is formed by the spatial association of a substitutional Cu^+ and an interstitial Cu^+.[35] Such a proposed center would have C_{3v} symmetry.

In ZnSe, the center corresponding to the B-Cu center in ZnS shows green luminescence. Measurements of optically detected electron spin resonance indicates that the center has the structure of a pair composed of a substitutional Cu^+ and an interstitial Cu^+.[36] It is clear that the B-Cu center in ZnS has the same type of structure and has C_{3v} symmetry. The B-Cu luminescence shows no spectral shift, indicating that it is due to an intra-center transitions. However, the nature of the initial state of the transition is not clear.

(d) Other luminescence characteristics
(i) Stimulation and quenching.   It has been known since the beginning of this century that Cu-activated green ZnS phosphors show distinct stimulation or quenching of the luminescence if irradiated by red or near-infrared (NIR) light while under ultraviolet excitation or during the luminescent decay following excitation. Whether stimulation or quenching occurs depends on the conditions of observation, i.e., the ratio of the intensity of the excitation light to the red or NIR light and on the temperature. Exposure with red or NIR light during decay usually results in stimulation first, changing to quenching as time

progresses. The spectra of the light producing stimulation and quenching have two peaks at 0.6–0.8 and 1.3 μm. It is seen from Figures 64 and 65 that these two peaks correspond to the induced absorption bands C and D, indicating that absorption by excited Cu activators (Cu^{2+} ions) is responsible for these phenomena.

By measuring time-resolved emission spectra and photoconductivity under irradiation with red to NIR light, the mechanisms for stimulation and quenching have been established.[37] Holes are created in the valence band by induced absorption of the C and D bands (in the case of the C band, only at room temperature). These holes move in the valence band and are trapped by unexcited Cu activators. If Al (or Cl) co-activators exist in the vicinity and have trapped electrons, green luminescence is produced immediately. In D-A type luminescence, the transition probability is larger when the intra-pair distance is smaller, so that this immediate luminescence process occurs. Under the stimulation irradiation of red to NIR light with ultraviolet excitation, the average value of the intra-pair distance for excited Cu-Al pairs becomes shorter, increasing the average transition probability and producing stimulated luminescence. On the other hand, quenching is caused by a process where holes created in the valence band are trapped by various nonradiative recombination centers, thus decreasing the effective number of holes.

(ii) Killer effect.   It has also been known since the 1920s that the luminescence intensity of ZnS phosphors is greatly reduced by contamination with very small amounts of the iron group elements Fe, Ni, and Co. Because of it, the iron group elements were called *killers* of luminescence. It follows that it is very important to remove iron group elements in the manufacturing processes of ZnS phosphors.

Figure 67 shows how contamination by Fe^{2+}, Ni^{2+}, and Co^{2+} reduces the intensity of the green luminescence in ZnS:Cu,Al.[38] The results for Mn^{2+}, which also belongs to the iron group, are also shown. In this case, an orange luminescence of Mn^{2+} is produced, as will be described in 2.7.4.2; the Mn^{2+} intensity is also shown in the figure.

The iron group ions have absorption bands in the visible region, and their spectra overlap the G-Cu luminescence spectrum. It is thought that resonance energy transfer from excited G-Cu centers to iron group ions takes place, reducing the G-Cu luminescence intensity and causing the killer effect. The overlap between the luminescence and the absorption spectra of iron group ions obtained using single crystals, and the decrease in the green luminescence intensity caused by energy transfer were calculated. The results are shown in Figure 67(b) by the dotted lines. The decrease of the luminescence intensity in Mn^{2+} is well described by this energy transfer effect. In the case of Fe^{2+}, Ni^{2+}, and Co^{2+}, the actual intensity decrease is considerably more than that which was calculated. As the reason for this disagreement, it is suggested that iron group ions create deep levels, and electrons and holes recombine nonradiatively via these levels.

(iii) Concentration quenching and luminescence saturation.   In ZnS:Cu,Al phosphors, if the concentration of activators and co-activators is increased to obtain brighter phosphors, concentration quenching results. The optimum concentration is Cu: 1.2×10^{-4} mol/mol and Al: 2×10^{-4}. These phosphors show, when used in CRTs, the phenomenon of luminescence intensity saturation when the current density is raised, as shown in Figure 68.[39] A very similar phenomenon occurs in ZnS:Ag,Al.[40] Luminescence saturation phenomena present serious problems in the practical use of these phosphors for CRT purposes.

The cause of the concentration quenching and luminescence saturation is not well understood, but the nonradiative Auger effect is thought to play an important role.[38] In this effect, excited Cu-Al pairs are annihilated nonradiatively, their energy is transferred

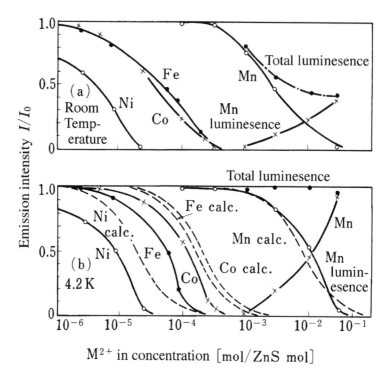

Figure 67 Dependence of the G-Cu luminescence intensity (solid curves) in ZnS:Cu,Al(M) phosphors on the concentration of Mn^{2+} (Mn^{2+}, Fe^{2+}, Ni^{2+}, and Co^{2+}) at (a) room temperature and (b) 4.2K. In the case of Mn^{2+}, the intensity of the Mn^{2+} orange luminescence and the total of the intensities of the G-Cu and Mn^{2+} luminescence are also shown. Dotted curves in (b) are calculated assuming the killer effect due to resonance energy transfer. (From Tabei, M., Shionoya, S., and Ohmatsu, H., *Jpn. J. Appl. Phys.*, 14, 240, 1975. With permission.)

to unexcited Cu activators, and electrons are raised into the conduction band. In this way, the excitation energy migrates in the lattice, and is dissipated in nonradiative recombination traps.

2.7.4.2 *Luminescence of transition metal ions*

Divalent transition metal ions with a $3d^n$ electron configurations have ionic radii close to those of the cations of IIb-VIb compounds. Therefore, these transition metal ions can be easily introduced into IIb-VIb compounds; the optical properties of such ions have been investigated in detail. Optical transitions taking place within the ions can be interpreted by crystal field theory (see 2.2), assuming T_d symmetry in ZB lattices. Besides intra-ion absorption, various charge-transfer absorptions, such as the valence band $\rightarrow$ ion and ion $\rightarrow$ the conduction band, are observed. Luminescence from charge-transfer transitions is also observed in some cases. The absorption band A of ZnS:Cu,Al in Figure 64 is an example of this kind of absorption.

$\underline{Mn^{2+}}$—The orange luminescence of ZnS:Mn^{2+} has been known since early days, and is important in applications in electroluminescence. The emission spectrum has a peak at 585 nm at room temperature with a halfwidth of 0.2 eV.

Absorption spectra of a ZnS:Mn^{2+} single crystal are shown in Figure 69.[41] The ground state of Mn^{2+} is $^6A_1$ (originating from the free ion state ^6S). Absorption bands at 535, 495, 460, 425, and 385 nm correspond to transitions from the ground state to $^4T_1(^4G)$, $^4T_2(^4G)$, $(^4A_1,^4E)(^4G)$, $^4T_2(^4D)$, and $^4E(^4D)$ excited states, respectively. Luminescence is produced from

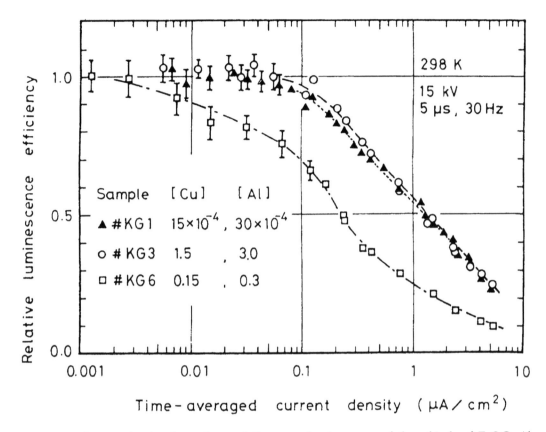

Figure 68 Current density dependence of the green luminescence of three kinds of ZnS:Cu,Al (W type) phosphors at room temperature. Excitation was made by a pulsed 15-kV electron beam with 5-µs duration and a repetition frequency of 30 Hz. (From Kawai, H., Kuboniwa, S., and Hoshina, T., *Jpn. J. Appl. Phys.*, 13, 1593, 1974. With permission.)

the lowest excited state $^4T_1(^4G)$, and a sharp zero-phonon line common to the absorption and the emission is observed at 558.94 nm.[42]

The location of the Mn^{2+} energy levels in relation to the energy bands of the ZnS host is important in discussing the excitation mechanism for Mn^{2+}. From measurements of X-ray photoelectron spectroscopy, it has been found that the Mn^{2+} ground state is located 3 eV below the top of the ZnS valence band.[43] In more recent measurements using resonant synchrotron radiation, the locations of the Mn^{2+} ground state in ZnS, ZnSe, and ZnTe are 3.5, 3.6, and 3.8 eV, respectively, below the top of the valence band.[44]

$\underline{Fe^{2+} \text{ and } Fe^{3+}}$ — Iron in ZnS is usually divalent. Two kinds of infrared luminescence due to intra-ion transitions of Fe^{2+} are known. One appears at 3.3 to 4.2 µm and is due to the transition from the lowest excited state $^5T_2(^5D)$ to the ground state $^5E(^5D)$ of Fe^{2+}.[45] The other is composed of two bands at 971 nm and 1.43 µm, which are due to transitions from a higher excited state $^3T_1(^3H)$ to the ground state ^5E and the lowest excited state $^5T_2$, respectively.[46] In the killer effect of Fe^{2+} mentioned in Section 3.7.4.1, excitation energy is transferred from the G-Cu center to Fe^{2+} and is down-converted to infrared luminescence or is dissipated nonradiatively.

It has also been known since early days that iron in ZnS emits red luminescence at 660 nm. This luminescence is due to a pair-type transition, and is considered to be caused by a transition from a Cl^{2-} state (Cl^- donor trapping and electron) to an Fe^{3+} state

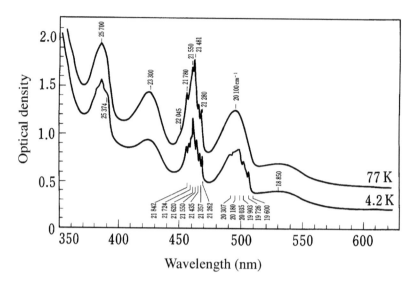

Figure 69 Absorption spectra of a ZnS:Mn^{2+} (4.02 mol%) (ZB type) single crystal with 1.34-mm thickness at 4.2 and 77K. (From McClure, D.S., *J. Chem. Phys.*, 39, 2850, 1963. With permission.)

(photoionized Fe^{2+}).[47] In this process, luminescence at 980 nm is also produced from intra-ion transitions in Fe^{3+}.[48]

Cu^{2+}—Copper in ZnS is usually monovalent. As mentioned in 2.7.4.1(c)(iii), when ZnS:Cu,Al phosphors are excited, an ionization process Cu$^+$ → Cu^{2+} takes place and luminescence at 1.44 μm, due to an intra-ion transition from $^2T_2(^2D)$ to $^2E(^2D)$ of Cu^{2+}, is produced (see Figure 64).

ZnO:Cu exhibits green luminescence in a band spectrum with the peak at 510 nm and a broad halfwidth of 0.4 eV. The zero-phonon line was observed at 2.8590 eV.[49] In ZnO the Cu^{2+} level is located 0.2 eV below the bottom of the conduction band, which is different from the case of ZnS. Luminescence is considered to be produced by the recombination of an electron trapped at the Cu^{2+} level (Cu$^+$ state) with a hole in the valence band.

2.7.4.3 Luminescence of rare-earth ions

The luminescence of trivalent rare-earth ions in IIb-VIb compounds has also been investigated in fair detail. The introduction of trivalent rare-earth ions with high concentrations into IIb-VIb compounds is difficult, because the valence is different from the host cations and the chemical properties are also quite different. This is one of the reasons that bright luminescence from trivalent rare-earth ions in ZnS is rather difficult to obtain.

Exceptionally bright and highly efficient luminescence has been observed in ZnS:Tm^{3+}.[50] This phosphor shows bright blue luminescence under cathode-ray excitation. The spectrum is shown in Figure 70; the strongest line at 487 nm is due to the $^1G_4 \to {}^3H_6$ transition (see 2.3).[51] Beside this line, there are weak lines at 645 nm ($^1G_4 \to {}^3F_4$) and 775–800 nm ($^1G_4 \to {}^3H_5$). In the figure, the spectra of commercial ZnS:Ag,Cl phosphors (P22 and P11) used in color CRTs are also shown for comparison. The energy efficiency of ZnS:Tm^{3+} under cathode-ray excitation is 0.216 W/W and is very high; this value should be compared with 0.23 W/W obtained in ZnS:Ag,Cl (P22).

2.7.5 ZnO phosphors

It has been known since the 1940s that by firing pure ZnO powders in a reducing atmosphere, a phosphor showing bright white-green luminescence can be prepared.[6] The

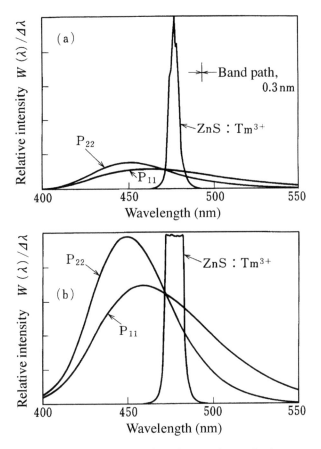

Figure 70 Emission spectra of a ZnS:Tm³⁺ phosphor under cathode-ray excitation. Spectra of ZnS:Ag,Cl phosphors (P22, P11) are also shown for comparison. (From Shrader, R.E., Larach, S., and Yocom, P.N., *J. Appl. Phys.*, 42, 4529, 1971. With permission.)

chemical formula of this phosphor is written customarily as ZnO:Zn because it is obtained in a reducing atmosphere without adding activators. The emission spectrum has a peak at 495 nm and is very broad, having a halfwidth of 0.4 eV. This phosphor shows conductivity, and hence is an exceedingly valuable phosphor for applications to vacuum fluorescent displays and field emission displays.

The origin of the luminescence center and the luminescence mechanism of ZnO:Zn phosphors are barely understood. Due to firing in reducing atmospheres, these phosphors contain excess zinc; its amount can be determined by colorimetric analysis and has been found to be in the 5–15 ppm range. This amount of excess zinc is almost proportional to the intensity of the white-green luminescence of these phosphors.[52] Therefore, it is clear that intertitial zinc or oxygen vacancy participates in the formation of the luminescence center, but nothing has been established with certainty. The decay constant of the emission is about 1 μs; the decay becomes faster with increased excitation intensity.[53] From this, the luminescence is thought to be due to a bimolecular type recombination.

References

1. Review of luminescence of IIb-VIb compounds:
   a. Shionoya, S., Luminescence of lattices of the ZnS type, in *Luminescence of Inorganic Solids*, Goldberg, P., Ed., Academic Press, New York, 1966, chap. 4.

b. Shionoya, S., *II–VI Semiconducting Compounds, 1967 International Conference*, Thomas, D.G., Ed., W. A. Benjamin, Inc., 1967, 1.

c. Shionoya, S., *J. Luminesc.*, 1/2, 17, 1970.

d. Curie, D. and Prener, J.S., Deep Center Luminescence, in *Physics and Cemistryn of II-VI Compounds*, Aven, M. and Prener, J.S., Ed., North-Holland Pub. Co., Amsterdam, 1967, chap. 4.

2. Arpiarian, N., *Proc. Int. Conf. Luminesc.*, Budapest, 1966, Szigeti, G., Ed., Akadémiai Kiadó, Budapest, 1968, 903.

3. Becquerel, E., *Compt. Rend. Acad. Sci.*, LXIII, 188, 1866.

4. Chelikowsky, J.R. and Cohen, M.L., *Phys. Rev.*, B14, 556, 1976.

5. Nakamura, S., Sakashita, T., Yoshimura, K., Yamada, Y., and Taguchi, T., *Jpn. J. Appl. Phys.*, 36, L491, 1997.

6. Shrader, R.E. and Leverenz, H.W., *J. Opt. Soc. Am.*, 37, 939, 1947.

7. Miyamoto, S., *Jpn. J. Appl. Phys.*, 17, 1129, 1978.

8. Klick, C.C., *J. Opt. Soc. Am.*, 37, 939, 1947.

9. Thomas, D.G., Gershenzon, M., and Trumbore, F.A., *Phys. Rev.*, 133, A269, 1964.

10. Henry, C.H., Faulkner, R.A., and Nassau, K., *Phys. Rev.*, 183, 708, 1968.

11. Reynolds, D.C. and Collins, T.C., *Phys. Rev.*, 188, 1267, 1969.

12. Dean, P.J. and Merz, J.L., *Phys. Rev.*, 178, 1310, 1969.

13. Merz, J.L., Nassau, K., and Siever, J.W., *Phys. Rev.*, B8, 1444, 1973.

14. Colbow, K., *Phys. Rev.*, 141, 742, 1966.

15. Bhargava, R.N., *J. Cryst. Growth*, 59, 15, 1982.

16. Leverenz, H.W., *An Introduction to Luminescence of Solids*, John Wiley & Sons, New York, 1950.

17. Kröger, F.A. and Hellingman, J.E., *Trans. Electrochem. Soc.*, 95, 68, 1949.

18. Kröger, F.A., Hellingman, J.E., and Smit, N.W., *Physica*, 15, 990, 1949.

19. Kröger, F.A. and Dikhoff, J.A.M., *Physica*, 16, 297, 1950.

20. Klasens, H.A., *J. Electrochem. Soc.*, 100, 72, 1953.

21. Prener, J.S. and Williams, F.E., *J. Electrochem. Soc.*, 103, 342, 1956.

22. van Gool, W., *Philips Res. Rept. Suppl.*, 3, 1, 1961.

23. Shionoya, S., Koda, T., Era, K., and Fujiwara, H., *J. Phys. Soc. Japan*, 19, 1157, 1964.

24. Koda, T. and Shionoya, S., *Phys. Rev. Lett.*, 11, 77, 1963; *Phys. Rev.*, 136, A541, 1964.

25. Kukimoto, H., Shionoya, S., Koda, T., and Hioki, R., *J. Phys. Chem. Solids*, 29, 935, 1968.

26. Prener, J.S. and Williams, F.E., *J. Chem. Phys.*, 25, 361, 1956.

27. Urabe, K. and Shionoya, S., *J. Phys. Soc. Japan*, 24, 543, 1968.

28. Urabe, K., Shionoya, S., and Suzuki, A., *J. Phys. Soc. Japan*, 25, 1611, 1968.

29. Shionoya, S., Urabe, K., Koda, T., Era, K., and Fujiwara, H., *J. Phys. Chem. Solids*, 27, 865, 1966.

30. Suzuki, A. and Shionoya, S., *J. Phys. Soc. Japan*, 31, 1719, 1971.

31. Shionoya, S., Kobayashi, Y., and Koda, T., *J. Phys. Soc. Japan*, 20, 2046, 1965; Suzuki, A. and Shionoya, S., *J. Phys. Soc. Japan*, 31, 1462, 1971.

32. Era, K., Shionoya, S., and Washizawa, Y., *J. Phys. Chem. Solids*, 29, 1827, 1968; Era, K., Shionoya, S., Washizawa, Y., and Ohmatsu, H., *J. Phys. Chem. Solids*, 29, 1843, 1968.

33. Suzuki, A. and Shionoya, S., *J. Phys. Soc. Japan*, 31, 1455, 1971.

34. Broser, I., Maier, H. and Schultz, H.J., *Phys. Rev.*, 140, A2135, 1965.

35. Blicks, H., Riehl, N., and Sizmann, R., *Z. Phys.*, 163, 594, 1961.

36. Patel, J.L., Davies, J.J., and Nicholls, J.E., *J. Phys.*, C14, 5545, 1981.

37. Tabei, M. and Shionoya, S., *J. Luminesc.*, 15, 201, 1977.

38. Tabei, M., Shionoya, S., and Ohmatsu, H., *Jpn. J. Appl. Phys.*, 14, 240, 1975.

39. Kawai, H., Kuboniwa, S., and Hoshina, T., *Jpn. J. Appl. Phys.*, 13, 1593, 1974.

40. Raue, R., Shiiki, M., Matsukiyo, H., Toyama, H., and Yamamoto, H., *J. Appl. Phys.*, 75, 481, 1994.

41. McClure, D.S., *J. Chem. Phys.*, 39, 2850, 1963.

42. Langer, D. and Ibuki, S., *Phys. Rev.*, 138, A809, 1965.

43. Langer, D., Helmer, J.C., and Weichert, N.H., *J. Luminesc.*, 1/2, 341, 1970.

44. Weidemann, R., Gumlich, H.-E., Kupsch, M., and Middelmann, H.-U., *Phys. Rev.*, B45, 1172, 1992.

45. Slack, G.A. and O'Meara, B.M., *Phys. Rev.*, 163, 335, 1967.
46. Skowrónski, M. and Lire, D., *J. Luminesc.*, 24/25, 253, 1981.
47. Jaszczyn-Kopec, P. and Lambert, B., *J. Luminesc.*, 10, 243, 1975.
48. Nelkowski, H., Pfützenreuter, O. and Schrittenlacher, W., *J. Luminesc.*, 20, 403, 1979.
49. Dingle, R., *Phys. Rev. Lett.*, 23, 579, 1969.
50. Shrader, R.E., Larach, S., and Yocom, P.N., *J. Appl. Phys.*, 42, 4529, 1971.
51. Charreire, Y. and Parche, P., *J. Electrochem. Soc.*, 130, 175, 1983.
52. Harada, T. and Shionoya, S., *Tech. Digest, Phosphor Res. Soc., 174th Meeting*, February 1979 (in Japanese).
53. Pfanel, A., *J. Electrochem. Soc.*, 109, 502, 1962.

Principal phosphor materials and their optical properties

Toshiya Yokogawa

Contents

2.8   *ZnSe and related luminescent materials*

The wide-bandgap ZnSe and related luminescent materials have attracted considerable recent attention because of the advent of blue-green lasers and light emitting diodes (LEDs). Since ZnSe is nearly lattice matched to GaAs (which is a high-quality substrate material), high-quality ZnSe can be grown. Furthermore, addition of S, Mg, or Cd to ZnSe leads to a ternary or quaternary alloy with a higher or lower bandgap, a property which is needed to fabricate heterostructure devices. Advanced crystal growth techniques such as metal organic vapor phase epitaxy (MOVPE) and molecular beam epitaxy (MBE) have made it possible to grow not only high-quality ZnSe but also lattice-matched ternary and quaternary alloys on (100) GaAs substrates. The success of these growth techniques at low temperatures has resulted in limiting the concentration of background impurities. The reduction of background impurities has allowed us to control the conductivity by the incorporation of shallow acceptors and donors. In this section, crystal growth techniques for ZnSe and related luminescent materials will be described first, and then application for ZnSe-based laser diodes will be discussed.

2.8.1   *MOVPE*

MOVPE growth of ZnSe involves the pyrolysis of a vapor-phase mixture of H_2Se and, most commonly, dimethylzinc (DMZn) or diethylzinc (DEZn). Free Zn atoms and Se molecules are formed and these species recombine on the hot substrate surface in an irreversible reaction to form ZnSe. Growth is carried out in a cold-wall reactor in flowing H_2 at atmospheric or low pressure. The substrate is heated to temperatures of 300 to 400°C, typically by radio frequency (RF) heating. Transport of the metal-organics to the growth zone is achieved by bubbling H_2 through the liquid sources, which are held in temperature-controlled containers.

When DMZn and H_2Se are used for the MOVPE growth, a premature reaction takes place, resulting in poor uniformity and poor surface morphology of the ZnSe epitaxial layers. This has been solved by using dialkyl zincs and dialkyl selenides (DMSe or DESe).[1] With these source materials, uniform growth of ZnSe epitaxial layers with smooth surface morphology has been achieved. However, the growth temperature has to be increased above 500°C for ZnSe.

Recently, it has been established that the growth at relatively lower temperature and the use of high-purity source materials are required to obtain high-quality ZnSe films. It has also been reported that photo-assisted MOVPE growth dramatically enhances the growth rate of ZnSe at temperatures as low as 350°C, reducing the optimum growth temperature.[2]

It was generally thought that the source materials of DMZn or DEZn typically contain 10 to 100 ppm chlorine impurity. The ZnSe epitaxial layers grown using such sources show strong bound-exciton emission due to chlorine donor impurities. On the other hand, ZnSe layers grown using high-purity DMZn showed dominant free-exciton emission in the low-temperature photoluminescence spectrum, as shown in Figure 71.[2] The DMZn used was purified so that the chlorine content was below the detection limit of 5 ppm.

Numerous attempts have been made to grow *p*-type ZnSe crystals using group I and V elements as acceptor dopants. However, there have been only a few attempts for the MOVPE growth of *p*-type ZnSe. With Li doping by MOVPE, high *p*-type conductivity (50 Ω^{-1} cm^{-1}) materials with carrier concentrations up to 10^{18} cm^{-3} have been demonstrated, although the very fast diffusion rate of Li dopants in ZnSe crystal results in poor controllability of carrier concentrations.[3] Nitrogen is thought to be a stable acceptor impurity. However, nitrogen doping in the MOVPE resulted in highly resistive ZnSe films. The nitrogen doping in MOVPE still experiences a problem with the low activation efficiency of acceptors due to hydrogen passivation.[4,5]

n-type doping elements for the MOVPE growth also has been investigated using Al and Ga to substitute for the Zn site and Cl, Br, and I for the Se site in the ZnSe lattice. It has been reported that iodine doping with ethyliodide or n-butyliodide results in a good controllability of carrier concentrations, which range from 10^{15} to 10^{19} cm^{-3}.[6,7]

2.8.2 MBE

Molecular beam epitaxy (MBE) is the growth of semiconductor films such as ZnSe by the impingement of directed atomic or molecular beams on a crystalline surface under ultra-high-vacuum (UHV) condition. Molecular beams of Zn and Se are generated in a resistively heated Knudsen cell within the growth chamber. GaAs substrates are usually used for the ZnSe growth. Modern II-VI MBE systems are generally a multichamber apparatus comprising a fast entry load-lock, a preparation chamber, and two growth chambers for II-VI and III-V films. Systems are of stainless steel construction pumped to UHV conditions. Base pressures of 10^{-11} to 10^{-10} Torr are normally attained. A major attraction of MBE is that the use of UHV conditions enables the incorporation of high-vacuum-based surface analytical and diagnostic techniques. Reflection high-energy electron diffraction (RHEED) is commonly employed to examine the substrate and the actual epitaxial film during

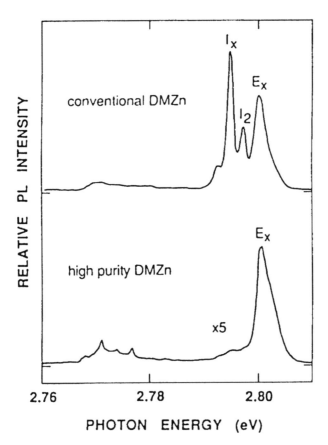

Figure 71 Photoluminescence spectra at 10K of ZnSe films grown using a conventional DMZn source in which about 15 ppm chlorine impurities are involved (upper trace), and using a purified source for which the chlorine content is below the detection limit of 5 ppm. The emission line labeled E_x is due to free excitons, and emission lines I_2 and I_x are due to excitons bound to neutral acceptors. (From Kukimoto, H., *J. Crystal Growth*, 101, 953, 1990. With permission.)

growth. A (quadrupole) mass spectrometer is essential for monitoring the gas composition in the MBE growth chamber.

Early ZnSe-based laser diodes show room-temperature and continuous wave (RTCW) lifetimes of the order of a minute because of degradation caused by extended crystalline defects such as stacking faults. Transmission electron microscopy (TEM) imaging indicates that the degradation originates from dislocation networks that developed in the quantum well region during lasing. The dislocation networks were produced by the stacking faults nucleated at the II-VI/GaAs interface and extending into the II-VI layer. To reduce the stacking fault density, incorporation of GaAs and ZnSe buffer layers and Zn treatment of the II-VI/GaAs interface were employed.[8] The lowest defect density films were reported to be obtained when the (2X4) As-stabilized GaAs surface was exposed to a Zn flux, which resulted in (2X4) to (1X4) surface reconstructions. This was then followed by the epitaxial growth of ZnSe. Stacking fault densities of 10^3 cm^{-2} or less were achieved under this growth condition.

2.8.3 *n-Type doping*

Group III atoms such as Al and Ga substituting in Zn sites and Group VII atoms such as Cl and I in Se sites are typical impurities producing *n*-type carriers in ZnSe crystals.

n-Type doping in ZnSe during MBE growth has been extensively studied. Ga impurities have often been used as a donor dopant although maximum carrier concentrations are limited to approximately 10^{17} cm^{-3}. The photoluminescence (PL) properties of Al- or Ga-doped layers shows a remarkable degradation of the band-edge emission when the carrier concentration exceeds 10^{17} cm^{-3}.[9] Cl impurities were also studied as a donor dopant at the Se site.[10] In Cl doping, *n*-type carrier concentration increases with the temperature of the ZnCl$_2$ cell. The maximum carrier concentration has been estimated to be 10^{19} cm^{-3}, resulting in a small resistivity of 10^{-3} Ωcm. Lately, Cl impurities have been used to fabricate blue-green laser diodes and light-emitting diodes because of advantages presented by controllability and crystalline quality.

2.8.4   p-Type doping

Group I atoms such as Li and Na at Zn sites and group V atoms such as N, P, and As at Se sites are typical impurities used to produce *p*-type carriers in ZnSe crystals. Net acceptor concentrations (N_A–N_D) of 10^{17} cm^{-3} have been achieved using Li doping at a growth temperature of 300°C.[11] Capacitance-voltage (C-V) profiling is usually used to measure N_A–N_D, which implies uncompensated acceptor concentration. When the Li impurity concentration (N_A) exceeds 10^{17} cm^{-3}, N_A–N_D decreases due to increased compensation. This compensation is thought to originate from increased concentration (N_D) of Li interstitial donors in heavily doped ZnSe. Lithium doping is also problematic in that lithium atoms can easily diffuse within the epitaxial layer.

Highly resistive ZnSe films have been grown with As and P doping. A first principles total energy calculation suggests that two neutral acceptors combine to form a new deep state that results in the high resistivity of As- and P-doped ZnSe.[12] Experimental results, which show *p*-type conduction is difficult in As- or P-doped ZnSe, are consistent with this proposed model.

An important breakthrough came with the development of a N$_2$ plasma source for MBE.[13,14] This technique employs a small helical-coil RF plasma chamber replacing the Knudsen cell in the MBE chamber. The active nitrogen species is thought to be either neutral, monoatomic N free radicals, or neutral, excited N$_2$ molecules. This technique has been used to achieve N_A–N_D = 3.4×10^{17} cm^{-3} and blue emission in LEDs. This advance was rapidly followed by the first ZnSe-based laser. Maximum net acceptor concentration has been limited to around 1×10^{18} cm^{-3} in ZnSe. At present, however, the nitrogen-plasma doping is the best way available to achieve *p*-type ZnSe and has been most frequently used to grow *p-n* junctions by MBE and to fabricate ZnSe-based laser diodes. N incorporation depends on the growth temperature and the plasma power. Increased N incorporation is found with low growth temperature and high RF power.

Photoluminescence (PL) spectra in lightly N-doped ZnSe layers with concentrations less than 10^{17} cm^{-3} show a neutral acceptor bound-exciton emission and a weak emission due to donor-acceptor pair (DAP) recombination. With increasing N concentration, up to 10^{18} cm^{-3}, DAP emission became dominant in the PL spectrum. This highly N-doped ZnSe shows a *p*-type conduction as confirmed by capacitance-voltage and Van der Pau measurements. From PL analyses of the excitonic and DAP emissions, the N-acceptor ionization energy was estimated to be about 100 meV, which is in good agreement with the result calculated with an effective mass approximation.

2.8.5   ZnSe-based blue-green laser diodes

ZnSe-based blue-green laser diodes have been studied intensively to be applied in next-generation, high-density optical disk memories and laser printers. Since the first demonstration of II-VI blue-green laser diodes,[15] further improvements in materials quality coupled with the use of wide bandgap ZnMgSSe quaternary alloys for improved electrical as well as optical confinement and the development of ohmic contacts to *p*-type layers have led to room-temperature (RT) CW operation of ZnSe-based laser diodes with very reduced threshold currents and voltages has been achieved.[16]

The first electrically injected ZnSe-based laser was obtained using ZnSSe cladding layers lattice-matched to the GaAs substrate and a ZnCdSe single quantum well surrounded by ZnSe waveguide layers. The band structure in the strained-layer $Zn_{0.82}Cd_{0.18}Se/ZnSe$ system was thought to be a type I quantum well structure with conduction and valence band offsets of $\Delta E_c = 230$ meV and $\Delta E_v = 50$ meV, respectively. According to a common anion rule, the conduction band offset is relatively larger than that of the valence band in this system. Optical and electrical confinement in this prototypical laser structure is quite weak due to the constraint in the device design by the large lattice mismatch between ZnSe and CdSe. The use of the lattice-matched quaternary ZnMgSSe allows greater refractive index and bandgap differences to be realized. The incorporation of Mg into the cladding layer improves the confinement factor, resulting in the RTCW operation of the II-VI lasers. Shorter-wavelength lasers with a ZnSe active layer have also been made possible.

A typical structure of the ZnCdSe/ZnSSe/ZnMgSSe separate-confinement heterostructure (SCH) laser is shown schematically in Figure 72.[16] The incorporation of GaAs:Si and ZnSe:Cl buffer layers and the Zn beam exposure on an As-stabilized surface of the GaAs buffer layer were employed to reduce stacking fault density. The stacking fault density of the laser structure was estimated to be 3×10^3 cm^{-2}. For the *p*- and *n*-$Zn_{1-x}Mg_xS_ySe_{1-y}$ cladding layers, designed for optical confinement, the Mg concentration was nominally x = 0.1 and the sulfur concentration y = 0.15. The Cd composition of 0.35 in the ZnCdSe active layer results in lasing wavelength $\lambda = 514.7$ nm. Low-resistance quasi-ohmic contact to *p*-ZnSe:N is usually achieved using heavily *p*-doped ZnTe:N and ZnSe/ZnTe multiquantum wells as an intermediate layer. The threshold current under CW operation was found to be 32 mA, corresponding to a threshold current density of 533 A cm^{-2}, for a laser diode with a stripe area of 600 μm × 10 μm and 70/95% high reflective coating. The threshold voltage was 11 V. Currently, the lifetime of laser diodes operating at a temperature of 20°C has been reported to be 101.5 hours, the longest for ZnSe-based laser diodes.[17]

The spectacular progress in edge-emitting lasers has stimulated exploration of more advanced designs such as the vertical-cavity surface-emitting lasers (VCSELs) operating in the blue-green region. VCSELs have recently attracted much attention because of their surface-normal operation, potential for extremely low threshold currents, and the ease with which they may be fabricated in closely spaced and two-dimensional arrays. These lasers are ideal for integration with other devices such as transistors for photonic switching applications. Output characteristics such as narrow divergence beams and operation in a single longitudinal mode, due to the large mode spacing of a short cavity, are additional advantages.

Blue-green VCSELs have experienced significant progress recently. For example, electrical pumped operation has been demonstrated at 77K.[18] The VCSEL structures used were consistent with a CdZnSe/ZnSe multiquantum-well (MQW) active layer, *n*- and *p*-ZnSe cladding layers, and two SiO_2/TiO_2 distributed Bragg reflectors (DBRs), as shown in Figure 73. The reflectivity of the SiO_2/TiO_2 dielectric mirrors was greater than 99%. The VCSEL devices were characterized at 77K under pulsed operation. A very low threshold current of 3 mA was obtained in the VCSEL. Single longitudinal mode operation was obtained at the lasing wavelength of 484 nm. Above the threshold, the far-field radiation angle was as narrow as 7°, which indicated the spatial coherence expected for VCSEL

Pd /Pt/Au electrode

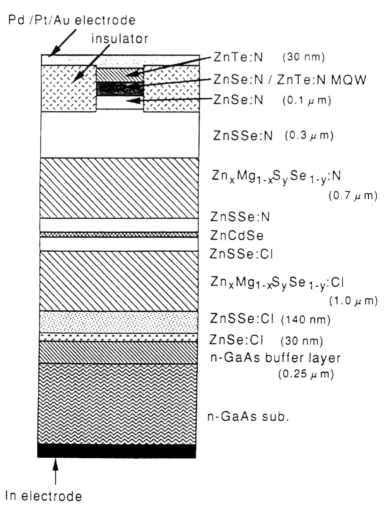

Figure 72 Schematic structure of ZnCdSe/ZnSSe/ZnMgSSe SCH lasers. (From Itoh, S., Nakayama, N., Matsumoto, S., et al., *Jpn. J. Appl. Phys.*, 33, L938, 1994. With permission.)

emission. This blue VCSEL opens the door for a broad range of new device applications for II-VI materials.

2.8.6 *ZnSe-based light-emitting diodes*

Further improved performance of ZnSe-based light-emitting diodes (LEDs) has also been demonstrated since the first demonstration of II-VI blue-green laser diodes. Recently, high-brightness ZnSe-based LEDs operating at peak wavelengths in the 489- to 514-nm range have been reported.[19] The LED consisted of a 3-μm thick *n*-type ZnSe:Cl layer, a 50- to 100-nm green-emitting active region of $ZnTe_{0.1}Se_{0.9}$, and a 1-μm thick *p*-type ZnSe:N layer grown by MBE on the ZnSe substrates. The devices produced 1.3 mW (10 mA, 3.2 V), peaking at 512 nm with an external efficiency of 5.3%. The emission spectrum of the LED was relatively broad (50 nm) due to emission from the ZnTeSe active region. The luminous performance of the device was 17 lm W^{-1} at 10 mA, which is comparable to the performance of super-bright red LEDs (650 nm) based on AlGaAs double heterostructures. The lifetime of the LEDs at RT has been reported to exceed 2000 hours, which is the longest lifetime of any ZnSe-LED device.

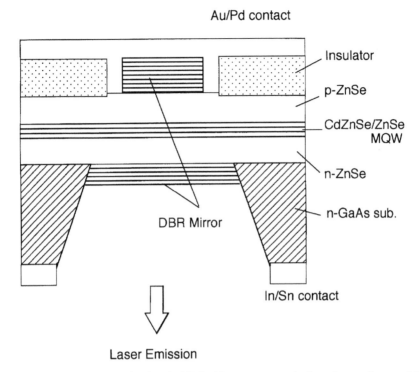

Au/Pd contact

Insulator

p-ZnSe

CdZnSe/ZnSe
MQW

n-ZnSe

n-GaAs sub.

DBR Mirror

In/Sn contact

Laser Emission

Figure 73 Schematic structure of CdZnSe/ZnSe blue-green vertical cavity surface emitting lasers. (From Yokogawa, T., Yoshii, S., Tsujimura, A., Sasai, Y., and Merz, J.L., *Jpn. J. Appl. Phys.*, 34, L751, 1995. With permission.)

High-brightness LEDs with a ZnCdSe quantum well have been demonstrated.[20] The LED consists of a ZnCdSe/ZnSSe multiquantum well and ZnMgSSe cladding layers grown on GaAs substrates. The devices produced 2.1 mW (20 mA, 3.9 V), and peaked at 512 nm with an external efficiency of 4.3%. A narrow spectral output of 10 nm has been obtained in the devices.

References

1. Mitsuhashi, H., Mitsuishi, I., Mizuta, M., and Kukimoto, H., *Jpn. J. Appl. Phys.*, 24, L578, 1985.
2. Kukimoto, H., *J. Cryst. Growth*, 101, 953, 1990.
3. Yasuda, T., Mitsuishi, I., and Kukimoto, H., *Appl. Phys. Lett.*, 52, 57, 1988.
4. Kamata, A., Mitsuhashi, H., and Fujita, H., *Appl. Phys. Lett.*, 63, 3353, 1993.
5. Wolk, J.A., Ager, III, J.W., Duxstad, K., Haller, J.E.E., Tasker, N.R., Dorman, D.R., and Olego, D.J., *Appl. Phys. Lett.*, 63, 2756, 1993.
6. Shibata, N., Ohki, A., and Katsui, A., *J. Cryst. Growth*, 93, 703, 1988.
7. Yasuda, T., Mitsuishi, I., and Kukimoto, H., *Appl. Phys. Lett.*, 52, 57, 1988.
8. Gunshor, R.L., Kolodziejski, L.A., Melloch, M.R., Vaziri, M., Choi, C., and Otsuka, N., *Appl. Phys. Lett.*, 50, 200, 1987.
9. Niina, T., Minato, T., and Yoneda, K., *Jpn. J. Appl. Phys.*, 21, L387, 1982.
10. Ohkawa, K., Mitsuyu, T., and Yamazaki, O., *J. Appl. Phys.*, 62, 3216, 1987.
11. DePuydt, J.M., Hasse, M.A, Cheng, H., and Potts, J.E., *Appl. Phys. Lett.*, 55, 1103, 1989.
12. Chadi, D.J. and Chang, K.L., *Appl. Phys. Lett.*, 55, 575, 1989.
13. Park, R.M., Troffer, M.B., Rouleau, C.M., DePuydt, J.M., and Haase, M.A., *Appl. Phys. Lett.*, 57, 2127, 1990.
14. Ohkawa, K., Karasawa, T., and Mitsuyu, T., *Jpn. J. Appl. Phys.*, 30, L152, 1991.
15. Hasse, M.A., Qiu, J., DePuydt, J.M., and Cheng, H., *Appl. Phys. Lett.*, 59, 1272, 1991.

16. Itoh, S., Nakayama, N., Matsumoto, S., Nagai, M., Nakano, K., Ozawa, M., Okuyama, H., Tomiya, S., Ohata, T., Ikeda, M., Ishibashi, A., and Mori, Y., *Jpn. J. Appl. Phys.*, 33, L938, 1994.
17. Taniguchi, S., Hino, T., Itoh, S., Nakano, K., Nakayama, N., Ishibashi, A., and Ikeda, M., *Electron. Lett.*, 32, 552, 1996.
18. Yokogawa, T., Yoshii, S., Tsujimura, A., Sasai, Y., and Merz, J.L., *Jpn. J. Appl. Phys.*, 34, L751, 1995.
19. Eason, D.B., Yu, Z., Hughes, W.C., Roland, W.H., Boney, C., Cook, Jr., J.W., Schetzina, J.F., Cantwell, G., and Harsch, W.C., *Appl. Phys. Lett.*, 66, 115, 1995.
20. *Nikkei Electronics*, 614, 20, 1994 (in Japanese).

Principal phosphor materials and their optical properties

Hiroshi Kukimoto

Contents

2.9   IIIb-Vb compounds

2.9.1   General overview

IIIb-Vb compounds, which consist of the group IIIb and Vb elements of the periodic table, include many important semiconductors such as GaP, GaAs, GaN, and InP. These materials are not used for phosphors in a polycrystalline form as is the case of IIb-VIb compounds, but they are utilized for many optoelectronic devices such as light-emitting diodes, semiconductor lasers, and photodiodes in a single crystalline form of thin films. IIIb-Vb compounds are to some extent similar to IIb-VIb compounds in terms of the nature of the atomic bond; more precisely, from a viewpoint of the nature of their ionic and covalent bonding, they are well situated between group IV elemental semiconductors and IIb-VIb compound semiconductors. Therefore, many similarities in the optical properties can be seen between these classes of compounds. In some cases, the optical properties due to impurities can be more clearly observed in IIIb-Vb compounds than in IIb-VIb compounds. Before moving on to the optical properties of typical IIIb-Vb compounds, an overview of the composition of this group is presented.

Typical characteristics of IIIb-Vb compounds are shown in Table 23. The materials composed of lighter elements tend to be more ionic than those composed of heavier elements. This trend is reflected in their crystal structure and energy gap; the wurtzite

Table 23 Properties of IIIb-Vb Compounds

Material	Crystal structure[a]	Lattice const. (Å) a	Lattice const. (Å) c	Density (g cm^{-3})	Melting point (°C)	Band structure[b]	Bandgap[c] (eV)	Effective mass[d] Electron	Effective mass[d] Hole	Mobility (cm^2 V^{-1} s^{-1}) Electron	Mobility Hole	Dielectric constant Static ε_0	Optical ε_∞	Refractive index[e]
BN	ZB	3.615	—	3.49		ID	7.2					7.1	4.5	2.12 (0.589)
	H	2.51	6.69	2.26		D	3.8					6.85 (⊥c), 5.09 (∥c)	4.95 (⊥c), 4.10 (∥c)	2.20 (0.05)
BP	ZB	4.538	—	2.97		ID	2.0					11	8.2	3.0~3.5
BAs	ZB	4.777	—	5.22	>2000	ID	1.6	1.2	0.51 (h), 0.2 (l)				10.2	
BSb	ZB	—	—											
AlN	W	3.111	4.980	3.26		D	6.20	0.29				8.5	4.8	2.25 (0.4)
AlP	ZB	5.467	—	2.40	1525	ID	2.45		0.63 (h), 0.20 (l)	80		9.8	7.5	2.99 (0.5)
AlAs	ZB	5.662	—	3.60	1740	ID	2.15	0.5, 1.56 (∥), 0.19 (⊥)	0.5 (h), 0.26 (l)	180	290	10.1	8.2	3.2 (0.56)
AlSb	ZB	6.136	—	4.26	1080	ID	1.63	0.39, 1.64 (∥), 0.23 (⊥)	0.5 (h), 0.11 (l)	200	400	12.0	10.2	3.45 (1.1)
GaN	W	3.189	5.185	6.10		D	3.39	0.22	0.8	1200		9.5 (⊥c), 10.4 (∥c)	5.4	2.00 (0.58)
GaP	ZB	5.451	—	4.13	1465	ID	2.27	0.25	0.67 (h), 0.17 (l)	300	100	11.0	9.1	5.19 (0.344)
GaAs	ZB	5.653	—	5.32	1238	D	1.43	0.0665	0.475 (h), 0.087 (l)	8500	400	12.9	10.9	3.66 (0.8)
GaSb	ZB	6.096	—	5.61	712	D	0.70	0.042	0.32 (h), 0.045 (l)	4000	1400	15.7	14.4	3.82 (1.8)
InN	W	3.533	5.693	6.88	~1200	D	0.6~0.7	0.04		4000		15.0	6.3	
InP	ZB	5.869	—	4.79	1070	D	1.34	0.079	0.45 (h), 0.12 (l)	4600	650	12.6	9.6	3.33 (1.0)
InAs	ZB	6.058	—	5.67	943	D	0.35	0.023	0.41 (h), 0.025 (l)	33000	460	15.2	12.3	3.52 (3.74)

Table 23 Properties of IIIb-Vb Compounds (continued)

Material	Crystal structure[a]	Lattice const. (Å) a	Lattice const. (Å) c	Density (g cm⁻³)	Melting point (°C)	Band structure[b]	Bandgap[c] (eV)	Effective mass[d] Electron	Effective mass[d] Hole	Mobility (cm² V⁻¹ s⁻¹) Electron	Mobility (cm² V⁻¹ s⁻¹) Hole	Dielectric constant Static ε₀	Dielectric constant Optical ε∞	Refractive index[e]
InSb	ZB	6.479	—	5.78	525	D	0.18	0.014	0.40 (h), 0.016 (l)	78000	750	16.8	15.7	4.00 (7.87)

[a] ZB: zinc-blende, H: hexagonal, W: wurtzite.
[b] D: direct type, ID: indirect type.
[c] At 300K.
[d] ∥, ⊥: parallel and perpendicular to the principal axis; h and l: heavy and light holes.
[e] Wavelength μm in parenthesis.

structure and wider gaps are prevalent in lighter materials, while the zinc-blende structure and narrower bandgaps occur in heavier materials. Furthermore, one should note that optical properties of these materials largely depend on the type of energy band structure, direct (D) or indirect (ID).

Light materials, including BN, BP, AlN, and GaN, have high melting points and wide bandgaps. In general, conductivity control of these materials has not been so easy. Recently, GaN and related alloys have become important materials for blue light emission as is described in Section 2.8.5. In contrast, heavy materials including AlSb, GaSb, InSb, and InAs have low melting points and narrow bandgaps. In addition, they features high mobility. These properties are suited for light-emitting devices and photodetectors operating in the infrared region.

AlP, AlAs, GaP, GaAs, and InP are located between the above two extremes. Their bandgaps range from the near-infrared to the visible light region, and these materials and related alloys of AlGaAs, GaPAs, GaInP, GaInAs, and GaAlPAs are key materials for the optoelectronic applications, which are described in Section 2.8.3.

2.9.2   GaP as luminescence material

2.9.2.1   Energy band structure

GaP is an indirect gap semiconductor with an energy band structure similar to that of Si, as illustrated in Figure 74. The bottom of the conduction band is located near the X point in momentum or wave vector (k) space, while the top of the valence band is found at the Γ point. Before and after the event of optical transition, momentum must be conserved for light and electron alike. Because the momentum of light is negligibly small, direct electron transitions must take place between bands at the same k values. Therefore, the probability of an intrinsic optical transition across the bandgap in GaP, i.e., between the X and Γ points, is inherently very low unless phonons participate in the transition. For the same reason, the optical transition probability associated with *shallow* donors and acceptors is also small. Nevertheless, GaP is an important material for practical light-emitting diodes. The reason for this is to be found in the following section.

2.9.2.2   Isoelectronic traps

The description of isoelectronic traps given in this section can also be found in 1.4 dealing with the fundamentals of luminescence in semiconductors. Considering nitrogen-doped GaP, the nitrogen (N) atom enters at the phosphorous (P) site of the GaP lattice. N and P atoms are isoelectronic with each other since they belong to the same Vb column of the periodic table and have the same number of valence electrons. Therefore, the N impurity in GaP would appear to be unable to bind electrons or holes to itself as do other common donor or acceptor impurities in semiconductors. Yet, the N atom in GaP does bind an electron because N is more attractive to electrons than P, owing to the nuclear charge of N being more exposed, i.e., due to large difference of electron negativity between N and P. Similarly, a Bi atom in GaP can bind a hole. These impurities in GaP are called *isoelectronic traps* (or centers).[2]

Since the trapped electron is localized around the N atom in real space, its wavefunction is spread considerably in momentum space. Such a situation is shown in Figure 75, where the electron density of an isoelectronic trap with a binding energy of 10 meV is compared to that of a shallow donor with a binding energy of 100 meV. One can clearly see that the amplitude of the electron wavefunction at the Γ point for the isoelectronic trap is about three orders of magnitude larger than that for the shallow donor.

Once an electron is bound to N by a short range potential, a hole can also be bound to the negatively charged center by the Coulombic potential, resulting in the formation

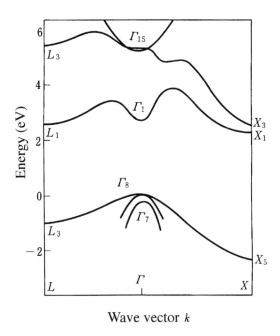

Figure 74 Energy band structure of GaP. Energies at 0K are: $X_1 - \Gamma_8 = (2.339 + 0.002)$ eV, $\Gamma_1 - \Gamma_8 = (2.878 + 0.002)$ eV. (From Cohen, M.L. and Bergstresser, T.K., *Phys. Rev.*, 141, 789, 1966. With permission.)

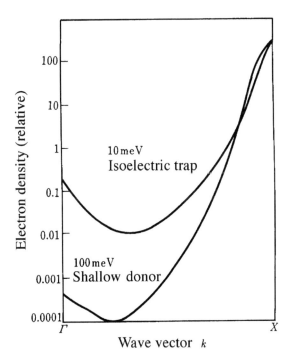

Figure 75 Electron density distributions for a 10-meV isoelectronic trap and a 100-meV shallow donor in GaP. (From Dean, P.J., *J. Luminesc.*, 1/2, 398, 1970. With permission.)

of an exciton bound to N. Thus, the radiative recombination of excitons bound to N in GaP takes place with high probability. The high efficiency of bound exciton recombination at N centers is further promoted by the fact that Auger recombination due to a third particle (electron or hole) cannot take place as it does in the case of exciton recombination at neutral donors or acceptors. Thus, N is responsible for the efficient luminescence observed in green GaP light-emitting diodes.

At high N concentrations, the luminescence due to excitons preferably bound to N-N pairs is observed.[4] The pair distribution in GaP occurs with different distances and leads to the spectrum shown in Figure 76. This type of luminescence is used for yellow-green GaP light-emitting diodes.

Slightly more complicated isoelectronic traps in GaP consist of nearest-neighbor donor-acceptor complexes of Zn-O, Cd-O, or Mg-O, and a triplex of Li-O-Li.[5,6] Each of these complexes can be regarded as being isoelectronic with one GaP molecule where eight valence electrons reside. Because of the highly localized nature of the O potential, these complexes can bind electrons and form bound excitons as is the case of N. The luminescence due to excitons bound to Zn-O is utilized for red GaP light-emitting diodes.

2.9.2.3 *Donor-acceptor pair emission*

A general concept and nature of donor-acceptor pair emission is described in 1.4. The important equations for the emission are repeated here. The transition energy $E(R)$ of a discrete pair with separation R is given by:

$$E(R) = E_g - (E_D + E_A) + e^2/4\pi\varepsilon R \tag{39}$$

where E_g is bandgap energy, E_D and E_A are the donor and acceptor binding energies, respectively, e is the electronic charge, and ε is the static dielectric constant. On the other hand, the transition probability $W(R)$ between a tightly bound electron (or hole) and a loosely bound hole (or electron) is approximately given by:

$$W(R) = W_0 \exp(-2R/R_B) \tag{40}$$

where W_0 is a constant and R_B is the Bohr radius of a loosely bound electron (or hole).

Since GaP is an indirect gap semiconductor with a low transition probability, emission from the remote pair can be easily saturated under high excitation conditions. This situation results in the observation of well-resolved, fine line structure in the luminescence spectra corresponding to various donor-acceptor pairs with discrete values of R. The spectrum as shown in Figure 24 in 1.4 is for the emission taking place between S donors substituting into the P sites and Si acceptors substituting into the Ga sites. For this type of emission (type I) in a zinc-blende structure, R can be expressed in terms of a shell number m as $R = (m/2)^{1/2}a_0$, where $m \neq 14, 30, 46,\ldots$. From this relation, it is possible to assign specific lines with corresponding R values. Once the value of R is determined, the observed energy can be plotted against R. Then, with extrapolation to $R = \infty$, $E_g - (E_D + E_A)$ can be determined. Fits of the simple expression Eq. 39 to some observed values are shown in Figure 77 as examples.

If either E_D or E_A is known through other experiments, the other is determined. The results obtained in this manner for emission spectra arising from various pair combinations of donors and acceptors in GaP are shown in Table 24.

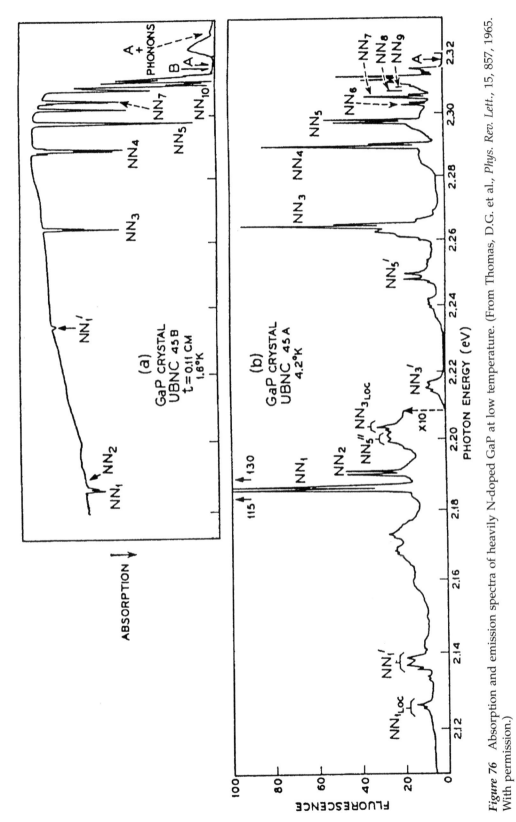

Figure 76 Absorption and emission spectra of heavily N-doped GaP at low temperature. (From Thomas, D.G. et al., *Phys. Rev. Lett.*, 15, 857, 1965. With permission.)

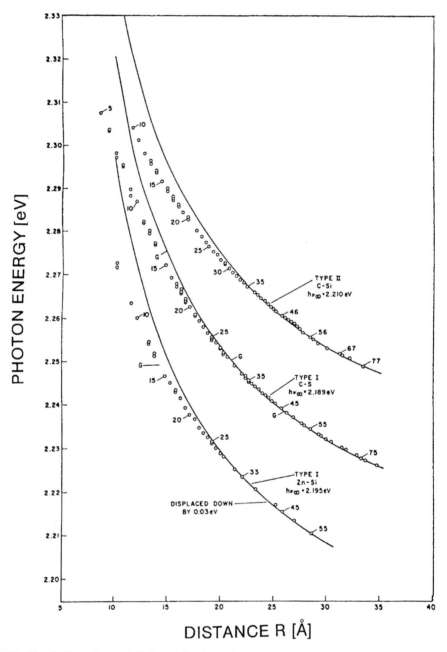

Figure 77 The fitting of type-I C-S and Zn-Si and of type-II C-Si pair spectra in GaP to Eq. 39. (From Dean, P.J., Frosch, C.J., and Henry, C.H., *J. Appl. Phys.*, 39, 5631, 1968. With permission.)

2.9.2.4 Application for light-emitting diodes

The characteristics of practical GaP light-emitting diodes are summarized in Table 25. One should note again that isoelectronic impurities of N, N-N pairs, and Zn-O are utilized for green, yellow, and red light-emitting diodes, respectively. Another thing to be noted is that pure-green diodes have also become available, where GaP without isoelectronic impurities is used. This has become possible by improving crystal quality in terms of decreasing the defects that act as nonradiative recombination centers. The emission is ascribed to the transition associated with free holes and donor bound electrons.

Table 24   Energy Depths of Donors and
Acceptors in GaP

Donor	E_D (meV)	Acceptor	E_A (meV)
Li(int. A)	58	C(P)	46.4
Sn(Ga)	69	Be(Ga)	48.7
Si(Ga)	82.1	Mg(Ga)	52.0
Li(int. B)	88.3	Zn(Ga)	61.7
Te(P)	89.8	Cd(Ga)	94.3
Se(P)	102.6	Si(P)	202
S(P)	104.2	Ge(P)	257
Ge(Ga)	201.5		
O(P)	896.0		

Note: P or Ga in parentheses indicates the lattice
sites to be substituted. Li occupies two different interstitial sites A and B.

Table 25   Properties of GaP Light-Emitting Diodes

Materials	Emission color	Peak wavelength (nm)	Ext. quantum efficiency (%)	Luminous efficiency (lm/W)
GaP:Zn,O	Red	700	4	0.8
GaP:NN	Yellow	590	0.2	0.9
GaP:N	Green	565	0.3	1.8
GaP	Pure green	555	0.2	1.4

References

1. Cohen, M.L. and Bergstresser, T.K., *Phys. Rev.*, 141, 789, 1966.
2. Thomas, D.G. and Hopfield, J.J., *Phys. Rev.*, 150, 680, 1966.
3. Dean, P.J., *J. Luminesc.*, 1/2, 398, 1970.
4. Thomas, D.G. et al., *Phys. Rev. Lett.*, 15, 857, 1965.
5. Henry, C.H., Dean, P.J., and Cuthbert, J.D., *Phys. Rev.*, 166, 754, 1968.
6. Dean, P.J. and Illegems, M., *J. Luminesc.*, 4, 201, 1971.
7. Dean, P.J., Frosch, C.J., and Henry, C.H., *J. Appl. Phys.*, 39, 5631, 1968.

Principal phosphor materials and their optical properties

Gen-ichi Hatakoshi

Contents

2.10   (Al,Ga,In)(P,As) alloys emitting visible luminescence

2.10.1   Bandgap energy

GaAlAs, GaAsP, InGaAsP, and InGaAlP are IIIb-Vb compound semiconductor materials used for devices in the visible wavelength region. Table 26 shows the compositional dependence of the bandgap energy E_g,[1–11] where E_g^Γ, E_g^X and E_g^L correspond to the distance between valence-band edge and conduction-band edge for Γ, X, and L valleys, respectively. Emission by direct transition occurs in a composition region, where the E_g^Γ value is smaller than that for E_g^X and E_g^L.

Lattice constants of alloys are determined by their composition and generally vary depending on the composition ratio. Therefore, the lattice constant of ternary alloys such as GaAlAs and GaAsP is determined uniquely by the bandgap energy value. In the case of GaAlAs, the compositional dependence of the lattice constant a is very small: for example, a = 5.653 Å for GaAs and a = 5.661 Å for AlAs.[1,2] Therefore, epitaxial layers of GaAlAs can be grown using a GaAs substrate. The change in the lattice constant of GaAsP is comparatively large; in this case, GaAs or GaP is used as a substrate, depending on the composition of the epitaxial layer.

In quaternary alloys such as InGaAsP and InGaAlP, the bandgap energy can be varied without altering the value of the lattice constant. The E_g value for InGaAlP[9–11] in Table 26 corresponds to the case where the alloy is lattice-matched to GaAs. This means that GaAs can be used as a substrate for crystal growth of InGaAlP alloys. InGaAsP can also be lattice-matched to GaAs, and visible light emission is obtained for this case. Such lattice

Table 26   Compositional Dependence of the Bandgap Energy

Material system	Bandgap energy (eV)	Direct-indirect transition point[a]	Ref.
$Ga_{1-x}Al_xAs$	$E_g^{\Gamma} = 1.425 + 1.155x + 0.370x^2$ $E_g^{X} = 1.911 + 0.005x + 0.245x^2$ $E_g^{L} = 1.734 + 0.574x + 0.055x^2$	$x_c = 0.4{-}0.45$ $E_g(x = 0.4) \sim 1.95$ eV	3
$GaAs_{1-x}P_x$	$E_g^{\Gamma} = 1.424 + 1.150x + 0.176x^2$ $E_g^{X} = 1.907 + 0.144x + 0.211x^2$	$x_c = 0.45{-}0.49$ $E_g(x = 0.49) \sim 2.03$ eV	2,4,5
	$E_g^{\Gamma} = 1.514 + 1.174x + 0.186x^2$ (77K) $E_g^{X} = 1.977 + 0.144x + 0.211x^2$ (77K) $E_g^{L} = 1.802 + 0.770x + 0.160x^2$ (77K)		5,6
$In_{1-x}Ga_xAs_yP_{1-y}$	$E_g^{\Gamma} = 1.35 + 0.668x - 1.068y$ $+ 0.758x^2 + 0.078y^2 - 0.069xy$ $- 0.322x^2y + 0.03xy^2$		7
$In_{0.5}(Ga_{1-x}Al_x)_{0.5}P$	$E_g^{\Gamma} = 1.91 + 0.59x$ $E_g^{X} = 2.26 + 0.09x$	$x_c = 0.6{-}0.7$ $E_g(x = 0.7) \sim 2.32$ eV	9–11

Note: Values at room temperature except as indicated.

[a] There is some discrepancy in the value for x_c and several values are reported.

matching with GaAs can be realized by selecting the composition ratio according to y ~ 0.5 for the $In_{1-y}(Ga_{1-x}Al_x)_yP$ system and x ~ (1 + y)/2.08 for the $In_{1-x}Ga_xAs_yP_{1-y}$ system,[8] respectively.

All materials described here have the zinc-blende structure. Band structures for $Ga_{1-x}Al_xAs$, $In_{0.5}(Ga_{1-x}Al_x)_{0.5}P$, and $GaAs_{1-x}P_x$ vary between direct transition and indirect transition types. In general, direct transition-type crystals have the advantages of high radiative efficiency and narrow emission spectrum.

2.10.2   *Crystal growth*

Thin-film crystals for optical devices using the aforementioned compound semiconductors are grown by liquid phase epitaxy (LPE), vapor phase epitaxy (VPE), and molecular beam epitaxy (MBE). LPE utilizes the recrystallization of the solute from a supersaturated solution. Conventional halogen-transport VPE is classified into hydride VPE and chloride VPE. Metalorganic chemical vapor deposition (MOCVD), another VPE method, uses metal organic compounds, such as trimethylgallium (TMGa) and trimethylindium (TMIn), as source gases for Group III materials. MBE is a type of ultra-high-vacuum deposition, where molecules or atoms of the constituent elements are supplied from solid sources or gas sources.

Attainable device structures for light-emitting diodes (LEDs) and semiconductor lasers depend on the method of crystal growth. For example, the growth aspect on a stepped or grooved substrate varies, depending on the method. In the LPE method, crystal growth proceeds so as to embed and level the groove. Such a characteristic feature has been utilized to obtain various structures of GaAlAs semiconductor lasers for practical use.[12,13] InGaAsP crystal can also grown by LPE. Transverse-mode stabilized structures for InGaAsP/GaAlAs semiconductor lasers oscillating in the 0.6-μm wavelength range have been grown by the LPE method.[14]

The problems with the LPE method arise from the difficulty to grow ultra-thin layers and to control the composition of epitaxial layers for some material systems. For example, the segregation coefficient (defined as the ratio of atoms incorporated from the liquid solution to those in the solid crystal) of Al is relatively large in the case of LPE growth for GaAlAs. This causes a gradual decrease in the Al amount in the solution, resulting in a graded composition structure for thick GaAlAs growth. The problem of segregation is even more serious for InGaAlP growth. It was very difficult to obtain high-quality InGaAlP crystals by the LPE methods because of the extremely large segregation coefficient of Al.[15]

The development of MBE and MOCVD techniques has enabled the production of high-quality, thin-film crystals for the InGaAlP systems.[9,16–18] The MBE and MOCVD methods have an advantage in that controlled ultra-thin layers, which can be applied to form multiquantum well (MQW) structures for light-emitting devices, can easily be obtained.

In order to realize a double heterostructure for semiconductor lasers and LEDs, *p*-type and *n*-type semiconductor crystals are required. In general, Group VI elements, such as Se and S, act as donors for the III-V system and thus are used as *n*-type dopants. Group II elements such as Zn, Mg, and Be behave as acceptors and are used as *p*-type dopants. Group IV dopants such as Si and Ge are amphoteric impurities. For example, when Si is substituted for a Group III site atom, it acts as a donor. On the contrary, it acts as an acceptor when substituted for a Group V site atom. The substitution site depends on the growth condition.

2.10.3 Characteristics of InGaAlP crystals grown by MOCVD

An attractive technique for MOCVD growth of InGaAlP material systems is growth on an off-angle substrate. This process is related to the formation of a natural superlattice.[19–21] In the InGaAlP system, the bandgap energy value depends on the growth condition, for example, on the growth temperature. This is attributed to the dependence of atomic ordering on the growth temperature. An ordered structure of an InGaAlP alloy is produced by the formation of a natural superlattice, where the Group III atoms are arranged systematically. It is known that a disordered alloy has a larger bandgap energy than that of an ordered

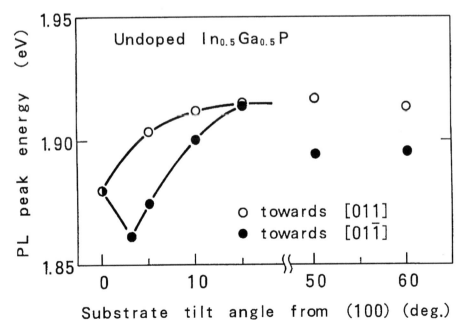

Figure 78 Photoluminescence (PL) peak energy of InGaP vs. substrate orientation. (From Suzuki, M., Nishikawa, Y., Ishikawa, M., and Kokubun, Y., *J. Crystal Growth*, 113, 127, 1991. With permission.)

alloy, and that the disordered state is enhanced by using an intentionally misoriented substrate. It follows that the bandgap energy of the InGaAlP crystal grown on a misoriented substrate has a larger bandgap energy than that grown on a (100)-oriented substrate.

Figure 78 shows the dependence of the bandgap energy, obtained by photoluminescence measurement, of InGaAlP alloys on the substrate orientation.[22] As shown in the figure, the bandgap energy increases with increasing substrate tilt angle away from the (100) plane toward the [011] direction.* This is considered to be due to the suppression of crystal ordering.

The off-angle substrate technique is utilized to fabricate short-wavelength InGaAlP lasers. In general, shortening of the oscillating wavelength or, equivalently, an increase in the bandgap energy is obtained by increasing the Al composition of the alloy as shown in Table 26. Introduction of off-angle substrates has the advantage of wavelength shortening while using a smaller Al content in the active layer. This is preferable because formation of undesirable nonradiative recombination centers arising from incorporation of oxygen impurities (which increase with increasing Al composition) is reduced. The

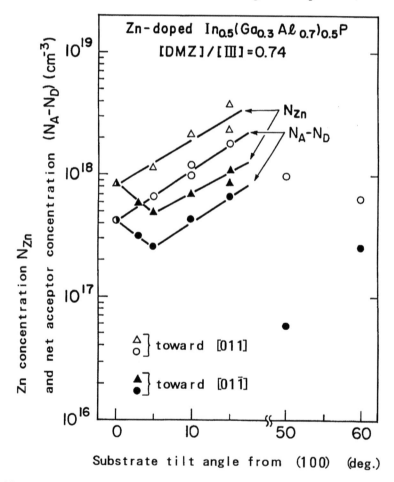

Figure 79 Net acceptor concentration and Zn concentration vs. substrate orientation. (From Suzuki, M., Nishikawa, Y., Ishikawa, M., and Kokubun, Y., *J. Crystal Growth*, 113, 127, 1991. With permission.)

* Conventionally, (*hkl*) and [*hkl*] represent a crystal plane and a crystal direction, respectively: e.g., (100) denotes a crystal plane normal to the [100] direction.

bandgap of the cladding layers can also be increased by using off-angle substrates. Thus, electron overflow can be effectively suppressed by creating a larger bandgap difference between the active and cladding layers in this way.

Another effect of off-angle substrates is to increase acceptor concentration in *p*-type layers. As shown in Figure 79, Zn incorporation and the net acceptor concentration strongly depend on the tilt angle of the substrate.[22] This dependence is similar to that for the PL peak energy shown in Figure 78. Both the Zn concentration and the net acceptor concentration increase with increasing tilt angle from (100) toward the [011] direction. High acceptor concentrations are preferable for *p*-type cladding layers because of their effect in reducing electron overflow from the active layer to the *p*-cladding layer,[23,24] due to the increase in the conduction-band heterobarrier height at the interface between the active and the *p*-cladding layers.

Electrical activity of *p*-type dopants depends on the effects of residual impurities such as hydrogen and oxygen and also upon growth conditions. Oxygen incorporation into InGaAlP crystals results in the electrical compensation of Zn acceptors. It also causes a nonradiative center due to the formation of deep levels. These phenomena are serious problems for light-emitting devices. Oxygen incorporation can be reduced by increasing the V:III ratio in MOCVD growth[25] and by the utilization of the off-angle technique.[26] An example of experimental results is shown in Figure 80.[26] The effect on oxygen reduction in off-angle substrate is remarkable, especially for high Al composition crystals, and is very useful for producing highly doped *p*-type cladding layers. High acceptor concentrations exceeding 1×10^{18} cm^{-3} have been reported for InAlP crystals fabricated by MOCVD growth on off-angle substrates.[27]

Experimental results showing improvements in the hetero-interface properties of quantum wells grown on misoriented substrates have been reported.[28] Full width at half maximum (FWHM) value of the PL spectra for InGaP/InGaAlP single quantum wells shows a strong dependence on the substrate misorientation. The FWHM value is found

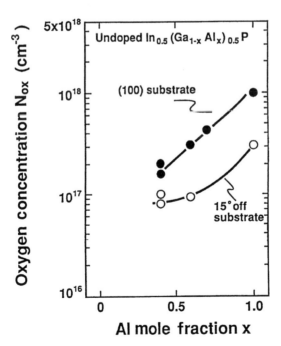

Figure 80 Oxygen concentration vs. Al mole fraction. (From Suzuki, M., Itaya, K., Nishikawa, Y., Sugawara, H., and Okajima, M., *J. Crystal Growth*, 133, 303, 1993. With permission.)

to decrease with increasing misorientation from the (100) toward the [011] direction. This result indicates that the interface smoothness and abruptness are improved by employing off-angle substrates.

A remarkable improvement in the temperature characteristics of InGaAlP lasers has been achieved by employing an off-angle technique. Short-wavelength and high-temperature operation have been reported for InGaAlP lasers grown on misoriented substrates.

2.10.4 Light-emitting devices

Semiconductor lasers and LEDs in the visible wavelength region are obtained using GaAlAs, GaAsP, InGaAsP, and InGaAlP systems. Figure 81 shows the available wave-

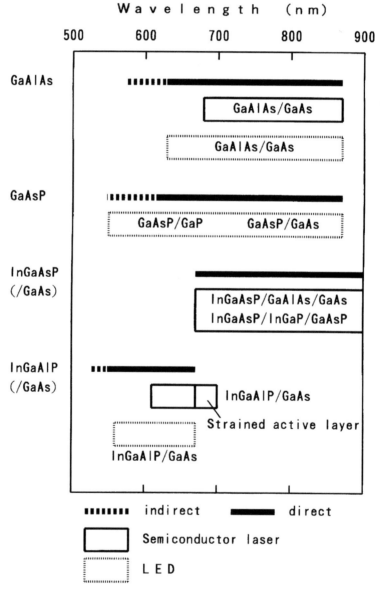

Figure 81 Available wavelength range for semiconductor lasers and LEDs. Constituent alloy systems are indicated by D/B or A/C/B, where D, A, C, and B denote the material systems for the double heterostructure, the active layer, the cladding layer, and the substrate, respectively.

length range for semiconductor lasers and LEDs. The wavelength range is restricted in the case of semiconductor lasers because the active layer is required to have a direct-transition-type band structure; here also, cladding layers with bandgap energies greater than that of the active layer are required in order to confine the injected carrier within the active layer. It is difficult, in general, to obtain shorter wavelength semiconductor lasers for a given material system because the bandgap difference between the active and the cladding layers decreases with shortening oscillation wavelength, resulting in a significant carrier overflow from the active layer. Visible-light oscillations in the 0.6-μm wavelength region have been realized for InGaAlP/GaAs,[16-18] GaAlAs/GaAs,[29] InGaAsP/GaAlAs/GaAs,[14] and InGaAsP/InGaP/GaAsP[30] systems.

As for LEDs, indirect-transition-type alloys can also be used for emission layers, and cladding layers are not necessarily required. Therefore, the possible wavelength range for LEDs is larger than that for semiconductor lasers. In general, high-brightness characteristics are obtained by using direct-transition alloys and by introducing a double heterostructure. The isoelectronic trap technique, which is effective in improving the emission efficiency of GaP LEDs, is also applicable to the GaAsP systems[5,6,31] in the indirect transition region. Nitrogen is used as the isoelectronic impurity. GaAsP:N LEDs show electroluminescence efficiencies of an order of magnitude higher than those without nitrogen doping.[31]

Examples of emission spectra for visible-light LEDs are shown in Figure 82. GaAlAs[32] and InGaAlP[33] alloys have direct transition band structures and thus the LEDs with these alloys have higher brightness and narrower emission spectra, as shown in the figure.

Light-extraction efficiency of LEDs is affected by various factors, which can be controlled by device design.[34,35] Remarkable enhancement of light-extraction efficiency has been reported for InGaAlP LEDs by introducing current-spreading and current-blocking layers.[33,34] Introduction of DBR mirror[36] is effective for LEDs with absorbing substrates. High-power InGaAlP/Gap LEDs with chip reshaping[37] have also been reported. Other

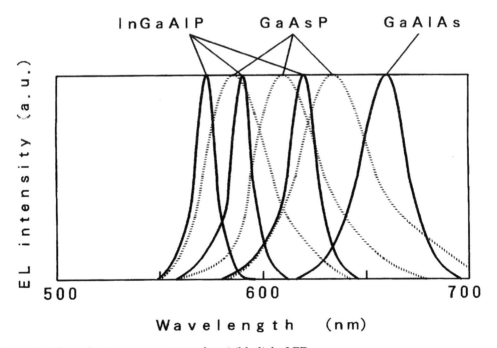

Figure 82 Electroluminescence spectra for visible-light LEDs.

approaches such as surface texture, resonant cavity structure, and photonic crystals have been investigated for improving the LED efficiency.[38]

References

1. Madelung, O., Ed., *Landolt-Börnstein, Numerical Data and Functional Relationships in Science and Technology*, III, 17 and 22a, Springer-Verlag, Berlin, 1982.
2. Casey, H.C., Jr. and Panish, M.B., *Heterostructure Lasers*, Academic Press, New York, 1978.
3. Lee, H.J., Juravel, L.Y., Woolley, J.C., and SpringThorpe, A.J., *Phys. Rev. B*, 21, 659, 1980.
4. Thompson, A.G., Cardona, M., Shaklee, K.L., and Woolley, J.C., *Phys. Rev.*, 146, 601, 1966.
5. Craford, M.G., Shaw, R.W., Herzog, A.H., and Groves, W.O., *J. Appl. Phys.*, 43, 4075, 1972.
6. Holonyak, N., Jr., Nelson, R.J., Coleman, J.J., Wright, P.D., Fin, D., Groves, W.O., and Keune, D.L., *J. Appl. Phys.*, 48, 1963, 1977.
7. Kuphal, E., *J. Cryst. Growth*, 67, 441, 1984.
8. Adachi, S., *J. Appl. Phys.*, 53, 8775, 1982.
9. Asahi, H., Kawamura, Y., and Nagai, H., *J. Appl. Phys.*, 53, 4928, 1982.
10. Honda, M., Ikeda, M., Mori, Y., Kaneko, K., and Watanabe, N., *Jpn. J. Appl. Phys.*, 24, L187, 1985.
11. Watanabe, M.O. and Ohba, Y., *Appl. Phys. Lett.*, 50, 906, 1987.
12. Aiki, K., Nakamura, M., Kuroda, T., Umeda, J., Ito, R., Chinone, N., and Maeda, M., *IEEE J. Quantum Electron.*, QE-14, 89, 1978.
13. Yamamoto, S., Hayashi, H., Yano, S., Sakurai, T., and Hijikata, T., *Appl. Phys. Lett.*, 40, 372, 1982.
14. Chong, T. and Kishino, K., *IEEE Photonics Tech. Lett.*, 2, 91, 1990.
15. Kazumura, M., Ohta, I., and Teramoto, I., *Jpn. J. Appl. Phys.*, 22, 654, 1983.
16. Kobayashi, K., Kawata, S., Gomyo, A., Hino, I., and Suzuki, T., *Electron. Lett.*, 21, 931, 1985.
17. Ikeda, M., Mori, Y., Sato, H., Kaneko, K., and Watanabe, N., *Appl. Phys. Lett.*, 47, 1027, 1985.
18. Ishikawa, M., Ohba, Y., Sugawara, H., Yamamoto, M., and Nakanisi, T., *Appl. Phys. Lett.*, 48, 207, 1986.
19. Suzuki, T., Gomyo, A., Iijima, S., Kobayashi, K., Kawata, S., Hino, I., and Yuasa, T., *Jpn. J. Appl. Phys.*, 27, 2098, 1988.
20. Nozaki, C., Ohba, Y., Sugawara, H., Yasuami, S., and Nakanisi, T., *J. Crystal Growth*, 93, 406, 1988.
21. Ueda, O., Takechi, M., and Komeno, J., *Appl. Phys. Lett.*, 54, 2312, 1989.
22. Suzuki, M., Nishikawa, Y., Ishikawa, M., and Kokubun, Y., *J. Crystal Growth*, 113, 127, 1991.
23. Hatakoshi, G., Itaya, K., Ishikawa, M., Okajima, M., and Uematsu, Y., *IEEE J. Quantum Electron.*, 27, 1476, 1991.
24. Hatakoshi, G., Nitta, K., Itaya, K., Nishikawa, Y., Ishikawa, M., and Okajima, M., *Jpn. J. Appl. Phys.*, 31, 501, 1992.
25. Nishikawa, Y., Suzuki, M., and Okajima, M., *Jpn. J. Appl. Phys.*, 32, 498, 1993.
26. Suzuki, M., Itaya, K., Nishikawa, Y., Sugawara, H., and Okajima, M., *J. Crystal Growth*, 133, 303, 1993.
27. Suzuki, M., Itaya, K., and Okajima, M., *Jpn. J. Appl. Phys.*, 33, 749, 1994.
28. Watanabe, M., Rennie, J., Okajima, M., and Hatakoshi, G., *Electron. Lett.*, 29, 250, 1993.
29. Yamamoto, S., Hayashi, H., Hayakawa, T., Miyauchi, N., Yano, S., and Hijikata, T., *Appl. Phys. Lett.*, 41, 796, 1982.
30. Usui, A., Matsumoto, T., Inai, M., Mito, I., Kobayashi, K., and Watanabe, H., *Jpn. J. Appl. Phys.*, 24, L163, 1985.
31. Craford, M.G. and Groves, W.O., *Proc. IEEE*, 61, 862, 1973.
32. Ishiguro, H., Sawa, K., Nagao, S., Yamanaka, H., and Koike, S., *Appl. Phys. Lett.*, 43, 1034, 1983.
33. Sugawara, H., Itaya, K., Nozaki, H., and Hatakoshi, G., *Appl. Phys. Lett.*, 61, 1775, 1992.
34. Hatakoshi, G. and Sugawara, H., *Display and Imaging*, 5, 101, 1997.
35. Hatakoshi, G., *10th Int. Display Workshop (IDW'03)*, Fukuoka, 1125, 2003.
36. Sugawara, H., Itaya, K., and Hatakoshi, G., *J. Appl. Phys.*, 74, 3189, 1993.
37. Krames, M.R., Ochiai-Holcomb, M., Hofler, G.E., Carter-Coman, C., Chen, E.I., Tan, I.-H., Grillot, P., Gardner, N.F., Chui, H.C., Huang, J.-W., Stockman, S.A., Kish, F.A., and Craford, M.G., *Appl. Phys. Lett.*, 75, 2365, 1999.
38. Issue on High-Efficiency Light-Emitting Diodes, *IEEE J. Sel. Top. Quantum Electron.*, 8, No. 2, 2002.

chapter two — section eleven

Principal phosphor materials and their optical properties

Kenichi Iga

Contents

2.11 (Al,Ga,In)(P,As) alloys emitting infrared luminescence

2.11.1 Compound semiconductors based on InP

Semiconductors for which bandgaps correspond to a long wavelength spectral region (1 to 1.6 μm) are important for optical fiber communication using silica fibers exhibiting extremely low loss and low dispersion, *infrared* imaging, lightwave sensing, etc. Figure 83 depicts a diagram of lattice constant vs. bandgap of several compound semiconductors based on InP, InAs, GaAs, GaN and AlAs, which can emit light in this infrared region.

    Semiconductor crystals for 1 to 1.6-μm wavelength emission. Ternary or quaternary semiconductor crystals are used since binary semiconductor crystals with 1 to 1.6-μm bandgaps are not available. Matching of lattice constants to substrates in crystal growth processes is important for fabricating semiconductor devices such as semiconductor lasers and light-emitting diodes (LEDs) with high current injection levels (>5 kA cm^{-2} μm^{-1}) or a high-output power density (>1 mW μm^{-2}) or for photodiodes used for low-noise detection of very weak optical signals.

    The bandgap of a specific quaternary crystal can be varied widely while completely maintaining the lattice match to a binary crystal used as a substrate, as shown in Figure 83. An example is Ga$_x$In$_{1-x}$As$_y$P$_{1-y}$, which utilizes InP (a = 5.8696 Å) as a substrate; the bandgap can be changed in the region of $0.7 \leq E_g \leq 1.35$ eV when the composition is adjusted along the vertical line. The corresponding emission wavelength ranges from 0.92 to 1.67 μm. The ternary materials lattice-matched to the InP substrate are Al$_{0.47}$In$_{0.53}$As and Ga$_{0.47}$In$_{0.53}$As.

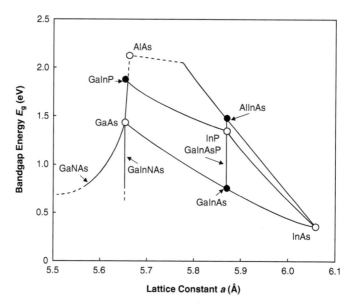

Figure 83 Diagram of lattice constant vs. bandgap for several compound semiconductors.

Possible compound crystals corresponding to light emission of 0.8 to 2 μm are as follows:

1. $Ga_xIn_{1-x}As_yP_{1-y}$ (InP): $0.92 < \lambda_g < 01.67$ (μm)
2. $(Ga_{1-x}Al_x)_yIn_{1-y}As$(InP): $0.83 < \lambda_g < 1.55$ (μm)
3. $Ga_{1-x}Al_xAs_ySb_{1-y}$(GaSb): $0.8 < \lambda_g < 1.7$ (μm)
4. $Ga_xIn_{1-x}As_ySb_{1-y}$(InAs): $1.68 < \lambda_g < 2$ (μm)
5. $Ga_xIn_{1-x}As_ySb_{1-y}$(GaSb): $1.8 < \lambda_g < 2$ (μm)
6. $Ga_xIn_{1-x}N_xAs_{1-x}$(GaAs): $1.1 < \lambda_g < 1.6$ (μm)

The binaries in the parentheses indicate the substrates to be used. Crystal growth of these materials is possible with a lattice mismatch ±0.1% or less. Among these, the heterostructure composed of $Ga_xIn_{1-x}As_yP_{1-x}$ and InP has been widely employed as a material for semiconductor lasers or photodiodes for lightwave systems.

The relationship between x, y, and the bandgap energy associated with $Ga_xIn_{1-x}As_yP_{1-y}$, which are lattice-matched to InP, can be expressed as follows.

$$x = \frac{0.466y}{1.03 - 0.03y} (0 \leq x \leq 1) \tag{41}$$

$$E_g(y) = 1.35 - 0.72y + 0.12y^2, \text{ (eV)} \tag{42}$$

which was phenomenologically obtained by Nahory et al.[1]

The values of x and y are no longer independent of one another, since the lattice constant must be adjusted so as to be matched to that of the InP substrate, 5.86875 Å. Consequently, the bandgap energy can be expressed by specifying the Ga or As contents. The band-structure parameters of GaInAsP/InP are summarized in Table 27.[2]

Longer-wavelength materials. Fluoride glass fibers have found use in long-distance optical communication in the 2- to 4-μm wavelength range. Signal loss in fluoride glass fibers is predicted to be one or two orders of magnitude lower than that for silica fibers. Also, this spectral band is important for LIDAR (Light Detection and Ranging) and optical

Table 27   The Band Structure Parameters of $Ga_xIn_{1-x}As_yP_{1-y}/InP$

Parameter	Dependence on the mole fractions x and y
Energy gap at zero doping	E_g [eV] = $1.35 - 0.72y + 0.12y^2$
Heavy-hole mass	$m^*_{hh}/m_0 = (1-y)[0.79x + 0.45(1-x)] + y[0.45x + 0.4(1-x)]$
Light-hole mass	$m^*_{lh}/m_0 = (1-y)[0.14x + 0.12(1-x)] + y[0.082x + 0.0261(1-x)]$
Dielectric constant	$\varepsilon = (1-y)[8.4x + 9.6(1-x)] + y[13.1x + 12.2(1-x)]$
Spin-orbit splitting	Δ [eV] = $0.11 - 0.31y + 0.09x^2$
Conduction-band mass	$m^*_e/m_0 = 0.080 - 0.039y$

From Agrawal, G.P. and Dutta, N.K., *Long-wavelength Semiconductor Lasers*, Van Nostrand Reinhold, New York, 1986, 85. With permission.

sensing. A potential material system to cover the wavelength range from 1.7 to 5 μm is GaInAsSb/AlGaAsSb.

2.11.2   Determination of GaInAsP/InP solid compositions

First, a review of the general concepts of crystal preparation for GaInAsP lattice-matched to InP, which has been commonly used in light-emitting devices. $Ga_xIn_{1-x}As_yP_{1-y}$ contains two controllable parameters, enabling independent adjustment of the lattice constant and the bandgap energy. The lattice constant $a(x,y)$ of $Ga_xIn_{1-x}As_yP_{1-y}$ is given as follows:

$$a(x,y) = a(GaAs)xy + a(GaP)x(1-y) + a(InAs)(1-x)y + a(InP)(1-x)(1-y) \qquad (43)$$

According to measurements by Nahory et al.,[1] the binary lattice constants are: $a(GaAs)$ = 5.653 Å, $a(GaP)$ = 5.4512 Å, $a(InAs)$ = 6.0590 Å, and $a(InP)$ = 5.8696 Å. The following equation is obtained by inserting this data into Eq. 43:

$$a(x,y) = 0.1894y - 0.4184x + 0.013xy + 5.8696 \; (\text{Å}) \qquad (44)$$

The relation between x and y, therefore, is given by the following equation, when the $a(x,y)$ coincides with the lattice constant of InP:

$$0.1894y - 0.4184x + 0.0130xy = 0 \qquad (45)$$

Usually, Eq. 45 is approximated as:

$$x = 0.467y \qquad (46)$$

According to the theory by Moon et al.[3] and experimental results, the relation between the bandgap energy and compositions x and y is given by:

$$E_g(x,y) = 1.35 + 0.672x - 1.091y + 0.758x^2 + 0.101y^2$$
$$-0.157xy - 0.312x^2y + 0.109xy^2 \qquad (47)$$

The bandgap energy calculated in terms of x and y using Eq. 47 agrees with the phenomenological results of Nahory et al.[1]

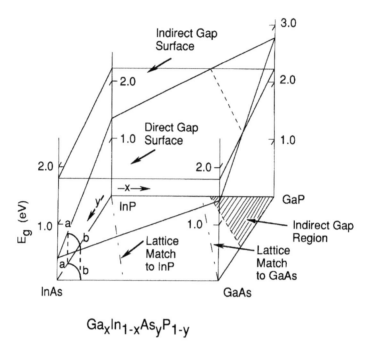

$$Ga_xIn_{1-x}As_yP_{1-y}$$

Figure 84 Bandgap energy vs. compositions x and y in $Ga_xIn_{1-x}As_yP_{1-y}$. (From Casey, H.C. and Panish, M.B., *Heterostructure Lasers*, Part B, Academic Press, New York, 1978. With permission.)

The bandgap energy vs. compositions x and y is illustrated in Figure 84.[4] With the aid of this figure, one can obtain the band structure of GaInAsP lattice-matched to InP for the entire set of allowed compositions of y. The bandgap of GaInAsP in the vicinity of GaP is seen to be indirect in the figure.

2.11.3 Crystal growth

Liquid phase epitaxy (LPE). In the case of liquid phase epitaxy, one has to determine the liquid composition of an In-rich melt in thermal equilibrium with the solid phase of the desired x and y compositions for $Ga_xIn_{1-x}As_yP_{1-y}$. The As composition y in the $Ga_xIn_{1-x}As_yP_{1-y}$ solid of the desired bandgap energy is given by Eq. 42 when its lattice constant is equal to that of InP. The Ga composition x is obtained by Eq. 46. In this way, the atomic fractions of Ga, As, and P in the In-rich melt that exists in equilibrium with the desired $Ga_xIn_{1-x}As_yP_{1-y}$ solid can be obtained. The actual weights of InP, InAs, and GaAs per gram of In can be estimated. The degree of lattice mismatching $|\Delta a/a|$ can be examined by X-ray diffraction and should be less than 0.05%.

Metal-organic chemical vapor deposition (MOCVD). In the metal-organic chemical vapor deposition (MOCVD) method, gas sources are used for growth of the structures.[5] To satisfy the lattice-match condition, the flow rates of trimethylindium and arsine (AsH_3) are fixed and the triethylgallium flow rate is adjusted. The phosphine (PH_3) flow rate is varied to obtain different compositions. Growth rates of InP and quaternary materials are about 2 μm/h, differing slightly for different alloy compositions. The compositions are calculated from the wavelength of the photoluminescence spectral peak intensities.

Chemical beam epitaxy (CBE). Trimethylindium and triethylgallium with H_2 carrier gas are used as Group III sources in chemical beam epitaxy (CBE) deposition.[6] Group V sources are pure AsH_3 and PH_3, which are precracked at 1000°C by a high-temperature cracking cell. Solid Si and Be are used as *n*-type and *p*-type dopants, respectively. The typical growth temperature is 500°C, which must be calibrated, for example, using the melting point of InSb (525°C). Typical growth rates for InP, GaInAsP (λ_g = 1.3 µm), and GaInAsP (λ_g = 1.55 µm) are 1.5, 3.8, and 4.2 µm/h, respectively.

Impurity doping control over wide ranges is one of the most important issues in the fabrication of optoelectronic devices. The advantages of using Be are that it is a well-behaved acceptor producing a shallow level above the valence band, and it can be incorporated into GaInAsP at a relatively high level (on the order of 10^{19} cm^{-3}).

The impurity levels of GaInAs grown by various epitaxial techniques are 3×10^{15} cm^{-3} by MBE, 8×10^{15} cm^{-3} by MOCVD, and 5×10^{14} cm^{-3} by CBE.

2.11.4   Applied devices

Semiconductor lasers emitting 1 to 1.6-µm wavelength.   The optical fiber made of silica glass exhibits a very low transmission loss, i.e., 0.154 dB/km at 1.55 µm. The material dispersion of retractive index is minimum at the wavelength of 1.3 µm. These are advantageous for long-distance optical communications. Semiconductor lasers emitting 1.3-µm wavelength using lattice-matched GaInAsP/InP have been developed having low thresholds of about 10 mA and very long device lifetimes. The 1.3-µm wavelength system has been used since 1980 in public telephone networks and undersea cable systems.

In the 1990s, the 1.55-µm system was realized by taking the advantage of the minimum transmission loss. In this case, the linewidth of the light source must be very small, since the dispersion of the silica fiber is relatively large compared to that at 1.3 µm. Figure 85 exhibit an example of a single-mode laser structure that provides narrow linewidth even when modulated at high speed-signals.[7]

High-power semiconductor lasers emitting at 1.48 µm are employed as a pumping source for Er-doped optical fiber amplifier (EDFA). A surface-emitting laser operating at this wavelength is shown in Figure 86 and is expected to be used in long-wavelength networks and optical interconnects.[8]

For the purpose of substantially improving laser performance, quantum wells have been considered for use as the active region of semiconductor lasers. Figure 87 gives an example of quantum wire lasers employing a GaInAs/GaInAsP system that emits at 1.55 µm.[9]

Other optoelectronic devices.   The counterpart of semiconductor lasers is a photodetector that receives the transmitted optical signal. Photodiodes having high quantum efficiencies in wavelength 1.3 to 1.6 µm band employ the GaInAs ternary semiconductors lattice-matched to InP as well. This system provides low-noise and high-speed photodiodes, i.e., PIN diodes and avalanche photodiodes (APDs). Infrared (IR) detectors and CCDs are important for infrared imaging. Illumination by IR LEDs are useful for imaging as well. Eye-safe radiation in the 1.3- to 1.55-µm range is another important issue in IR imaging.

References

1. Nahory, R.E., Pollack, M.A., Johnstone, W.D., and Barnes, R.L., *Appl. Phys. Lett.*, 33, 659, 1978.
2. Agrawal, G.P. and Dutta, N.K., *Long-Wavelength Semiconductor Lasers*, Van Nostrand Reinhold, New York, 1986, 85.
3. Moon, R.L., Antypas, G.A., and James, L.W., *J. Electron. Mater.*, 3, 635, 1974.
4. Casey, H.C. and Panish, M.B., *Heterostructure Lasers*, Part B, Academic Press, New York, 1978.
5. Manasevit, H.M., *Appl. Phys. Lett.*, 12, 156, 1968.
6. Tsang, W.T., *IEEE J. Quant. Electron.*, QE-23, 936, 1987.

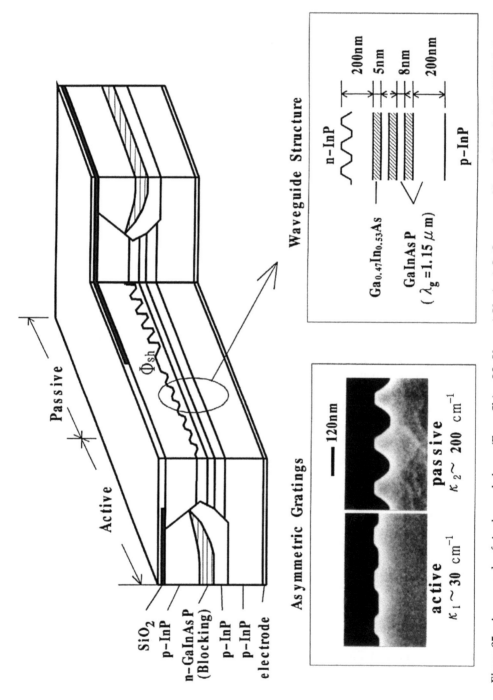

Figure 85   An example of single-mode laser. (From Shim, J.I., Komori, K., Arai, S., Suematsu, Y., and Somchai, R., *IEEE J. Quant. Electron.*, QE-27, 1736, 1991. With permission.)

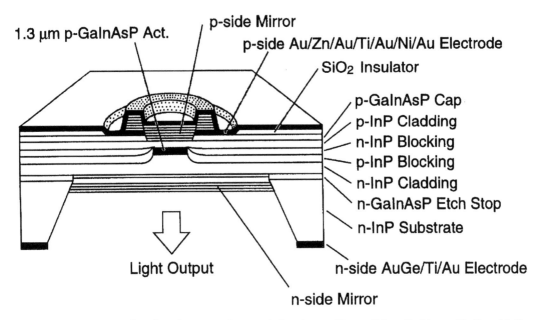

1.3 µm p-GaInAsP Act.
p-side Mirror
p-side Au/Zn/Au/Ti/Au/Ni/Au Electrode
SiO₂ Insulator
p-GaInAsP Cap
p-InP Cladding
n-InP Blocking
p-InP Blocking
n-InP Cladding
n-GaInAsP Etch Stop
n-InP Substrate
Light Output
n-side AuGe/Ti/Au Electrode
n-side Mirror

Figure 86 An example of 1.48 µm surface emitting laser. (From Baba, T., Yogo, Y., Suzuki, T., Koyama, F., and Iga, K., *IEICE Trans. Electronics.*, E76-C, 1423, 1993. With permission.)

7. Shim, J.I., Komori, K., Arai, S., Suematsu, Y., and Somchai, R., *IEEE J. Quant. Electron.*, QE-27, 1736, 1991.
8. Baba, T., Yogo, Y., Suzuki, T., Koyama, F., and Iga, K., *IEICE Trans. Electronics.*, E76-C, 1423, 1993.
9. Kudo, K., Nagashima, Y. Tamura, S., Arai, S., Huang, Y., and Suematsu, Y., *IEEE Photon. Technol. Lett.*, 5, 864, 1993.

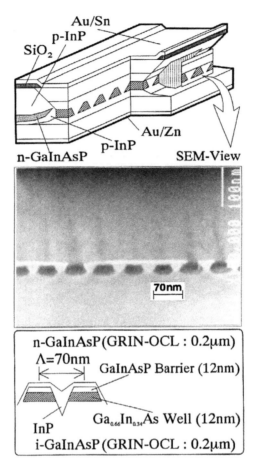

Figure 87   An example of quantum wire lasers employing GaInAs/GaInAsP system to emit 1.55 μm wavelength. (From Kudo, K., Nagashima, Y. Tamura, S., Arai, S., Huang, Y., and Suematsu, Y., *IEEE Photon. Lett.*, 5, 864, 1993. With permission.)

Principal phosphor materials and their optical properties

Shuji Nakamura

Contents

2.12   GaN and related luminescence materials

2.12.1   Introduction

GaN and related materials such as AlGaInN are III-V nitride compound semiconductors with the wurtzite crystal structure and an energy band structure that allow direct interband transitions which are suitable for light-emitting devices (LEDs). The bandgap energy of AlGaInN varies between 6.2 and 1.95 eV at room temperature, depending on its composition. Therefore, these III-V semiconductors are useful for light-emitting devices, especially in the short-wavelength regions. Among the AlGaInN systems, GaN has been most intensively studied. GaN has a bandgap energy of 3.4 eV at room temperature.

Recent research on III-V nitrides has paved the way for the realization of high-quality crystals of GaN, AlGaN, and GaInN, and of *p*-type conduction in GaN and AlGaN.[1,2] The mechanism of acceptor-compensation, which prevents obtaining low-resistivity *p*-type GaN and AlGaN, has been elucidated.[3] In Mg-doped *p*-type GaN, Mg acceptors are deactivated by atomic hydrogen that is produced from NH_3 gas used to provide nitrogen during GaN growth. After growth, thermal annealing in N_2 ambience can reactivate the Mg acceptors by removing the atomic hydrogen from the Mg-hydrogen complexes.[3] High-brightness blue GaInN/AlGaN LEDs have been fabricated on the basis of these results, and luminous intensities over 2 cd have been achieved.[4-6] Also, blue/green GaInN single-quantum-well (SQW) LEDs with a narrow spectrum width have been developed.[7,8] These LEDs are now commercially available. Furthermore, recently, bluish-purple laser light emission at room-temperature (RT) in GaInN/GaN/AlGaN-based heterostructure laser diodes (LDs) under pulsed currents[9-14] or continuous-wave (CW) operation was demonstrated.[15-17] Recent studies of (Al,Ga,In)N compound semiconductors are described in this section.

2.12.2   n-Type GaN

GaN films are usually grown on a sapphire substrate with (0001) orientation (c face) at temperatures around 1000°C by the metal-organic chemical vapor deposition (MOCVD) method. Trimethylgallium (TMG) and ammonia are used as Ga and N sources, respectively. The lattice constants along the *a*-axis of the sapphire and GaN are 4.758 and 3.189 Å, respectively. Therefore, the lattice-mismatch between the sapphire and the GaN is very large. The lattice constant along the *a*-axis of 6H-SiC is 3.08 Å, which is relatively close to that of GaN. However, the price of a SiC substrate is extraordinarily expensive to use for the practical growth of GaN. Therefore, at present, there are no alternative substrates to sapphire from considerations of price and high-temperature properties, even as the lattice mismatch is large. Grown GaN layers usually show *n*-type conduction without any intentional doping. The donors are probably native defects or residual impurities such as nitrogen vacancies or residual oxygen.

Recently, remarkable progress has been achieved in the crystal quality of GaN films by employing a new growth method using buffer layers. Carrier concentration and Hall mobility, with values of 1×10^{16} cm^{-3} and 600 cm^2 Vs^{-1} at room temperature, respectively, have been obtained by deposition of a thin GaN or AlN layer as a buffer before the growth of a GaN film.[18] In order to obtain *n*-type GaN with high carrier concentrations, Si or Ge is doped into GaN.[19] The carrier concentration can be varied between 1×10^{17} and 1×10^{20} cm^{-3} by Si doping. Figure 88 shows a typical photoluminescence (PL) spectra of Si-doped GaN films. In the spectra, relatively strong deep-level (DL) emission around 560 nm and the band-edge (BE) emission around 380 nm are observed. The intensity of DL emissions is always stronger than that of BE emissions in this range of Si concentrations.

2.12.3   p-Type GaN

Formerly, it was impossible to obtain a *p*-type GaN film due to the poor crystal quality of GaN films. Recently, Amano et al.[1] succeeded in obtaining *p*-type GaN films by means of Mg doping and low-energy electron-beam irradiation (LEEBI) treatment after growth. In 1992, Nakamura et al.[20] found that low-resistivity *p*-type GaN films are also obtained by post-thermal annealing in N$_2$ ambience of Mg-doped GaN films. The resistivity of as-grown films is 1×10^6 Ω·cm. When the temperature is raised to 400°C in a N$_2$ ambience for annealing, resistivity begins to decrease suddenly. After annealing at 700°C, the resistivity, hole carrier concentration and hole mobility become 2 Ω·cm, 3×10^{17} cm^{-3} and 10 cm^2 V·s^{-1}, respectively.

These changes of the resistivity of Mg-doped GaN films are explained by the hydogenation process model in which atomic hydrogen produced from NH$_3$ during the growth is assumed to be the origin of the acceptor compensation. If low-resistivity *p*-type GaN films, which are obtained by N$_2$-ambient thermal annealing or LEEBI treatment, are thermally annealed in NH$_3$ ambience at temperatures above 400°C, they show a resistivity as high as 1×10^6 Ω·cm. This resistivity is almost the same as that of as-grown Mg-doped GaN films. Therefore, these results indicate that the abrupt resistivity increase in NH$_3$-ambient thermal annealing at temperatures above 400°C is caused by the NH$_3$ gas itself.

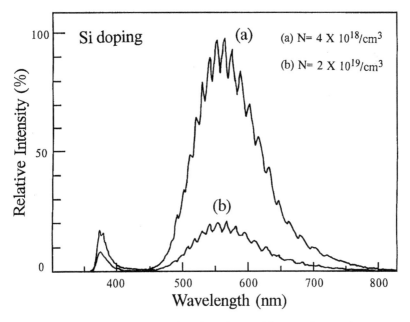

Figure 88   Room-temperature PL spectra of Si-doped GaN films. Both samples were grown under the same growth conditions but changing the flow rate of SiH_4. The carrier concentrations are (a) 4×10^{18} cm^{-3} and (b) 2×10^{19} cm^{-3}. (From Nakamura, S., Mukai, T., and Senoh, M., *Jpn. J. Appl. Phys.*, 31, 2883, 1992. With permission.)

Atomic hydrogen produced by the NH_3 dissociation at temperatures above 400°C is considered to be related to the acceptor compensation mechanism. A hydrogenation process whereby acceptor-H neutral complexes are formed during the growth of *p*-type GaN films has been proposed.[3] The formation of these complexes during film growth causes acceptor compensation. The N_2-ambient thermal annealing or LEEBI treatment after growth can reactivate the acceptors by removing atomic hydrogen from the neutral complexes. As a result, noncompensated acceptors are formed and low-resistivity *p*-type GaN films are obtained.

2.12.4   GaInN

The ternary III-V semiconductor compound, GaInN, is one of the candidates for blue to blue-green emitting LEDs, because its bandgap varies from 1.95 to 3.4 eV depending on the indium mole fraction. It was very difficult to grow high-quality single crystal GaInN films due to the high dissociation pressure of GaInN at the growth temperature. Recently, this difficulty has been overcome by means of the two-flow (TF)-MOCVD method,[21] and high-quality GaInN films have been obtained. Figure 89 shows the results of room-temperature PL measurements of high-quality GaInN films grown by this method. A strong sharp peak is observed at 400 nm in (a) and at 438 nm in (b). These spectra are due to BE emission of GaInN films because they have a very narrow halfwidth (about 70 meV).

Figure 90 shows the bandgap energy ($E_g(X)$) of $Ga_{(1-X)}In_XN$ films estimated from PL spectra at room temperature as a function of the indium mole fraction X.[22] The indium mole fraction of the GaInN films was determined by the measurements of the difference of the X-ray diffraction peak positions between GaInN and GaN films. Osamura et al.[23] showed that $E_g(X)$ in ternary alloys of $Ga_{(1-X)}In_XN$ has the following parabolic dependence on the molar fraction X:

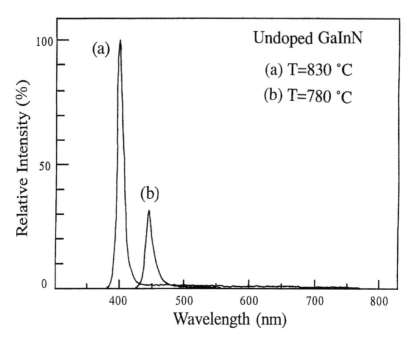

Figure 89 Room-temperature PL spectra of the GaInN films. Both samples were grown on GaN films under the same growth conditions but changing the growth temperature: (a) 830°C and (b) 780°C. (From Nakamura, S. and Mukai, T., *Jpn. J. Appl. Phys.*, 31, L1457, 1992. With permission.)

$$E_g(X) = (1-X)E_{g,\text{GaN}} + XE_{g,\text{InN}} - bX(1-X) \qquad (48)$$

where $E_{g,\text{GaN}}$ is 3.40 eV, $E_{g,\text{InN}}$ is 1.95 eV, and the bowing parameter b is 1.00 eV. The calculated curve is shown by the solid line in the figure. Here, the bowing parameter, which is also called nonlinear parameter, shows downward deviation of the bandgap energy of ternary compounds compared to the linear relation between the bandgap energy of binary compounds, that is, from $(1-X)E_{g,\text{GaN}} + XE_{g,\text{InN}}$.

Figure 91 shows a typical room-temperature PL spectrum of a Zn-doped GaInN film.[22] It has two peaks. The shorter wavelength peak is due to BE emission of GaInN, and the longer wavelength peak is due to a Zn-related emission and has a large halfwidth (about 70 nm, i.e., about 430 meV).

2.12.5 GaInN/AlGaN LED

Figure 92 shows the structure of a GaInN/AlGaN double-heterostructure (DH) LED fabricated by Nakamura et al.[4-6,22] In this LED, Si and Zn are co-doped into the GaInN active layer in order to obtain a high output power. Zn-doped GaInN is used as the active layer to obtain strong blue emission, as shown in Figure 91. Mg-doped GaInN does not show strong blue emission, in contrast to the Zn-doped films.

Figure 93 shows the electroluminescence (EL) spectra of this system with forward currents of 0.1, 1, and 20 mA.[4-6,22] The typical peak wavelength and halfwidth are 450 and 70 nm, respectively, at 20 mA. The peak wavelength shifts to shorter wavelengths with increasing forward current. This blue shift suggests that the luminescence is dominated by the donor-acceptor (DA) pair recombination mechanism in the GaInN active layer co-doped with Si and Zn. At 20 mA, a narrower, higher-energy peak emerges around 385 nm. This peak is due to band-to-band recombination in the GaInN active layer. This peak is

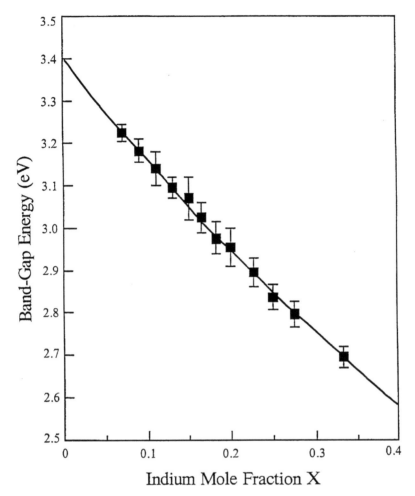

Figure 90 Bandgap energy of $Ga_{(1-X)}In_XN$ films as a function of the indium mole fraction X. (From Nakamura, S., *Jpn. J. Opt.*, 23, 701, 1994. With permission.)

resolved at injection levels where the intensity of impurity-related recombination luminescence is saturated. The output power of the GaInN/AlGaN DH blue LEDs is 1.5 mW at 10 mA, 3 mW at 20 mA, and 4.8 mW at 40 mA. The external quantum efficiency is 5.4% at 20 mA.[22] The typical on-axis luminous intensity with 15° conical viewing angle is 2.5 cd at 20 mA when the forward voltage is 3.6 V at 20 mA.

2.12.6 GaInN single-quantum-well (SQW) LEDs

High-brightness blue and blue-green GaInN/AlGaN DH LEDs with a luminous intensity of 2 cd have been fabricated and are now commercially available, as mentioned above.[4-6,22] In order to obtain blue and blue-green emission centers in these GaInN/AlGaN DH LEDs, the GaInN active layer was doped with Zn. Although these GaInN/AlGaN DH LEDs produced high-power light output in the blue and blue-green regions with a broad emission spectrum (FWHM = 70 nm), green or yellow LEDs with peak wavelengths longer than 500 nm have not been fabricated.[6] The longest peak wavelength of the EL of GaInN/AlGaN DH LEDs achieved thus far has been observed at 500 nm (blue-green) because the crystal quality of the GaInN active layer of DH LEDs deteriorates when the indium mole fraction is increased to obtain green band-edge

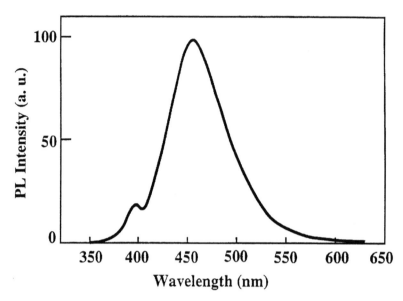

Figure 91 Room-temperature PL spectrum of a Zn-doped $Ga_{0.95}In_{0.05}N$ film. (From Nakamura, S., *Jpn. J. Opt.*, 23, 701, 1994. With permission.)

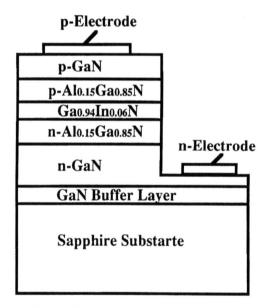

Figure 92 Structure of the GaInN/AlGaN double-heterostructure blue LED. (From Nakamura, S., *Jpn. J. Opt.*, 23, 701, 1994. With permission.)

emission.[6] Quantum-well (QW) LEDs with thin GaInN active layers (about 30 Å) fabricated to obtain high-power emission from blue to yellow with a narrow emission spectrum[7,8] are described below.

The green GaInN SQW LED device structures (Figure 94) consist of a 300-Å GaN buffer layer grown at low temperature (550°C), a 4-µm-thick layer of *n*-type GaN:Si, a 30-Å-thick active layer of undoped $Ga_{0.55}In_{0.45}N$, a 1000-Å-thick layer of *p*-type $Al_{0.2}Ga_{0.8}N$:Mg, and a 0.5-µm-thick layer of *p*-type GaN:Mg. This is the SQW structure.

Figure 95 shows the typical EL of the blue, green, and yellow SQW LEDs containing different indium mole fractions of the GaInN layer, all at a forward current of 20 mA. The

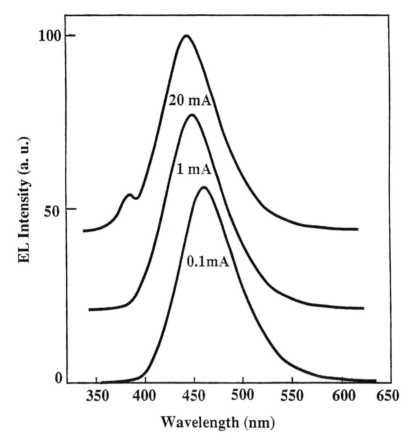

Figure 93   Electroluminescence spectra of a GaInN/AlGaN double-heterostructure blue LED. (From Nakamura, S., *Jpn. J. Opt.*, 23, 701, 1994. With permission.)

GaInN green SQW LEDs

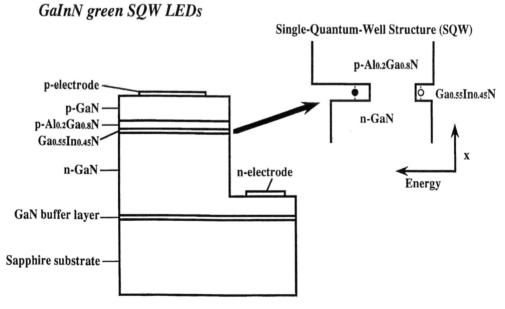

Figure 94   The structure of green SQW LED. (From Nakamura, S., Senoh, M., Iwasa, N., Nagahama, S., Yamada, T., and Mukai, T., *Jpn. J. Appl. Phys. Lett.*, 34, L1332, 1995. With permission.)

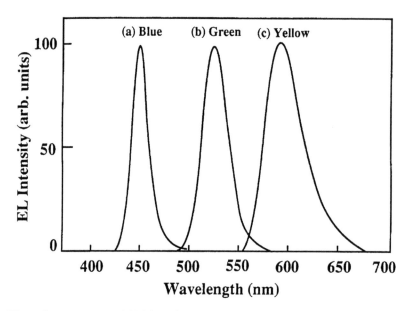

Figure 95 Electroluminescence of (a) blue, (b) green, and (c) yellow SQW LEDs at a forward current of 20 mA. (From Nakamura, S., Senoh, M., Iwasa, N., and Nagahama, S., *Jpn. J. Appl. Phys.*, 34, L797, 1995. With permission.)

peak wavelength and the FWHM of the typical blue SQW LEDs are 450 and 20 nm, respectively; of the green 525 and 30 nm; and of the yellow 600 and 50 nm, respectively. When the peak wavelength becomes longer, the FWHM of the EL spectra increases, probably due to the inhomogeneities in the GaInN layer or due to strain between well and barrier layers of the SQW caused by lattice mismatch and differences in the thermal expansion coefficients.

At 20 mA, the output power and the external quantum efficiency of the blue SQW LEDs are 5 mW and 9.1%, respectively. Those of the green SQW LEDs are 3 mW and 6.3%, respectively. A typical on-axis luminous intensity of the green SQW LEDs with a 10° cone viewing angle is 10 cd at 20 mA. These values of output power, external quantum efficiency, and luminous intensity of blue and green SQW LEDs are more than 100 times higher than those of conventional blue SiC and green GaP LEDs. By combining these high-power and high-brightness blue GaInN SQW, green GaInN SQW, and red AlGaAs LEDs, many kinds of applications such as LED full-color displays and LED white lamps for use in place of light bulbs or fluorescent lamps are now possible. These devices have the characteristics of high reliability, high durability, and low energy consumption.

Figure 96 is a chromaticity diagram in which the positions of the blue and green GaInN SQW LEDs are shown. The chromaticity coordinates of commercially available green GaP LEDs, green AlGaInP LEDs, and red AlGaAs LEDs are also shown. The color range of light emitted by a full-color LED lamp in the chromaticity diagram is shown as the region inside each triangle, which is drawn by connecting the positions of three primary color LED lamps. Three color ranges (triangles) are shown for differences only in the green LED (green GaInN SQW, green GaP, and green AlGaInP LEDs). In this figure, the color range of lamps composed of a blue GaInN SQW LED, a green GaInN SQW LED, and a red AlGaAs LED is the widest. This means that the GaInN blue and green SQW LEDs show much better color and color purity in comparison with other blue and green LEDs. Using these blue and green SQW LEDs together with LEDs made of AlGaAs, more realistic LED full color displays have been demonstrated.

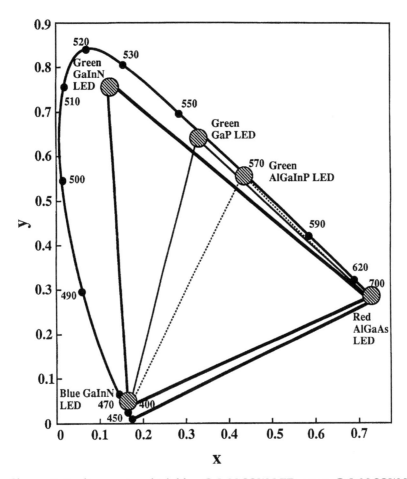

Figure 96 Chromaticity diagram in which blue GaInN SQW LED, green GaInN SQW LED, green GaP LED, green AlGaInP LED, and red AlGaAs LED are shown. (From Nakamura, S., Senoh, M., Iwasa, N., Nagahama, S., Yamada, T., and Mukai, T., *Jpn. J. Appl. Phys. Lett.*, 34, L1332, 1995. With permission.)

2.12.7   GaInN multiquantum-well (MQW) LDs

The structure of the GaInN MQW LDs is shown in Figure 97. The GaInN MQW LD device consists of a 300-Å-thick GaN buffer layer grown at a low temperature of 550°C, a 3-μm-thick layer of n-type GaN:Si, a 0.1-μm-thick layer of n-type $Ga_{0.95}In_{0.05}N$:Si, a 0.5-μm-thick layer of n-type $Al_{0.08}Ga_{0.92}N$:Si, and a 0.1-μm-thick layer of n-type GaN:Si. At this point, the MQW structure consists of four 35-Å-thick undoped $Ga_{0.85}In_{0.15}N$ well layers by 70-Å-thick undoped $Ga_{0.98}In_{0.02}N$ barrier layers. The four well layers form the gain medium. The heterostructure is then capped with a 200-Å-thick layer of p-type $Al_{0.2}Ga_{0.8}N$:Mg, a 0.1-μm-thick layer of p-type GaN:Mg, a 0.5-μm-thick layer of p-type $Al_{0.08}Ga_{0.92}N$:Mg, and a 0.5-μm-thick layer of p-type GaN:Mg. The n-type and p-type GaN layers are used for light-guiding, while the n-type and p-type $Al_{0.08}Ga_{0.92}N$ layers act as cladding for confinement of the carriers and the light from the active region.

Figure 98 shows typical voltage-current (V-I) characteristics and the light output power (L) per coated facet of the LD as a function of the forward DC current at RT. No stimulated emission was observed up to a threshold current of 80 mA, corresponding to a current density of 3.6 kA cm⁻², as shown in Figure 98. The operating voltage at the threshold was 5.5 V.

Ridge-waveguide purplish-blue InGaN MQW LDs

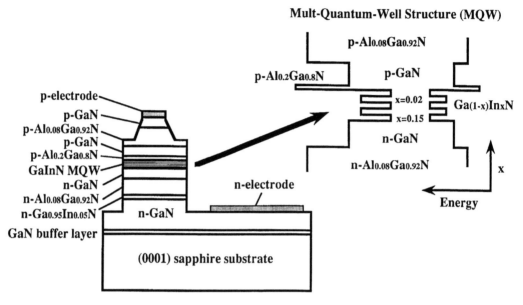

Figure 97 The structure of the GaInN MQW LDs. (From Nakamura, S., Senoh, M., Nagahama, S., Iwasa, N., Yamada, T., Matsushita, T., Sugimoto, Y., and Kiyoku, H., Presented at the *9th Annual Meeting of IEEE Lasers and Electro-Optics Society*, Boston, PD1.1, Nov. 18-21, 1996. With permission.)

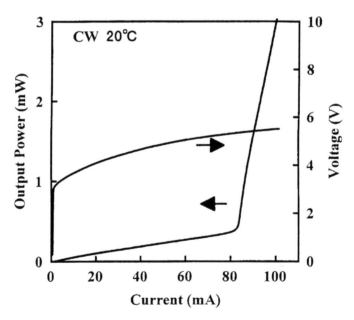

Figure 98 Typical light output power (*L*)-current (*I*) and voltage (*V*)-current (*I*) characteristics of GaInN MQW LDs measured under CW operation at RT. (From Nakamura, S., Senoh, M., Nagahama, S., Iwasa, N., Yamada, T., Matsushita, T., Sugimoto, Y., and Kiyoku, H., Presented at the *9th Annual Meeting of IEEE Lasers and Electro-Optics Society*, Boston, PD1.1, Nov. 18-21, 1996. With permission.)

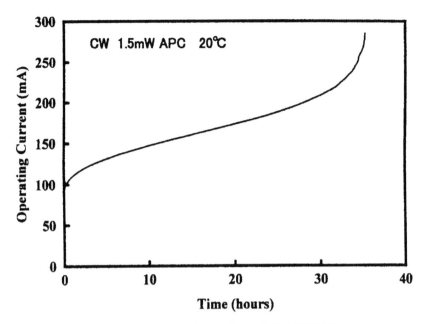

Figure 99 Operating current as a function of time for GaInN MQW LDs under a constant output power of 1.5 mW per facet controlled using an autopower controller. The LD was operated under DC at RT. (From Nakamura, S., presented at *Materials Research Society Fall Meeting*, Boston, N1.1, Dec. 2-6, 1996. With permission.)

Figure 99 shows the results of a lifetime test of CW-operated LDs carried out at RT, in which the operating current is shown as a function of time, keeping output power constant at 1.5 mW per facet using an autopower controller (APC). The operating current gradually increases due to the increase in the threshold current from its initial value; the current then increases sharply after 35 hours. This short lifetime is probably due to heating resulting from the high operating currents and voltages. Short-circuiting of the LDs occurred after the 35 hours, as mentioned above.

Figure 100 shows emission spectra of GaInN MQW LDs with various operating current under RT CW operation. The threshold current and voltage of this LD were 160 mA and 6.7 V, respectively. The threshold current density was 7.3 kA cm^{-2}. At a current of 156 mA, many longitudinal modes are observed with a mode separation of 0.042 nm; this separation is smaller than the calculated value of 0.05 nm, probably due to refractive index changes from the value used (2.54) in the calculation. Periodic subband emissions are observed with a peak separation of about 0.025 nm ($\Delta E = 2$ meV). The origin of these subbands has not yet been identified. On increasing the forward current from 156 to 186 mA, the laser emission becomes single mode and shows mode hopping of the peak wavelength toward higher energy; the peak emission is at the center of each subband emission. Figure 101 shows the peak wavelength of the laser emission as a function of the operating current under RT CW operation. A gradual increase of the peak wavelength is observed, probably due to bandgap narrowing of the active layer caused by the temperature increase. At certain currents, large mode hopping of the peak wavelength toward higher energy is observed with increasing operating current.

The delay time of the laser emission of the LDs as a function of the operating current was measured under pulsed current modulation using the method described in Reference 14 in order to estimate the carrier lifetime (τ_s). From this measurement, τ_s was estimated to be 10 ns, which is larger than previous estimates of 3.2 ns.[14] The threshold

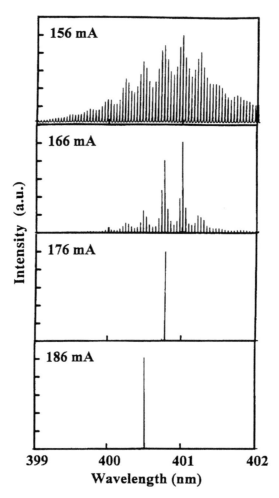

Figure 100 Emission spectra of GaInN MQW LDs with various operating currents under RT CW operation. (From Nakamura, S., presented at *Materials Research Society Fall Meeting*, Boston, N1.1, Dec. 2-6, 1996. With permission.)

carrier density (n_{th}) was estimated to be 2×10^{20} cm^{-3} for a threshold current density of 3.6 kA cm^{-2}, and an active layer thickness of 140 Å.[14] The thickness of the active layer was determined as 140 Å, assuming that the injected carriers were confined in the GaInN well layers. Other typical values are $\tau_s = 3$ ns, $J_{th} = 1$ kA cm^{-2}, and $n_{th} = 2 \times 10^{18}$ cm^{-3} for AlGaAs lasers and $n_{th} = 1 \times 10^{18}$ cm^{-3} for AlGaInP lasers. In comparison with other more conventional lasers, n_{th} in our structure is relatively large (two orders of magnitude higher), probably due to the large density of states of carriers resulting from their large effective masses.[14]

The Stokes' shift or energy differences between excitation and emission in GaInN MQW LDs can be as large as 100 to 250 meV at RT.[24-26] This means that the energy depth of the localized state of the carriers is 100 to 250 meV in these devices. Both the spontaneous emission and the stimulated emission of the LDs originates from these deep localized energy states.[24-26] Using high-resolution, cross-sectional transmission electron microscopy (TEM), a periodic indium composition fluctuation was observed in the LDs, probably caused by GaInN phase separation during growth.[25,26] Based on these results, the laser emission is thought to originate from GaInN quantum dot-like states formed in these structures. The many periodic subband emissions observed probably result from transitions between the subband energy levels of the GaInN quantum dots formed from

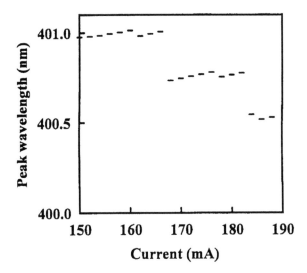

Figure 101   Peak wavelength of the emission spectra of the GaInN MQW LDs as a function of the operating current under RT CW operation. (From Nakamura, S., Characteristics of GaInN multi-quantum-well-structure laser diodes, presented at *Materials Research Society Fall Meeting*, Boston, N1.1, Dec. 2-6, 1996. With permission.)

In-rich regions in the GaInN well layers. The size of the GaInN dots is estimated to be approximately 35 Å from the high-resolution, cross-sectional TEM pictures.[25,26] It is difficult to control the size of GaInN dots that form in adjacent In-rich and -poor regions. The energy separation of each subband emission in Figure 100 is only about 2 meV, which is considered to be relatively small in comparison with the energy difference between the n = 1 and n = 2 subband energy transitions of other more controlled quantum dots. These periodic subband energy levels are probably caused between n = 1 subband levels of quantum dots with different dot sizes.

2.12.8   Summary

Superbright blue and green GaInN SQW LEDs have been developed and commercialized. By combining high-power, high-brightness blue GaInN SQW LEDs, green GaInN SQW LEDs, and red AlGaAs LEDs, many kinds of applications, such as LED full-color displays and LED white lamps for use in place of light bulbs or fluorescent lamps, are now possible. These devices have the characteristics of high reliability, high durability, and low energy consumption. RT CW operation of bluish-purple GaInN MQW LDs has been demonstrated recently with a lifetime of 35 hours. The carrier lifetime and the threshold carrier density were estimated to be 10 ns and 2×10^{20} cm^{-3}, respectively. The emission spectra of GaInN MQW LDs under CW operation at RT showed periodic subband emissions with an energy separation of 2 meV. These periodic subband emissions are probably due to the transitions between the subband energy levels of quantum dots formed from In-rich regions in the GaInN well layers. Further improvement in the lifetime of the LDs can be obtained by reducing the threshold current and voltage. The advances in this technology have been rapid in the past decade. Progress attained has been reviewed in a number of places.[27]

References

1. Amano, H., Kito, M., Hiramatsu, K., and Akasaki, I., *Jpn. J. Appl. Phys.*, 28, L2112, 1989.
2. Strite, S., Lin, M.E., and Morkoç, H., *Thin Solid Films*, 231, 197, 1993.

3. Nakamura, S., Iwasa, N., Senoh, M., and Mukai, T., *Jpn. J. Appl. Phys.*, 31, 1258, 1992.
4. Nakamura, S., *Nikkei Electronics Asia*, 3, 65, 1994.
5. Nakamura, S., Mukai, T., and Senoh, M., *Appl. Phys. Lett.*, 64, 1687, 1994.
6. Nakamura, S., Mukai, T., and Senoh, M., *J. Appl. Phys.*, 76, 8189, 1994.
7. Nakamura, S., Senoh, M., Iwasa, N., and Nagahama, S., *Jpn. J. Appl. Phys.*, 34, L797, 1995.
8. Nakamura, S., Senoh, M., Iwasa, N., Nagahama, S., Yamada, T., and Mukai, T., *Jpn. J. Appl. Phys. Lett.*, 34, L1332, 1995.
9. Nakamura, S., Senoh, M., Nagahama, S., Iwasa, N., Yamada, T., Matsushita, T., Kiyoku, H., and Sugimoto, Y., *Jpn. J. Appl. Phys.*, 35, L74, 1996.
10. Nakamura, S., Senoh, M., Nagahama, S., Iwasa, N., Yamada, T., Matsushita, T., Kiyoku, H., and Sugimoto, Y., *Jpn. J. Appl. Phys.*, 35, L217, 1996.
11. Nakamura, S., Senoh, M., Nagahama, S., Iwasa, N., Yamada, T., Matsushita, T., Kiyoku, H., and Sugimoto, Y., *Appl. Phys. Lett.*, 68, 2105, 1996.
12. Nakamura, S., Senoh, M., Nagahama, S., Iwasa, N., Yamada, T., Matsushita, T., Kiyoku, H., and Sugimoto, Y., *Appl. Phys. Lett.*, 68, 3269, 1996.
13. Nakamura, S., Senoh, M., Nagahama, S., Iwasa, N., Yamada, T., Matsushita, T., Sugimoto, Y., and Kiyoku, H., *Appl. Phys. Lett.*, 69, 1477, 1996.
14. Nakamura, S., Senoh, M., Nagahama, S., Iwasa, N., Yamada, T., Matsushita, T., Sugimoto, Y., and Kiyoku, H., *Appl. Phys. Lett.*, 69, 1568, 1996.
15. Nakamura, S., Senoh, M., Nagahama, S., Iwasa, N., Yamada, T., Matsushita, T., Sugimoto, Y., and Kiyoku, H., *Appl. Phys. Lett.*, 69, 3034, 1996.
16. Nakamura, S., Senoh, M., Nagahama, S., Iwasa, N., Yamada, T., Matsushita, T., Sugimoto, Y., and Kiyoku, H., First room-temperature continuous-wave operation of GaInN multi-quantum-well-structure laser diodes, presented at *9th Annual Meeting of IEEE Lasers and Electro-Optics Society*, Boston, PD1.1, Nov. 18-21, 1996.
17. Nakamura, S., Characteristics of GaInN multi-quantum-well-structure laser diodes, presented at *Materials Research Society Fall Meeting*, Boston, N1.1, Dec. 2-6, 1996.
18. Nakamura, S., *Jpn. J. Appl. Phys.*, 30, L1705, 1991.
19. Nakamura, S., Mukai, T., and Senoh, M., *Jpn. J. Appl. Phys.*, 31, 2883, 1992.
20. Nakamura, S., Mukai, T., Senoh, M., and Iwasa, N., *Jpn. J. Appl. Phys.*, 31, L139, 1992.
21. Nakamura, S. and Mukai, T., *Jpn. J. Appl. Phys.*, 31, L1457, 1992.
22. Nakamura, S., *Jpn. J. Opt.*, 23, 701, 1994.
23. Osamura, K., Naka, S., and Murakami, Y., *J. Appl. Phys.*, 46, 3432, 1975.
24. Chichibu, S., Azuhata, T., Sota, T., and Nakamura, S., *Appl. Phys. Lett.*, 69, 4188, 1996.
25. Narukawa, Y., Kawakami, Y., Fuzita, Sz., Fujita, Sg., and Nakamura, S., *Phys. Rev.*, B55, 1938R, 1997.
26. Narukawa, Y., Kawakami, Y., Funato, M., Fujita, Sz., Fujita, Sg., and Nakamura, S., Role of self-formed GaInN quantum dots for the exciton localization in the purple laser diodes emitting at 420 nm, *Appl. Phys. Lett.*, 70, 981, 1996.
27. Nakamura, S., and Fasol, G., *The Blue Laser Diode*, Springer - Verlag, Berlin, 2000.

chapter two — section thirteen

Fundamentals of luminescence

Hiroyuki Matsunami

Contents

2.13   Silicon carbide (SiC) as a luminescence material

2.13.1   Polytypes

Silicon carbide (SiC) is the oldest semiconductor known as a luminescence material. This material shows polytypism arising from different stacking possibilities. In hexagonal close packing of the Si-C pair, the positions of the pair in the first and second layers are uniquely determined (A and B) as shown in Figure 102(a). However, in the third layer, there are two possibilities, either A or C as shown. In the former case, the stacking order becomes ABAB..., giving a wurtzite (hexagonal) structure, and the latter becomes ABCABC..., giving a zinc-blende (cubic) structure.

In SiC crystals, there can exist various combinations of these two structures, which give different stacking orders called polytypes. Among the many polytypes, 3C-, 6H-, and 4H-SiC appear frequently: these structures are shown in Figures 102(b)-(d) together with 2H-SiC (Figure 102(e)). Here, the number indicates the period of stacking order and the letter gives its crystal structure: C = cubic, H = hexagonal, R = rhombohedral. Since the position of each atom has a different configuration of nearest-neighbor atoms, the sites are crystallographically different; that is, they have cubic or hexagonal site symmetry. Hence, when an impurity atom substitutes into the position of Si or C, it gives rise to different energy levels depending upon the number of inequivalent sites present in the material. In 3C-SiC and 2H-SiC, only one cubic or one hexagonal site exists, respectively,

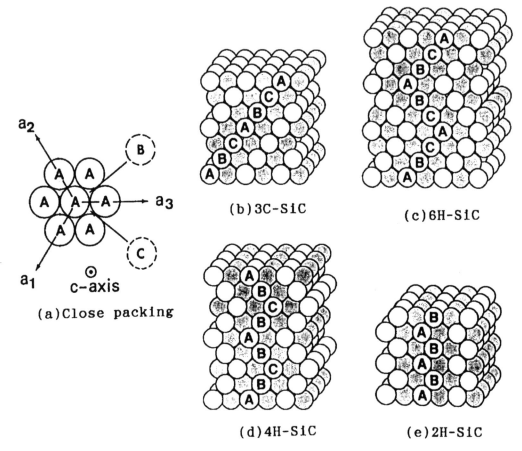

Figure 102 Position of Si-C pair in typical SiC polytypes. (a) Close packing of equal spheres (Si-C pair), (b) 3C-SiC, (c) 6H-SiC, (d) 4H-SiC, and (e) 2H-SiC.

whereas in 6H-SiC there exist one hexagonal and two cubic sites and in 4H-SiC one hexagonal and one cubic sites.

2.13.2   Band structure and optical absorption

Figure 103 shows the absorption spectra of different polytypes of SiC at 4.2K.[1] The spectra contain shoulder features related to phonon-assisted transitions, which are characteristics of indirect band structures. In the figure, the positions of the exciton bandgaps are shown. In Table 28, the values of exciton bandgaps and exciton binding energies are tabulated.[1] The characteristics near the fundamental absorption edge have quite similar structure for all the polytypes except 2H-SiC. This is due to the similarity of the phonons involved in optical absorption in the different polytypes.

2.13.3   Luminescence[1,2]

Since SiC has an indirect band structure, strong luminescence can be expected from the recombination of either bound excitons or donor-acceptor pairs.

2.13.3.1   Luminescence from excitons

Figure 104 depicts the photoluminescence spectrum from excitons bound at N donors in 3C-SiC.[1] From the energy difference between the exciton bandgap and the peak energy

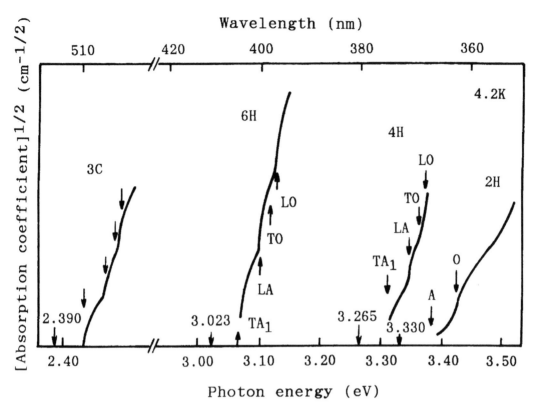

Figure 103 Absorption spectra for typical SiC polytypes. Exciton bandgap is shown for each polytype. (From Choyke, W.J., *Mater. Res. Bull.*, 4, S141-S152, 1969. With permission.)

Table 28 Bandgap Energies in Typical Polytypes of S:C

	E_{GX} (eV) 4.2K	E_{exc} (meV)	Conduction band minimum
3C (Zinc-blende)	2.390 (ID)[a]	13.5[b]	X[e]
6H	3.023 (ID)[a]	78[c]	U[f]
4H	3.265 (ID)[a]	20[d]	M[g]
2H (Wurtzite)	3.330 (ID)[a]	?	K[e]

Note: E_{GX}: Exciton bandgap, E_{exc}: Exciton binding energy.

ID: indirect band structure.

X, U, M, K: position in Brillouin zone.

[a] Choyke, W.J., *Mater. Res. Bull.*, 4, S141, 1969.

[b] Nedzvetskii, D.S. et al., *Sov. Phys. - Semicon.*, 2, 914, 1969.

[c] Sankin, V.I., *Sov. Phys. Solid State.*, 17, 1191, 1975.

[d] Dubrovskii, G.B. et al., *Sov. Phys. Solid State.*, 17, 1847, 1976.

[e] Herman, F. et al., *Mater. Res. Bull.*, 4, S167, 1969.

[f] Choyke, W.J., unpublished result, 1995.

[g] Patrick, L. et al., *Phys. Rev.*, 137, A1515, 1965.

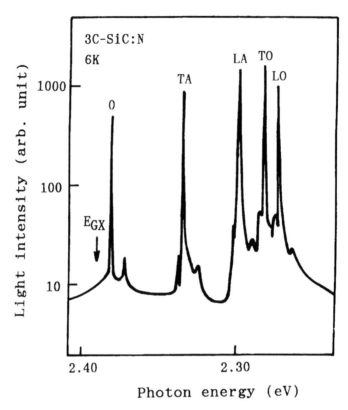

Figure 104 Photoluminescence spectrum of excitons bound at N donors in 3C-SiC. E_{GX} indicates the exciton bandgap. (0: zero phonon; TA: transverse acoustic; LA: longitudinal acoustic; TO: transverse optic; and LO: longitudinal optic). (From Choyke, W.J., *Mater. Res. Bull.*, 4, S141-S152, 1969. With permission.)

corresponding to the zero-phonon line, the exciton binding energy for N donors is estimated to be 10 meV. Since the resolution of peak energies is much better than that in the absorption spectra, the exact value of phonon energies can be obtained from the photoluminescence spectra.

In the photoluminescence spectrum of 6H-SiC, there exists a zero-phonon peak due to the recombination of excitons bound at N donors substituted into hexagonal C sites and two zero-phonon peaks due to those located in cubic C sites.[3] Since the energy levels of N donors in inequivalent (hexagonal, cubic) sites are different, the photoluminescence peaks have different energies.

2.13.3.2 *Luminescence from donor-acceptor pairs*

In SiC, N atoms belonging to the fifth column of the periodic table work as donors, and B, Al, and Ga in the third column work as acceptors. When donors and acceptors are simultaneously incorporated in a crystal, electrons bound at donors and holes at acceptors can create a pair due to the Coulombic force between electrons and holes. This interaction leads to strong photoluminescence through recombination and is known as donor-acceptor pair luminescence.

Figure 105 shows the photoluminescence spectrum from N-Al donor-acceptor pair recombination in 3C-SiC at 1.8K.[4] This gives a peculiar structure showing the recombination of electrons and holes in donor-acceptor pairs of type 2 with N donors replacing C and Al replacing Si. From a detailed analysis of this peculiar structure, the value of 310 meV is obtained for the sum of $E_D(N)$ and $E_A(Al)$, where $E_D(N)$ is the N-donor level

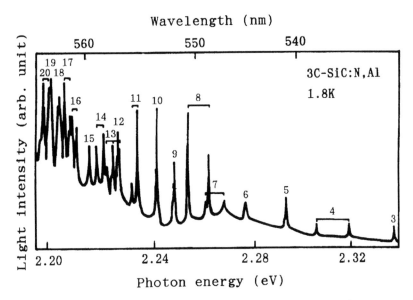

Figure 105 Photoluminescence spectrum of N donor-Al acceptor pair recombination in 3C-SiC. The number for each peak indicates the order of distance between donor and acceptor. (From Choyke, W.J. and Patrick, L., *Phys Rev.*, B2, 4959-4965, 1970. With permission.)

and E_A(Al) the Al-acceptor level. At 77K or higher, the spectrum changes to that due to the recombination of free electrons and holes bound at Al-acceptors (free-to-acceptor recombination) because of thermal excitation of electrons bound at N-donors to the conduction band. From the spectrum, the value of E_A(Al) can be determined precisely. Based on these studies, the values of E_A(Al) = 257 meV and E_D(N) = 53 meV were obtained.[2] From a similar analysis, the B- and Ga-acceptor levels can also be determined.

In most SiC polytypes, except for 3C-SiC and 2H-SiC, there are inequivalent sites, and impurities substituting into those sites give rise to different energy levels. Thus, spectra of donor-acceptor pair recombination and free-to-acceptor recombination can become complicated. As examples, donor-acceptor pair recombination spectra in 6H-SiC at 4.2K are shown in Figure 106(a) and free-to-acceptor recombination spectra at 77K in Figure 106(b).[5] Although the energy levels are different for different acceptors (B, Al, and Ga), the shapes of spectra are quite similar when the abscissa is shifted by an energy of the order of 0.05 eV, as shown in the figure.

The B series (peaks denoted as B) in the spectra show donor-acceptor pair luminescence for N donors in hexagonal C sites and Al acceptors, and the C series (peaks denoted as C) arising from N donors in cubic C sites and Al acceptors. Here, the energy levels of Al acceptors are thought to be very similar, whether they are in hexagonal or cubic Si sites. The subscripts in the figure are defined as follows: 0 implies a zero-phonon peak and LO implies peaks involving longitudinal optical phonons. Peaks A indicate free-to-acceptor recombination: A^a and A^b are due to acceptors substituting into hexagonal and cubic Si sites, respectively.

Since there are three different sites for donors and acceptors, respectively, in 6H-SiC, analysis for the peculiar structure observed in the spectra becomes very difficult. The photon energy, $h\nu(R)$, from donor-acceptor pair luminescence is given by $h\nu(R) \sim R^6 \exp(-4\pi c R^3/3)$, where R is the distance between a donor and an acceptor and c is larger of the donor or acceptor concentrations. By curve fitting of the above relation to the spectra, the value of $E_D + E_A$ can be obtained.[5] Since the value of E_A is calculated from free-to-acceptor recombination as in Figure 106(b), E_D can also be determined. Although

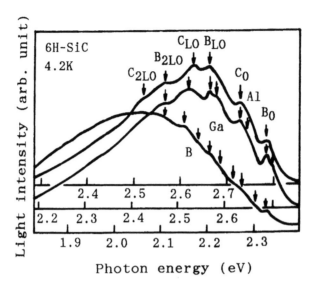

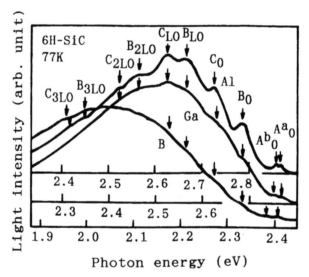

Figure 106 Photoluminescence spectra of (a) donor-acceptor pair recombination at 4.2K and (b) free-to-acceptor recombination at 77K in 6H-SiC doped with B, Al, and Ga. A_0: free-to-Al acceptor peak, (b) B_0: N-donor(hexagonal site)-Al acceptor, (c) C_0: N-donor(cubic site)-Al acceptor. LO indicates longitudinal phonon. (From Ikeda, M., Matsunami, H., and Tanaka, T., *Phys. Rev.*, B22, 2842-2854, 1980. With permission.)

one hexagonal site and two cubic sites exist in 6H-SiC, the difference between the energies for the two cubic sites seems to be very small. Curve fitting was carried out by assuming that the luminescence intensity related to cubic sites is two times larger than that related to hexagonal sites. The calculated energy levels of impurities are given in Table 29. In the table, the results of different polytypes are also shown.[5] In each polytype, the ratio between the acceptor energy levels for cubic and hexagonal sites is very small, whereas that of donor energy levels is large.

2.13.3.3 Other luminescence centers

In addition to the above luminescence centers, luminescence due to defects produced by ion implantation[2] and due to the localized centers such as Ti[2] have been reported.

Table 29 Energy Levels of Donor and Acceptors

		Energy level (meV)			
		Donor		Acceptor	
Polytype	Site	N	Al	Ga	B
3C-SiC	C	56.5	254	343	735
6H-SiC	C	155	249	333	723
	H	100	239	317	698
4H-SiC	C	124	191	267	647
	H	66			

From Ikeda, M., et al., *Phys. Rev.*, B22, 2842, 1980. With permission.

2.13.4   Crystal growth and doping

Crystals of SiC have been grown by the so-called Acheson method, in which a mixture of SiO_2 and C is heated to about 2000°C. To grow pure single crystals, the powdered SiC crystal mixture is sublimed in a specially designed crucible by the Lely method. Recent large-diameter (approximately 2-inch diameter) single crystal boules have been produced by a modified Lely method utilizing a SiC seed in the sublimation growth.

On those single crystals, epitaxial growth has been carried out by either liquid phase epitaxy (LPE) or vapor phase epitaxy (VPE). In LPE, molten Si in a graphite crucible is used as a melt in which a SiC substrate is dipped into.[6] In VPE, chemical vapor deposition (CVD) with SiH^4 and C_3H_8 has been widely used. To get a high-quality epitaxial layer at low temperatures, step-controlled epitaxy is used, which utilizes step-flow growth on off-oriented SiC substrates.[7]

Doping with third column elements as donors or fifth column elements as acceptors can be done easily through both in LPE and VPE.

2.13.5   Light-emitting diodes

Earlier, yellow light-emitting diodes (LEDs) of 6H-SiC utilizing N-B donor-acceptor pair luminescence were demonstrated; they were later replaced by $GaAs_{1-x}P_x$:N yellow LEDs. Blue LEDs of 6H-SiC p-n junction utilizing N-Al donor-acceptor pair luminescence are usually made by LPE[6] or VPE[7] methods. The mechanism for electroluminescence through injection of carriers was clarified by Ikeda et al.[8] A typical spectrum of blue LEDs is shown in Figure 107.[9] The spectral peak is located at 470 nm with a width of 70 nm for a forward current I_F of 20 mA (0.3×0.3 mm²). The diode consists of LPE-grown Al-doped p-SiC/N-doped n-SiC/n-6H-SiC substrate. LEDs are fabricated with a p-side down structure, and the light comes through the n-SiC. The maximum external quantum efficiency is 0.023% ($I_F = 5$ mA). Since the blue LEDs utilize N-Al donor-acceptor pair luminescence in n-type epilayers, the brightness increases with incorporation of Al, and it exceeds 20 mCd ($I_F = 20$ mA).

References

1. Choyke, W.J., *Mater. Res. Bull.*, 4, S141-S152, 1969.
2. Choyke, W.J. and Patrick, L., *Silicon Carbide—1973*, Marshall, R.C., Faust, J.W., and Ryan, C.E., Eds., University of South Carolina Press, 1974, 261-283.
3. Choyke, W.J. and Patrick, L., *Phys. Rev.*, 127, 1868-1877, 1962.
4. Choyke, W.J. and Patrick, L., *Phys. Rev.*, B2, 4959-4965, 1970.
5. Ikeda, M., Matsunami, H., and Tanaka, T., *Phys. Rev.*, B22, 2842-2854, 1980.
6. Matsunami, H., Ikeda, M., Suzuki, A., and Tanaka, T., *IEEE Trans. Elec. Devices*, ED-24, 958-961, 1977.

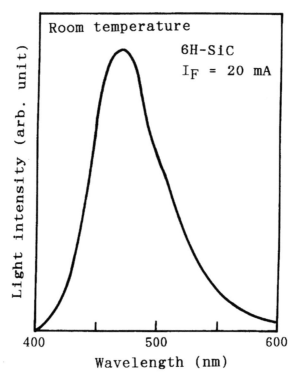

Figure 107   A typical spectrum of bright blue LEDs of 6H-SiC. (From Matsushita, Y., Koga, K., Ueda, Y., and Yamaguchi, T., *Oyobuturi*, 60, 159-162, 1991 (in Japanese).)

7. Shibahara, K., Kuroda, N., Nishino, S., and Matsunami, H., *Jpn. J. Appl. Phys.*, 26, L1815-L1817, 1987.
8. Ikeda, M., Hayakawa, T., Yamagiwa, S., Matsunami, H., and Tanaka, T., *J. Appl. Phys.*, 50, 8215-8225, 1979.
9. Matsushita, Y., Koga, K., Ueda, Y., and Yamaguchi, T., *Oyobuturi*, 60, 159-162, 1991, (in Japanese).

chapter two — section fourteen

Fundamentals of luminescence

Rong-Jun Xie, Naoto Hirosaki, and Mamoru Mitomo

Contents

2.14 Oxynitride phosphors

2.14.1 Introduction

Inorganic phosphors are composed of a host lattice doped with a small amount of impurity ions that activate luminescence. Most of these materials are oxides, sulfides, fluorides, halides, and oxysulfides doped with transition metal ions or rare-earth ions. Recently, with the advent of solid-state lighting technologies as well as the development of plasma and field emission display panels, a great number of traditional phosphors cannot meet the requirements for new applications, for example: (1) excitation by near-ultraviolet (UV) or visible light; (2) efficient emission of appropriate colors; and (3) survival at adverse environments. Therefore, novel phosphors with superior luminescent properties are being sought using new host materials.

The integration of nitrogen (N) in silicates or aluminosilicates produces a wide range of complex structures with increased flexibility compared to the oxosilicates, and thus a new class of materials, nitridosilicates, nitridoaluminosilicates, and sialons, are obtained.[1] These novel luminescent materials—the oxynitride phosphors—have been synthesized by doping with appropriate amounts of rare-earth activators.[2-20] The rare earths doped in the oxynitride phosphors usually enter into interstitial sites and are coordinated by (O, N) ions located at various distances. For those rare earths (i.e., Eu^{2+} and Ce^{3+}) emitting from their 5d excited state, which is strongly affected by the crystal-field environment (e.g., covalency, coordination, bond length, crystal-field strength), appropriate emission colors can be obtained by carefully selecting the host lattice. Due to a higher charge of N^{3-} compared with that of O^{2-} and because of the nephelauxetic effect (high covalency), the crystal-field splitting of the 5d levels of rare earths is larger and the center of gravity of the 5d states is shifted to low energy (i.e., longer wavelength) in these oxynitride compounds. Furthermore, the Stokes shift becomes smaller in a more rigid lattice, which results when more N^{3-} is incorporated. This will result in more versatile luminescent properties of oxynitride phosphors, increasing their range of applications. In this section, the characteristic features and potential applications of rare-earth-doped nitride phosphors are described.

2.14.2   Overview of oxynitride phosphors

Table 30 lists oxynitride phosphors reported in the literature in recent years. The host lattice of these phosphors is based on nitridosilicates, oxonitridosilicates, or oxonitrido-aluminosilicates, which are derived from silicates by formal exchanges of O and Si by N and Al, respectively. The structure of these host lattices is built on highly condensed networks constructed from the corner-sharing (Si, Al)–(O, N) tetrahedra. The degree of condensation of the network structures (i.e., the molar ratio Si:X > 1:2, with X = O, N) is higher than the maximum value for oxosilicates ($1:4 \leq$ Si:O $\leq 1:2$).[21] Consequently, these highly condensed materials exhibit high chemical and thermal stabilities. Moreover, the structural variabilities of this class of materials provide a significant extension of conventional silicate chemistry, forming a large family of Si–Al–O–N multiternary compounds.

Table 30   Emission Color and Crystal Structure of Oxynitride Phosphors

Phosphor	Emission color	Crystal structure	References
Y-Si-O-N:Ce^{3+}	Blue	—	[3]
$BaAl_{11}O_{16}N$:Eu^{2+}	Blue	β-Alumina	[2,4]
JEM:Ce^{3+}	Blue	Orthorhombic	[19]
$SrSiAl_2O_3N_2$:Eu^{2+}	Blue-green	Orthorhombic	[14]
$SrSi_5AlO_2N_7$:Eu^{2+}	Blue-green	Orthorhombic	[14]
$BaSi_2O_2N_2$:Eu^{2+}	Blue-green	Monoclinic	[18]
α-SiAlON:Yb^{2+}	Green	Hexagonal	[15]
β-SiAlON:Eu^{2+}	Green	Hexagonal	[17]
$MYSi_4N_7$:Eu^{2+} (M = Sr, Ba)	Green	Hexagonal	[12]
$MSi_2O_2N_2$:Eu^{2+} (M = Ca, Sr)	Green-yellow	Monoclinic	[18]
α-SiAlON:Eu^{2+}	Yellow-orange	Hexagonal	[7,8,10,11]
$LaSi_3N_5$:Eu^{2+}	Red	Orthorhombic	[6]
$LaEuSi_2N_3O_2$	Red	Orthorhombic	[6]
$Ca_2Si_5N_8$:Eu^{2+}	Red	Monoclinic	[5]
$M_2Si_5N_8$:Eu^{2+} (M = Sr, Ba)	Red	Orthorhombic	[5]
$CaAlSiN_3$:Eu^{2+}	Red	Orthorhombic	[20]

The most usual approaches for synthesizing oxynitride phosphors are solid-state reactions and gas-reduction–nitridation. The solid-state reaction involves the reaction among chemical components including metals, nitride, and oxide starting powders at high temperatures (1400–2000°C) under an N_2 atmosphere. The nitridation reaction is generally performed in an alumina boat containing the oxide precursor powder loaded inside an alumina/quartz tube through which NH_3 or NH_3–CH_4 gas flows at appropriate rates at high temperatures (600–1500°C). The NH_3 or NH_3–CH_4 gas acts as both a reducing and nitridation agent.

2.14.3   Characteristics of typical oxynitride phosphors

2.14.3.1 $LaAl(Si_{6-z}Al_z)N_{10-z}O_z$:$Ce^{3+}$ (z = 1)

Crystal structure. The $LaAl(Si_{6-z}Al_z)N_{10-z}O_z$ (JEM) phase was identified in the preparation of La-stabilized α-SiAlON materials.[22] It has an orthorhombic structure (space group *Pbcn*) with a = 9.4303, b = 9.7689, and c = 8.9386 Å. The Al atoms and the (Si, Al) atoms are tetrahedrally coordinated by the (N, O) atoms, yielding an $Al(Si, Al)_6(N, O)_{10}^{3-}$ network. The La atoms are located in the tunnels extending along the [001] direction and are irregularly coordinated by seven (N, O) atoms at an average distance of 2.70 Å.

Luminescence characteristics. As shown in Figure 108, the emission spectrum of JEM:Ce^{3+} displays a broad band with the peak located at 475 nm under 368-nm excitation.[19] The emission efficiency (external quantum efficiency) is about 55% when excited at 368 nm. This blue phosphor has a broad excitation spectrum, extending from the UV to the visible range. When the concentration of Ce^{3+} or the z value increases, both the excitation and emission spectra are red shifted.

Preparation. The starting materials for JEM are Si_3N_4, AlN, Al_2O_3, La_2O_3, and CeO_2. The powder phosphor is synthesized by heating the powder mixture at 1800–1900°C for 2 h under 1.0 MPa N_2.

2.14.3.2   β-SiAlON:Eu^{2+}

Crystal structure. The structure of β-SiAlON is derived from β-Si_3N_4 by substitution of Al–O by Si–N, and its chemical composition can be written as $Si_{6-z}Al_zO_zN_{8-z}$ (z represents

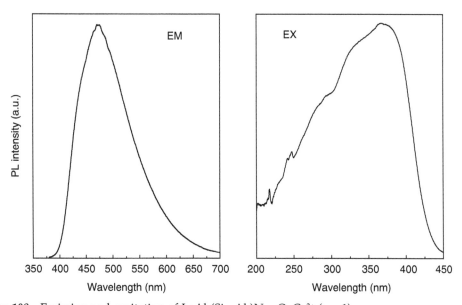

Figure 108   Emission and excitation of $LaAl(Si_{6-z}Al_z)N_{10-z}O_z$:$Ce^{3+}$ (z = 1).

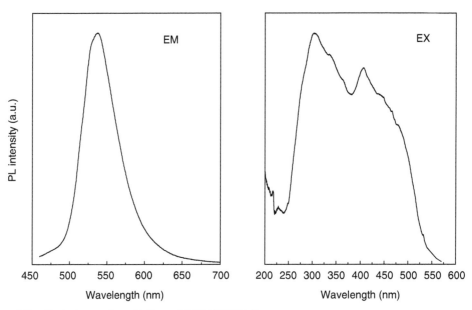

Figure 109   Emission and excitation of β-SiAlON:Eu²⁺.

the number of Al–O pairs substituting for Si–N pairs and $0 < z \leq 4.2$).[23] β-SiAlON has a hexagonal crystal structure and the $P6_3$ space group. In this structure, there are continuous channels parallel to the *c* direction.

Luminescence characteristics. The β-SiAlON:Eu²⁺ phosphor gives intense green emission with the peak located at 538 nm,[17] as seen in Figure 109. The broad emission spectrum has a full width of half maximum of 55 nm. Two well-resolved broad bands centered at 303 and 400 nm are observed in the excitation spectrum. The broad excitation range enables the β-SiAlON:Eu²⁺ phosphor to emit strongly under near UV (390–410 nm) or blue-light excitation (450–470 nm). This green phosphor has a chromaticity coordinates of $x = 0.31$ and $y = 0.60$. The external quantum efficiency is about 41% when excited at 405 nm.

Preparation. Starting from Si_3N_4, AlN, Al_2O_3, and Eu_2O_3, the β-SiAlON:Eu²⁺ phosphor is synthesized at 1800–2000°C for 2 h under 1.0 MPa N_2. An Eu concentration of <1.0% is used.

2.14.3.3 $MSi_2O_2N_2$:Eu²⁺ (M = Ca, Sr, Ba)

Crystal structure. All $MSi_2O_2N_2$ compounds crystallize in a monoclinic lattice with different space groups and lattice parameters for M = Ca, Sr, Ba: $CaSi_2O_2N_2$, $P2_1/C$, $a = 15.036$, $b = 15.450$, $c = 6.851$ Å; $SrSi_2O_2N_2$, $P2_1/M$, $a = 11.320$, $b = 14.107$, $c = 7.736$ Å; $BaSi_2O_2N_2$, P_2/M, $a = 14.070$, $b = 7.276$, $c = 13.181$ Å.[18,24] A nitrogen-rich phase $MSi_2O_{2-\delta}N_{2+2/3\delta}$ (M = Ca, Sr, $\delta > 0$) has been identified, suggesting that some modifications of $MSi_2O_2N_2$ (M = Ca, Sr) exist depending on the synthesis temperature.

Luminescence characteristics. All $MSi_2O_2N_2$:Eu²⁺ phosphors have a broad-band emission spectrum with different full widths at half maximum: $CaSi_2O_2N$:Eu²⁺, 97 nm; $SrSi_2O_2N$:Eu²⁺, 82 nm; and $BaSi_2O_2N$:Eu²⁺, 35 nm (see Figure 110). $CaSi_2O_2N$:6%Eu²⁺ shows a yellowish emission with a maximum at 562 nm. $SrSi_2O_2N$:6%Eu²⁺ emits green color with a maximum at 543 nm, and $BaSi_2O_2N$:6%Eu²⁺ yields a blue-green emission with a peak at 491 nm. The excitation spectrum of $CaSi_2O_2N$:6%Eu²⁺ shows a flat broad band covering the 300–450 nm range, whereas two well-resolved broad bands centered at 300 and 450 nm are seen in $SrSi_2O_2N$:6%Eu²⁺ and $BaSi_2O_2N$:6%Eu²⁺, respectively.

Preparation. The $MSi_2O_2N_2$:Eu²⁺ phosphors are synthesized by heating the powder mixture of Si_3N_4, SiO_2 and alkaline-earth carbonates at 1600°C under 0.5 MPa N_2.

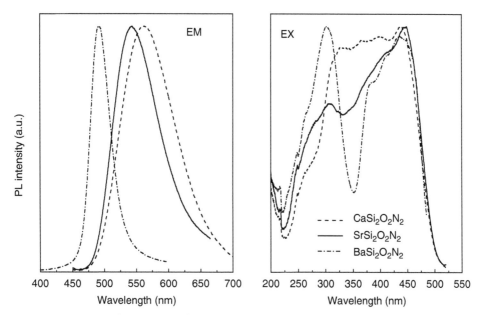

Figure 110 Emission and excitation of $MSi_2O_2N_2:Eu^{2+}$ (M = Ca, Sr, Ba).

2.14.3.4 α-SiAlON:Eu²⁺

Crystal structure. α-SiAlON is isostructural to α-Si_3N_4. It has a hexagonal crystal structure and the *P*31*c* space group. The α-SiAlON unit cell content, consisting of four "Si_3N_4" units, can be given in a general formula $M_xSi_{12-m-n}Al_{m+n}O_nN_{16-n}$ (*x* is the solubility of the metal M).[25,26] In the α-SiAlON structure, $m+n$ (Si–N) bonds are replaced by m (Al–N) bonds and n (Al–O) bonds; the charge discrepancy caused by the substitution is compensated by the introduction of M cations including Li^+, Mg^{2+}, Ca^{2+}, Y^{3+}, and lanthanides. The M cations occupy the interstitial sites in the α-SiAlON lattice and are coordinated by seven (N, O) anions at three different M-(N, O) distances.

Luminescence characteristics. α-SiAlON:Eu²⁺ phosphors give green-yellow, yellow, or yellow-orange emissions with peaks located in the range of 565–603 nm,[7,8,10,11] as shown in Figure 111. The broad-band emission spectrum covers from 500 to 750 nm with the full width of half maximum of 94 nm. The excitation spectrum of Eu²⁺ in α-SiAlON has two broad bands with peaks at 300 and 420 nm, respectively. The external quantum efficiency of the α-SiAlON:Eu²⁺ phosphor with optimal composition is about 58% when excited at 450 nm. By tailoring the composition of the host lattice and controlling the concentration of Eu²⁺, the emission color of α-SiAlON can be tuned through a wide range.

Preparation. The Ca-α-SiAlON:Eu²⁺ phosphor is synthesized by solid-state reactions. The powder mixture of Si_3N_4, AlN, $CaCO_3$, and Eu_2O_3 is fired at 1600–1800°C for 2 h under 0.5 MPa N_2. The gas-reduction–nitridation method is also used to prepare α-SiAlON:Eu²⁺ phosphor.[16] It is synthesized from the $CaO–Al_2O_3–SiO_2$ system, by using an $NH_3–CH_4$ gas mixture as a reduction–nitridation agent. The Eu concentration in α-SiAlON phosphors varies from 0.5 to 10%.

2.14.3.5 $M_2Si_5N_8:Eu^{2+}$ (M = Ca, Sr, Ba)

Crystal structure. $Ca_2Si_5N_8$ has a monoclinic crystal system with the space group of *Cc*, whereas both $Sr_2Si_5N_8$ and $Ba_2Si_5N_8$ have an orthorhombic lattice with the space group of $Pmn2_1$.[27,28] The local coordination in the structures is quite similar for these ternary alkaline-earth silicon nitrides, half of the nitrogen atoms connecting two Si neighbors and the other half have three Si neighbors. Each Ca atom in $Ca_2Si_5N_8$ is coordinated by seven nitrogen

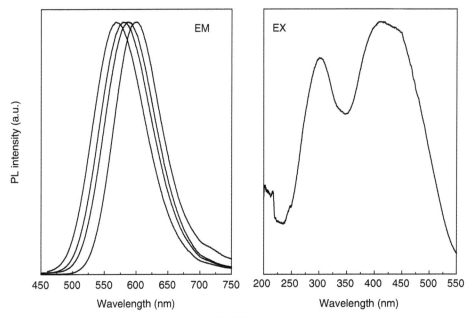

Figure 111 Emission and excitation of α-SiAlON:Eu²⁺.

atoms, whereas Sr in $Sr_2Si_5N_8$ and Ba in $Ba_2Si_5N_8$ are coordinated by eight or nine nitrogen atoms.

Luminescence characteristics. $M_2Si_5N_8$:Eu²⁺ (M = Ca, Sr, Ba) phosphors give orange-red or red emission, as shown in Figure 112. A single, broad emission band is centered at 623, 640, and 650 nm for $Ca_2Si_5N_8$, $Sr_2Si_5N_8$, and $Ba_2Si_5N_8$, respectively. A red shift in the emission wavelength is observed with increasing the ionic size of alkaline-earth metals. The excitation spectrum resembles each other, indicating the chemical environment of Eu²⁺ in these materials is very similar. The excitation spectrum extensively shifts to longer wavelengths, with the peak located at 450 nm for all samples.

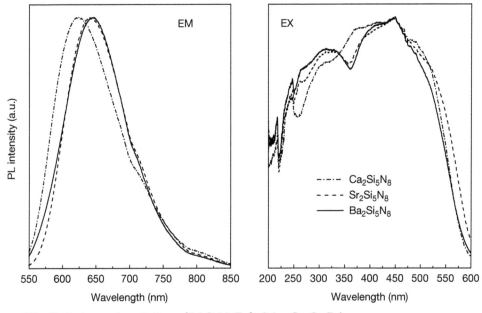

Figure 112 Emission and excitation of $M_2Si_5N_8$:Eu²⁺ (M = Ca, Sr, Ba).

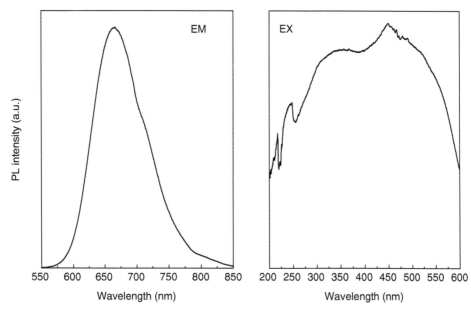

Figure 113   Emission and excitation of $CaAlSiN_3:Eu^{2+}$.

Preparation. The ternary alkaline-earth silicon nitrides are either synthesized by firing the powder mixture of Si_3N_4, M_3N_2, and EuN at 1600–1800°C under 0.5 MPa N_2 or prepared by the reactions of metallic alkaline-earths with silicon diimide at 1550–1650°C under nitrogen atmosphere.[5,27,28]

2.14.3.6 $CaAlSiN_3:Eu^{2+}$

Crystal structure. $CaAlSiN_3$ has an orthorhombic crystal structure and the space group of $Cmc2_1$, the unit cell parameter being $a = 9.8007$, $b = 5.6497$, and $c = 5.0627$ Å.[20] The Ca atoms are found in the tunnels surrounded by six corner-sharing tetrahedra of $(Al, Si)N_4$.

Luminescence characteristics. $CaAlSiN_3:Eu^{2+}$ is a red phosphor. The luminescence spectra are given in Figure 113. The excitation spectrum is extremely broad, ranging from 250 to 550 nm. Again a broad emission band centered at 650 nm is observed when excited at 450 nm. The chromaticity coordinates of red phosphor are $x = 0.66$ and $y = 0.33$. This phosphor has an external quantum efficiency as high as 86% under 450 nm excitation. The emission spectrum is red shifted with increasing Eu^{2+} concentrations.

Preparation. The $CaAlSiN_3:Eu^{2+}$ phosphor was synthesized by firing a powder mixture of Si_3N_4, AlN, Ca_3N_2, and EuN at 1600–1800°C for 2 h under 0.5 MPa N_2.

2.14.4   *Applications of oxynitride phosphors*

As shown in the previous section, oxynitride phosphors emit efficiently under UV and visible-light irradiation. This correlates well with the emission wavelengths of the UV chips or blue light-emitting diode (LED) chips, making their use as down-conversion phosphors in white LEDs feasible.

We have proposed that yellow α-SiAlON phosphors could be used to generate warm white light when combined with a blue LED. The first white LED lamp was reported by Sakuma et al. using an orange-yellow α-SiAlON:Eu^{2+} and a blue LED chip.[29] It emits warm white light with the color temperature of 2800 K. To obtain white LED lamps with high color rendering index, additional phosphors such as green and red phosphors are used. Sakuma et al. have reported white LEDs with various color temperatures and a color rendering index of >80 using β-SiAlON:Eu^{2+} (green), α-SiAlON:Eu^{2+} (yellow), and

CaAlSiN$_3$:Eu^{2+} (red) phosphors.[30] Mueller-Mach et al. have used (Ca,Sr,Ba)Si$_2$O$_2$N$_2$:Eu^{2+} (yellow-green) and (Ca,Sr,Ba)$_2$Si$_5$N$_8$:Eu^{2+} (orange-red) phosphors to fabricate highly efficient white LEDs.[31]

References

1. Schnick, W., *Inter. J. Inorg. Mater.*, 3, 1267, 2001.
2. Jansen, S.R., de Hann, J.W., van de Ven, L.J.M., Hanssen, R., Hintzen, H.T., and Metselaar, R., *Chem. Mater.*, 9, 1516, 1997.
3. van Krevel, J.W.H., Hintzen, H.T., Metselaar, R., and Meijerink, A., *J. Alloy Compd*, 268, 272, 1998.
4. Jansen, S.R., Migchel, J.M., Hintzen, H.T., and Metselaar, R., *J. Electrochem. Soc.*, 146, 800, 1999.
5. Höppe, H.A., Lutz, H., Morys, P., Schnick, W., and Seilmeier, A., *J. Phys. Chem. Solids*, 61, 2001, 2000.
6. Uheda, K., Takizawa, H., Endo, T., Yamane, H., Shimada, M., Wanf, C.M., and Mitomo, M., *J. Lum.*, 87–89, 867, 2000.
7. van Krevel, J.W.H., van Rutten, J.W.T., Mandal, H., Hintzen, H.T., and Metselaar, R., *J. Solid State Chem.*, 165, 19, 2002.
8. Xie, R.-J., Mitomo, M., Uheda, K., Xu, F.F., and Akimune, Y., *J. Am. Ceram. Soc.*, 85, 1229, 2002.
9. Xie, R.-J., Hirosaki, N., Mitomo, M., Yamamoto, Y., Suehiro, T., and Ohashi, N., *J. Am. Ceram. Soc.*, 87, 1368, 2004.
10. Xie, R.-J., Hirosaki, N., Mitomo, M., Yamamoto, Y., Suehiro, T., and Sakuma, K., *J. Phys. Chem. B*, 108, 12027, 2004.
11. Xie, R.-J., Hirosaki, N., Sakuma, K., Yamamoto, Y., and Mitomo, M., *App. Phys. Lett.*, 84, 5404, 2004.
12. Li, Y.Q., Fang, C.M., de With, G., and Hintzen, H.T., *J. Solid State Chem.*, 177, 4687, 2004.
13. Xie, R.-J., Hirosaki, N., Mitomo, M., Suehiro, T., Xin, X., and Tanaka, H., *J. Am. Ceram. Soc.*, 88, 2883, 2005.
14. Xie, R.-J., Hirosaki, N., Yamamoto, Y., Suehiro, T., Mitomo, M., and Sakuma, K., *Jpn. J. Ceram. Soc.*, 113, 462, 2005.
15. Xie, R.-J., Hirosaki, N., Mitomo, M., Uheda, K., Suehiro, T., Xin, X., Yamamoto, Y., and Sekiguchi, T., *J. Phys. Chem. B*, 109, 9490, 2005.
16. Suehiro, T., Hirosaki, N., Xie, R.-J., and Mitomo, M., *Chem. Mater.*, 17, 308, 2005.
17. Hirosaki, N., Xie, R.-J., Kimoto, K., Sekiguchi, T., Yamamoto, Y., Suehiro, T., and Mitomo, M., *App. Phys. Lett.*, 86, 211905, 2005.
18. Li, Y.Q., Delsing, C.A., de With, G., and Hintzen, H.T., *Chem. Mater.*, 17, 3242, 2005.
19. Hirosaki, N., Xie, R.-J., Yamamoto, Y., and Suehiro, T., Presented at *the 66th Autumn Annual Meeting of the Japan Society of Applied Physics* (Abstract No. 7ak6), Tokusima, Sept. 7–11, 2005.
20. Uheda, K., Hirosaki, N., Yamamoto, H., Yamane, H., Yamamoto, Y., Inami, W., and Tsuda, K., Presented at *the 206th Annual Meeting of the Electrochemical Society* (Abstract No. 2073), Honolulu, Oct. 3–8, 2004.
21. Schnick, W. and Huppertz, H., *Chem. Eur. J.*, 3, 679, 1997.
22. Grins, J., Shen, Z., Nygren, M., and Eskrtom, T., *J. Mater. Chem.*, 5, 2001, 1995.
23. Oyama, Y., and Kamigaito, O., *Jpn. J. Appl. Phys.*, 10, 1637, 1971.
24. Höppe, H.A., Stadler, F., Oeckler, O., and Schnick, W., *Angew. Chem. Int. Ed.*, 43, 5540, 2004.
25. Hampshire, S., Park, H.K., Thompson, D.P., and Jack, K.H., *Nature* (London), 274, 31, 1978.
26. Cao, G.Z. and Metselaar, R., *Chem. Mater.*, 3, 242, 1991.
27. Schlieper, T. and Schnick, W., *Z. Anorg. Allg. Chem.*, 621, 1037, 1995.
28. Schlieper, T. and Schnick, W., *Z. Anorg. Allg. Chem.*, 621, 1380, 1995.
29. Sakuma, K., Omichi, K., Kimura, N., Ohashi, M., Tanaka, D., Hirosaki, N., Yamamoto, Y., Xie, R.-J., and Suehiro, T., *Opt. Lett.*, 29, 2001, 2004.
30. Sakuma, K., Hirosaki, N., Kimura, N., Ohashi, M., Xie, R.-J., Yamamoto, Y., Suehiro, T., Asano, K., and Tanaka, D., *IEICE Trans. Electron.*, Vol.E88-C, 2005 (in press).
31. Mueller-Mach, R., Mueller, G., Krames, M.R., Höppe, H.A., Stadler, F., Schnick, W., Juestel, T., and Schmidt, P., *Phys. Stat. Sol. (a)* 202, 1727, 2005.

Index

Subject

Milton Keynes UK
Ingram Content Group UK Ltd.
UKHW052020071024
449327UK00027B/2356